KU-197-969

UTOPIA'S GARDEN

E. C. SPARY

UTOPIA'S GARDEN

French Natural History from Old Regime to Revolution

The University of Chicago Press
Chicago and London

E. C. Spary is senior researcher at the Max Planck Institute for the History of Science, Berlin. She is coeditor (with Nicholas Jardine and James A. Secord) of *Cultures of Natural History.*

The University of Chicago Press, Chicago 60637
The University of Chicago Press, Ltd., London

Printed in the United States of America

09 08 07 06 05 04 03 02 01 00   1 2 3 4 5

ISBN: 0-226-76862-7 (cloth)
ISBN: 0-226-76863-5 (paper)

Library of Congress Cataloging-in-Publication Data

Spary, E. C. (Emma C.)
Utopia's garden : French natural history from Old Regime to Revolution / E. C. Spary.
p. cm.
Includes bibliographical references (p.).
ISBN 0-226-76862-7 (alk. paper) — ISBN 0-226-76863-5 (alk. paper)
1. Natural history—France—History—Eighteenth century. 2. Muséum national d'histoire naturelle (France)—History—Eighteenth century. I. Title.

QH147.S62 2000
508.44′09′033—dc21

00-029897

⊗ The paper used in this publication meets the minimum requirements of the American National Standard for Information Sciences—Permanence of Paper for Printed Library Materials, ANSI Z39.48-1992.

For Paul and the A.B.

CONTENTS

ILLUSTRATIONS

ACKNOWLEDGMENTS

This book has taken me longer to complete than I like to dwell upon. As a result, my debts to those who have read it, commented on it, or simply tolerated it, surpass my space limits. But some debts are too large to be ignored. It could be said that my greatest debt is to the Department of History and Philosophy of Science in Cambridge. Various members of staff put up with initial tentatives toward this finished product, most notably Nick Jardine, supervisor, colleague, and friend; at different times I particularly benefited from the advice and suggestions of Jim Secord, Simon Schaffer, and Andrew Cunningham. I should also like to extend particular thanks to Laurence Brockliss, Janet Browne, Michael Dettelbach, Marina Frasca Spada, Michael Hagner, Jeff Hughes, Rob Iliffe, Shelley Innes, Adrian Johns, Ludmilla Jordanova, Julian Martin, Jim Moore, Edouard Pommier, Marsha Richmond, Harriet Ritvo, Anne Secord, Phillip Sloan, Peter Stevens, Frances Willmoth, Alison Winter, and many seminar and conference audiences whose suggestions have directed my interpretations at different times.

All translations in the work are my own, except where stated. I owe thanks to Frances Willmoth for translating passages from the *Genera plantarum* of Antoine-Laurent de Jussieu for me. I am also indebted to Debbie Feast, for laboring beyond the call of duty in preparing the maps and graphs for the book; and to many at the Press, notably Dave Aftandilian, Marty Hertzel, and Joann Hoy.

My thanks also go to the staff of numerous libraries: those of the

Whipple Library and the University Library, Cambridge; the British Library, London; the Bibliothèque Centrale du Muséum National d'Histoire Naturelle, Paris; the Archives Nationales, Paris; and the Bibliothèque Nationale, Paris. Newnham College, Cambridge; the British Academy; the Arnold Gerstenberg Trust; and the Spary family funded the initial stages of preparation of the book; the Royal Society; Girton College, Cambridge; the University of Warwick; and the Max-Planck-Institut für Wissenschaftsgeschichte, Berlin, have contributed toward its completion.

Without the support, in so many different ways, of Paul S. White, it is hard to know how this book would have reached completion. Martin, Stella, and Georgina Spary have supported me despite a healthy skepticism about the whole enterprise. Friends in Cambridge and elsewhere, many of whom are named above, have kept me on the right track. Finally, my warm recognition must also go to Susan Abrams, not just for her support for the whole enterprise but for her enlightened approach to that peculiar field, history of science. All deserve more than printed words by way of thanks.

ABBREVIATIONS

Actes CSP: François-Victor-Alphonse Aulard, ed. *Recueil des actes du Comité de Salut public, avec la correspondance officielle des représentants en mission et le registre du Conseil exécutif provisoire.* 28 vols. Paris: Imprimerie Nationale, 1889–1951.

AN: Archives Nationales, Paris.

AP: Jerôme Mavidal et al., eds. *Archives parlementaires de 1789 à 1860: Recueil complet des débats législatifs et politiques des chambres françaises.* Series 1, vols. 1–87 [1787–1794]. Paris: Imprimerie Nationale, 1867–1969.

BCMHN: Bibliothèque Centrale du Muséum d'Histoire Naturelle, Paris.

CIPCN: James Guillaume, ed. *Procès-verbaux du Comité d'Instruction publique de la Convention nationale.* 6 vols. Paris: Imprimerie Nationale, 1891.

Correspondance: Henri Nadault de Buffon, ed. *Buffon: Correspondance générale recueillie et annotée par H. Nadault de Buffon.* 2 vols. Geneva: Slatkine, 1971

La Décade philosophique: La Décade Politique, Philosophique, et Littéraire; par une Société de Républicains. Vols. 1–6. Paris Bureau de la Décade, years II–IV/1793–1796.

DSB: Charles Coulston Gillispie. *Dictionary of Scientific Biography.* 16 vols. New York: Charles Scribner, 1970–1980.

Époques: Jacques Roger, ed. *Buffon: Les époques de la nature: Édition critique.* Mémoires du Muséum National d'Histoire Naturelle, series C, Sciences de la terre, 10. Paris: Muséum National d'Histoire Naturelle, 1988.

"Etât": [A. Thouin and Trécourt]. "Etât de la Correspondance de M. A. Thouin." BCMHN, MS 314, circa 1791.

HARS: "Histoire de l'Académie Royale des Sciences," in *Histoire et Mémoires de l'Académie Royale des Sciences,* 1760–1788 (Paris: Imprimerie Royale, 1766–1791), and *Histoire et Mémoires de l'Académie des Sciences,* 1789 (Paris: Du Pont, 1793).

Histoire naturelle: Georges-Louis Leclerc, comte de Buffon, Louis-Jean-Marie Daubenton, Philibert Guéneau de Montbeillard, and Gabriel-Léopold-Charles-Aimé Bexon. *Histoire naturelle, générale et particulière, avec la description du cabinet du Roi.* 36 vols. Paris: Imprimerie Royale, 1749–1789. The subdivisions of this work are: Buffon [with Daubenton], *Histoire générale et des Quadrupèdes,* 15 vols., 1749–1767; Buffon [with Guéneau de Montbeillard and Bexon], *Histoire des Oiseaux,* 9 vols., 1770–1783; Buffon [with Bexon], *Histoire des Minéraux,* 5 vols., 1783–1788; and *Suppléments,* 7 vols., 1774–1789.

HSRM: Histoire de la société royale de médecine . . . Avec les mémoires de médecine et de Physique médicale . . . Tirés des Registres de cette Société. 9 vols. Paris: Philippe-Denys Pierres, 1779–1790.

Journal d'Histoire naturelle: Jean-Baptiste-Pierre-Antoine de Monet de Lamarck et al., eds. *Choix de mémoires sur divers sujets d'histoire naturelle par MM. Lamarck, Bruguière, Olivier, Hauy et Pelletier. Formant les Collections du Journal d'Histoire Naturelle.* Paris: Imprimerie du Cercle Social, 1792. *Journal d'Histoire Naturelle, rédigé par MM. Lamarck, Bruguière, Olivier, Hauy et Pelletier.* Paris: Directeurs du Cercle Social, year II [1794].

MARS: "Mémoires de l'Académie Royale des Sciences," in *Histoire et Mémoires de l'Académie Royale des Sciences,* 1760–1788 (Paris: Imprimerie Royale, 1766–1791), and *Histoire et Mémoires de l'Académie des Sciences,* 1789 (Paris: Du Pont, 1793).

NBU: Nouvelle biographie universelle. 46 vols. Paris: Firmin Didot, 1853–1866.

The form "*HSRM* 1776/1779" denotes the volume of memoirs for the year 1776, published in 1779. The same applies for the abbreviations *HARS* and *MARS*.

INTRODUCTION

Eighteenth-century observers variously described the Jardin du Roi in Paris. For some the renowned natural history establishment was "one of the most agreeable and richest of the Universe," "the precious cabinet of the king" and a center of national utility, but for others it appeared "purely decorative," too large, too small, or disorganized.[1] While Parisian natural historical practice was far from being restricted to the institutional setting of the Jardin du Roi, for many that establishment and the naturalists attached to it, most notably its intendant Georges-Louis Leclerc de Buffon, would come to exemplify the place of natural history in France.[2] The Jardin had originally been founded as a royal physic garden by Louis XIII, in 1626. By the mid–eighteenth century it was still at the edge of the metropolis, bordering the banks of the river Seine in the unfashionable and inaccessible Faubourg Saint-Victor, set between a hackney-carriage company and some tenanted lands belonging to the Abbaye de Saint-Victor. But the establishment was trans-

1 AN, AJ/15/503, piece 123: André Thouin, "Mémoire" to Georges-Louis Leclerc de Buffon; Antoine-Joseph Dezallier d'Argenville, *La Conchyliologie, ou Histoire Naturelle des Coquilles de mer, d'eau douce, terrestres et fossiles, avec un Traité de la Zoomorphose, ou représentation des Animaux qui les habitent,* 2 vols., ed. Jacques Favanne de Montcervelle and Guillaume Favanne de Montcervelle (Paris: De Bure, 1780), 199; BCMHN, MS 1459: Félix Vicq d'Azyr, plan for an academy of agriculture; BCMHN, MS 457: "Second Raport fait par la Commission des Travaux publics au Comité de Salut public relativement au Jardin des Plantes," thermidor year II/July 1794; Dorinda Outram, *Georges Cuvier: Vocation, Science, and Authority in Post-Revolutionary France* (Manchester: Manchester University Press, 1984), chapter 8.

2 Biographical details of all the Jardin's employees are in the appendix.

formed during the nearly fifty years of Buffon's tenure, which began in 1739. By the time of the intendant's death in 1788, the fame of his popular work, the *Histoire naturelle, générale et particulière, avec la description du Cabinet du Roi,* the high public profile of the establishment, and the resources that had been poured into it by royal ministers, all conspired to ensure that the Jardin was a principal site for the making of French natural historical knowledge.

The establishment's fame also attracted critics: enemies of Buffon's natural historical approach or those with competing agendas for running the foremost natural history establishment in Paris. Buffon's bitter relations with other European naturalists such as René-Antoine Ferchault de Réaumur or Carolus Linnaeus are well documented, and Buffon's natural historical project itself was framed against versions of natural history which claimed to show God's law from the design of the living world.[3] In one respect natural history at the Jardin can be seen as a local project, specific to one city in one part of Europe, practiced by and of interest to a small fraction of the French population in the eighteenth century. But just over a year after Buffon's death came the fall of the Bastille, and on June 10, 1793, the Jardin du Roi would be transformed into a Muséum d'Histoire Naturelle by decree of the Convention. By the century's end, that Muséum possessed a privileged position in France—indeed, in Europe—and so did the naturalists who worked there. This book seeks to show how the new Muséum came to acquire that unique place.

In retrospect, the "Golden Age" of French natural history has been deemed to have occurred in the first three decades of the nineteenth century.[4] Substantial works have focused on these years of glory and preeminence, but few have considered the period immediately preceding that peak of fame, in which the Muséum was fashioned into one of the leading scientific institutions in the world.[5] A single detailed study of the "last days of the Jardin du Roi" by Ernest-Théodore Hamy, itself over a century old, has long done duty as background for ac-

3 See Pascal Duris, *Linné et la France, 1780–1850* (Geneva: Droz, 1993), chapter 2; Frans A. Stafleu, *Linnaeus and the Linnaeans: The Spreading of Their Ideas in Systematic Botany* (Utrecht: Oosthoek, 1971), chapter 9; Charles Coulston Gillispie, *Science and Polity in France at the End of the Old Regime* (Princeton: Princeton University Press, 1980); Jean Torlais, *Un esprit encyclopédique en dehors de 'l'Encyclopédie': Réaumur: D'après des documents inédits,* 2d ed. (Paris: Albert Blanchard, 1961); Virginia P. Dawson, *Nature's Enigma: The Problem of the Polyp in the Letters of Bonnet, Trembley, and Réaumur* (Philadelphia: American Philosophical Society, 1987).

4 Camille Limoges, "The Development of the Muséum d'Histoire Naturelle of Paris, c. 1800–1914," in *The Organisation of Science and Technology in France, 1808–1914,* ed. Robert Fox and George Weisz (Cambridge: Cambridge University Press, 1980), 212.

5 Outram, *Cuvier;* Pietro Corsi, *The Age of Lamarck: Evolutionary Theories in France, 1790–1830,* revised ed., trans. Jonathan Mandelbaum (Berkeley: University of California Press, 1988).

counts concerned with the nineteenth century.[6] There are several reasons for this long neglect. The foundation of a new institution coincided with that great break in the history of the eighteenth century in France, the Revolution. Such termini have generally proved very inviting for historians, although naturalists' lives refused to obey these obvious boundaries. The staff of the Jardin du Roi, for the most part, became the staff of the Muséum d'Histoire Naturelle, even if this was not accomplished without some changes in their areas of expertise. How could a royal establishment survive the extremes of the French Revolution, when so many other institutions of the sciences closed?[7] Under what conditions was its reform possible? Such questions are fundamental to our understanding of the meaning of natural history, its practitioners, and its social place in the rapidly changing world of late-eighteenth-century France. If different accounts of the establishment could represent it in such different ways between 1750 and 1795, it is interesting to examine how the establishment's meaning was stabilized, its boundaries and the order of its specimens fixed. That stabilizing work was the basis on which the Muséum's postrevolutionary existence was assured. My account will explore the boundaries of the intellectual, political, and physical space of the Jardin as they were constructed through the associations and negotiations of the naturalists who worked there.[8]

6 Ernest-Théodore Hamy, "Les derniers jours du Jardin du Roi et la fondation du Muséum d'Histoire naturelle," in *Centenaire de la fondation du Muséum d'Histoire naturelle, 10 juin 1793–10 juin 1893: Volume commémoratif publié par les professeurs du Muséum* (Paris: Imprimerie Nationale, 1893), 1–162. More recent studies of the late-eighteenth-century Jardin are in Gillispie, *Science and Polity;* Yves Laissus, "Le Jardin du Roi," in *Enseignement et diffusion des sciences en France au XVIIIe siècle,* ed. René Taton (Paris: Hermann, 1986), 287–341; Charles Coulston Gillispie, "The *Encyclopédie* and the Jacobin Philosophy of Science: A Study in Ideas and Consequences," in *Critical Problems in the History of Science,* ed. Marshall Clagett (New York: Madison, 1959), 255–289; Joseph Fayet, *La Révolution française et la science, 1789–1795* (Paris: Marcel Rivière, 1960). But see Jean-Marc Drouin's call for further explorations of the institutionalization of natural history in eighteenth-century France, in "L'histoire naturelle: Problèmes scientifiques et engouement mondain," in *Nature, environnement, et paysage: L'héritage du XVIIIe siècle: Guide de recherche archivistique et bibliographique,* ed. Andrée Corvol and Isabelle Richefort (Paris: L'Harmattan, 1995), 26–27.

7 Roger Hahn, *The Anatomy of a Scientific Institution: The Paris Academy of Sciences, 1666–1803* (Berkeley: University of California Press, 1971); Janis Langins, *La République avait besoin de savants: Les débuts de l'École polytechnique: L'École centrale des travaux publiques et les cours révolutionnaires de l'an III* (Paris: Belin, 1987); Maurice P. Crosland, *Science under Control: The French Academy of Sciences, 1795–1914* (Cambridge: Cambridge University Press, 1992); Jean Dhombres and Nicole Dhombres, *Naissance d'un nouveau pouvoir: Sciences et savants en France, 1793–1824* (Paris: Bibliothèque Historique Payot, 1989).

8 For a different agenda in studying the Muséum as place, see Dorinda Outram, "New Spaces in Natural History," in *Cultures of Natural History,* ed. Nicholas Jardine, James A. Secord, and E. C. Spary (Cambridge: Cambridge University Press, 1996), 249–265. Elsewhere I have outlined other possible ways of being a naturalist in late-eighteenth-century France. See E. C. Spary, "The

Numerous secondary sources have explored the various eighteenth-century uses of nature as a new source of authority, superior to all temporal and spiritual powers. As such it would serve first in French critiques of the clergy and monarchy and afterward to overthrow them.[9] Medicine, natural history, and agriculture were arguably central to the generation of a new political language in the years prior to the French Revolution. Many political terms, such as *constitution, regénération,* and *corruption,* were simultaneously medical, natural historical, and agricultural, and Revolutionary reformers presented themselves as facilitators of the formation of a new society in France based upon natural laws. How, then, can we hope to comprehend the history of the Revolution itself if we possess only a partial history of the natural? The Jardin's demonstrator of botany, Antoine-Laurent de Jussieu, claimed to reflect the natural order in his new classificatory method, published in his book *Genera plantarum* of 1789. The use of the term *natural* here, as always, was not unproblematic. In the early nineteenth century de Jussieu's method was portrayed as reflecting a "natural order," the solution to the extended debate which had divided eighteenth-century naturalists over the relative value of classificatory systems and the very possibility of capturing natural relations between living beings. Such divisions became ever more embittered after the death of the principal claimant to the taxonomic throne, the Swedish naturalist Carolus Linnaeus, in 1778. My aim is not, however, to retrace the steps of these disputes, so thoroughly explored in the secondary literature.[10] Nor is my primary goal to address the principal assumptions of the substantial literature

'Nature' of Enlightenment," in *The Sciences in Enlightened Europe,* ed. William Clark, Jan V. Golinski, and Simon Schaffer (Chicago: University of Chicago Press, 1999), 272–304.

9 This claim, which has virtually become a commonplace of revolutionary history, is explored in R. Lenoble, *Histoire de l'idée de nature* (Paris: Albin Michel, 1969), chapter 6; J. Viard, *Le tiers espace: Essai sur la nature* (Paris: Méridiens Klincksieck, 1990); see also the substantial literature on natural law, especially Ernst Bloch, *Natural Law and Human Dignity,* trans. D. Schmidt (Cambridge: MIT Press, 1986); Otto Gierke, *Natural Law and the Theory of Society, 1500–1800,* trans. E. Barker (Boston: Beacon Press, 1957); Leo Strauss, *Natural Right and History* (Chicago: University of Chicago Press, 1953).

10 Henri Daudin, *De Linné à Jussieu: Méthode de la classification et idée de série en botanique et zoologie, 1740–1790,* facsimile reprint (Paris: Belles-Lettres, 1988); id., *Cuvier et Lamarck: Les classes zoologiques et l'idée de série animale, 1790–1830,* 2 vols., facsimile reprint (Paris: Belles-Lettres, 1988); Stafleu, *Linnaeus and the Linnaeans;* George H. M. Lawrence, ed., *Adanson: The Bicentennial of Michel Adanson's Familles des Plantes,* 2 vols. (Pittsburgh: Hunt Botanical Library and Carnegie Institute of Technology, 1963–1964); Duris, *Linné et la France;* Peter F. Stevens, *The Development of Biological Systematics: Antoine-Laurent de Jussieu, Nature, and the Natural System* (New York: Columbia University Press, 1994); Giulio Barsanti, "Linné et Buffon: Deux images differentes de la nature et de l'histoire naturelle," *Studies on Voltaire and the Eighteenth Century* 216 (1983): 306–307.

on the idea of nature in the eighteenth century.[11] Instead, I wish to show how an account of the culture within which eighteenth-century naturalists worked is central to a history of their practice. The Jardin's naturalists themselves downplayed the importance of classification: Louis-Jean-Marie Daubenton, for example, made programmatic statements calling for more attention to be paid to knowledge of the cultivation and properties of natural productions and less to nomenclature and classification.[12] In this respect my account addresses issues which were of particular importance to the natural historical concerns of those working at the Jardin.

Recent developments in the historiography of the sciences have notably focused attention on the problems of scientific practice and away from earlier concerns with an internalist history of science. It is difficult to find a homogeneous, fixed, readily identifiable body of theory, universally agreed to constitute "natural history" by contemporaries. Certainly, nothing like modern biology is identifiable with the projects which eighteenth-century individuals in general, or the Jardin's staff in particular, characterized as natural history. Moreover, to understand "natural history" solely as a set of theoretical debates would be to strip it of all historical meaning, since contemporaries portrayed the pursuit of natural history as inseparable from its social implications and practical uses. When eighteenth-century naturalists described *histoire naturelle* in programmatic statements, they might include within the discipline the practices of classifying, collecting, writing, experimenting, cultivating, and preserving; subjects as diverse as man, the fresh-

11 Clarence J. Glacken, *Traces on the Rhodian Shore: Nature and Culture in Western Thought from Ancient Times to the End of the Eighteenth Century* (Berkeley: University of California Press, 1967), 500–713; Lenoble, *Idée de nature;* Viard, *Tiers espace;* Jean Ehrard, *L'idée de nature en France dans la première moitié du XVIIIe siècle* (reprint, Paris: Albin Michel, 1994); D. G. Charlton, *New Images of the Natural in France: A Study in European Cultural History, 1750–1800* (Cambridge: Cambridge University Press, 1984); Robert Mauzi, *L'idée du bonheur au XVIIIe siècle* (reprint, Paris: Albin Michel, 1994); Lester G. Crocker, *Nature and Culture: Ethical Thought in the French Enlightenment* (Baltimore: Johns Hopkins University Press, 1963); Daniel Mornet, *Le sentiment de la nature en France de J.-J. Rousseau à Bernardin de Saint-Pierre: Essai sur les rapports de la littérature et des moeurs* (Paris: Hachette, 1907); id., *Les sciences de la nature en France au XVIIIe siècle: Un chapitre de l'histoire des idées* (New York: Franklin, 1971); Georges Gusdorf, *Dieu, la nature, l'homme au siècle des Lumières* (Paris: Payot, 1972). For more recent literature which explicitly acknowledges the social function of different symbolic uses of "nature," see Isabelle Richefort, "Métaphores et représentations de la nature sous la révolution," in *Nature, environnement, et paysage,* ed. Corvol and Richefort, 3–17; Andrée Corvol, ed., *La nature en Révolution, 1750–1800* (Paris: L'Harmattan, 1993).

12 Louis-Jean-Marie Daubenton, "Botanique," in *Encyclopédie, ou Dictionnaire raisonné des Sciences, des arts et des métiers,* ed. Denis Diderot and Jean Le Rond d'Alembert (Paris: Briasson, 1751), 2:340–345.

water polyp, and the potato. This diversity was precisely what allowed natural history to be shaped to cater for the needs of different regimes in a troubled and turbulent period. The eclectic quality of natural history has, not surprisingly, attracted less attention within current models of the social study of science, in which the experimental sciences have received the full benefit of historical attention.[13] But the insights of this literature can nevertheless be fruitfully employed in the history of natural history. While experimentalists of the eighteenth century often portrayed nature as artful, coquettish, modestly covering her innermost secrets, naturalists proudly announced that they perceived nature in herself, through a relationship of unmediated sensibility. What they thereby disguised was a vast enterprise: denying the validity of certain observations of nature in favor of others, determining who was to be trusted, they accumulated, ordered, displayed, managed, preserved, wrote, solicited. These issues of witnessing, managing, and accumulating have been addressed in the work of Bruno Latour, Steven Shapin, Simon Schaffer, and others.[14] In exploring the particular methodological issues raised by a history of natural history, I shall accordingly draw upon these recent studies of scientific practice developed for the experimental sciences, as well as the rapidly growing fields of material history and museology.[15] I shall also examine the role of the institution as a

13 For a selection of the principal approaches to experimental practice, many of which have now become classics in the field, see Bruno Latour and Steve Woolgar, *Laboratory Life: The Social Construction of Scientific Facts* (London: Sage, 1979); Karin D. Knorr-Cetina and Michael Mulkay, eds., *Science Observed: Perspectives on the Social Study of Science* (London: Sage, 1983); Simon Schaffer and Steven Shapin, *Leviathan and the Airpump: Hobbes, Boyle, and the Experimental Life* (Princeton: Princeton University Press, 1985); John Law, ed., *Power, Action, and Belief: A New Sociology of Knowledge?* (London: Routledge and Kegan Paul, 1986); Bruno Latour, *Science in Action: How to Follow Scientists and Engineers through Society* (Milton Keynes: Open University Press, 1987); Steve Woolgar, *Science: The Very Idea* (Chichester: Ellis Horwood, 1988); David Gooding, *Experiment and the Making of Meaning: Human Agency in Scientific Observation and Experiment* (Dordrecht: Kluwer Academic, 1990); Andrew Webster, *Science, Technology, and Society: New Directions* (Basingstoke: Macmillan, 1991); Andrew Pickering, ed., *Science as Practice and Culture* (Chicago: University of Chicago Press, 1992); Michael Lynch, *Scientific Practice and Ordinary Action: Ethnomethodology and Social Studies of Science* (Cambridge: Cambridge University Press, 1993); Harry Collins and Trevor Pinch, *The Golem: What Everyone Should Know about Science* (Cambridge: Cambridge University Press, 1993); Christian Licoppe, *La formation de la pratique scientifique: Le discours de l'expérience en France et en Angleterre, 1630–1820* (Paris: Éditions La Découverte, 1996); Barry Barnes, David Bloor, and John Henry, *Scientific Knowledge* (London: Athlone, 1996).

14 Steven Shapin, *A Social History of Truth: Civility and Science in Seventeenth-Century England* (Chicago: University of Chicago Press, 1994); Schaffer and Shapin, *Leviathan and the Airpump;* Bruno Latour, "Visualization and Cognition: Thinking with Eyes and Hands," *Knowledge and Society* 6 (1986): 1–40.

15 On material history see, in particular, Susan M. Pearce, ed., *Museum Studies in Material Culture* (Leicester: Leicester University Press, 1989); id., *Interpreting Objects and Collections* (Lon-

system of powers, a model created by Michel Foucault for other social practices, but not extended to natural history in his treatment of that discipline in *The Order of Things.*[16]

The history of the institution is a subject which has attracted little attention in recent methodological studies. At the outset of this introduction, I showed how one institution could be understood in very different ways by contemporaries. Such disagreements indicate the importance of studying the institution as a system of powers; how is the authority of an impersonal entity such as an institution created and maintained? We tend to think of the institution as identified with its geographical situation, as recognizable in terms of some boundary between outside and inside. In science studies, too, stress has been laid upon the importance of the isolation or insulation of the spaces of knowledge-making from the outside world, as places within which knowledge can be made, apparently uncontaminated by "external" factors such as politics, gender, or commerce; as places which draw their authority in large part from this apparent isolation of both setting and practitioner from normal life; as places whose access is strictly policed and within which behavior is highly ritualized and controlled. They are spaces within which, in the first instance, regulations can be made: for example, about proper modes of conduct, about what constitutes knowledge, and about the distinction between experimental fact and experimental error. Special rules are to operate there, and, accordingly, the multiple entanglements of everyday life can be ignored, expelled, or suppressed. Only once these rules are securely in place can the phenomena of ordinary life be processed, harnessed, or controlled effectively. In effect, then, this is a history about scientific consent rather than controversy—about the importance of accounts of scientific agreement and unity in defending the broader self-presentation of

don: Routledge, 1994); id., *Objects of Knowledge* (London: Athlone, 1990); Peter Vergo, ed., *The New Museology* (London: Reaktion Books, 1989); John Elsner and Roger Cardinal, eds., *The Cultures of Collecting* (London: Reaktion Books, 1994); George W. Stocking Jr., ed., *Objects and Others: Essays on Museums and Material Culture* (Madison: University of Wisconsin Press, 1985).

16 Michel Foucault, *The Order of Things: An Archaeology of the Human Sciences,* trans. Alan Sheridan (London: Tavistock, 1985); compare id., *Madness and Civilisation: A History of Insanity in the Age of Reason,* trans. Richard Howard (London: Tavistock, 1967); id., *The Birth of the Clinic: An Archaeology of Medical Perception,* trans. Alan Sheridan (London: Tavistock, 1973); id., *Discipline and Punish: The Birth of the Prison,* trans. Alan Sheridan (London: Penguin, 1991). Foucault's model of history is analyzed in detail in the essays in Jan Goldstein, ed., *Foucault and the Writing of History* (Oxford: Blackwell, 1994). The museum has attracted attention as an example of a Foucaldian disciplinary institution, however; see Tony Bennett, *The Birth of the Museum: History, Theory, Politics* (London: Routledge, 1995); Eilean Hooper-Greenhill, "The Museum in the Disciplinary Society," in *Museum Studies in Material Culture,* ed. Pearce, 61–72.

scientific practitioners since the seventeenth century as distanced from the world.[17]

In order to function as a site for constructing agreement, the institution's social "walls" must first have been built; in other words, its role, purpose, value, and legitimacy must have been agreed upon by different sides. The form and legitimation of that agreement, which is what constitutes and enables the institution as both social and material reality, may pass unquestioned for long periods, subject only to an ongoing process of maintenance and reinforcement. But in times of social change—the French Revolution being a good example—such forms and legitimations are overtly defended, challenged, or rejected. In developing a cultural history which is founded upon a new model of interpretation, Roger Chartier has opened the way for this study, in a sense, since my concern too is with the multiple processes of appropriation and negotiation which constituted the institution as a *single* entity to which individuals could refer, even though they presented different accounts of the existing and future nature, extent, and purpose of the institution.[18]

The practice of natural history itself reflected the concern of eighteenth-century naturalists to explore the processes of political, physical, and moral preservation and improvement. The involvement of scientific and medical practitioners in countless projects for improvement, conceived on a national scale, led them to portray themselves as able managers of the lower social levels. The successes of naturalists' large-scale enterprises during this period would reflect contemporary interest among the educated elite in the projects, publications, and places of natural history. By virtue of the form that natural historical practice (as opposed to other forms of scientific practice) took, the day-to-day activities of naturalists involved them in complex networks which served to create particular forms of order within the worlds they inhabited. Naturalists' orders passed between the natural and the social spheres that they themselves designated.[19] By appealing to the powers operating in one sphere, they could legitimate the work they performed to transform the other, and vice versa, before contemporaries prone to

17 See Schaffer and Shapin, *Leviathan and the Airpump;* Latour and Woolgar, *Laboratory Life,* 68–69; Adi Ophir and Steven Shapin, "The Place of Knowledge: A Methodological Survey," *Science in Context* 4 (1991): 3–22.

18 Roger Chartier, *Cultural History: Between Practices and Representations,* trans. Lydia G. Cochrane (Cambridge: Polity Press, 1988); id., "Texts, Printings, Readings," in *The New Cultural History,* ed. Lynn Hunt (Berkeley: University of California Press, 1989), 156–175.

19 See Bruno Latour, *We Have Never Been Modern,* trans. Catherine Porter (New York: Harvester Wheatsheaf, 1993).

claiming that the ideal social order must mirror the natural. In a series of maneuvers, naturalists invented literary, material, and bodily technologies to domesticate rulers, noble patrons, other naturalists, the poor, useful animals and plants, or specimens.[20]

Eighteenth-century naturalists thus sought to change French society through their scientific practice. But as that society changed, so the social space within which it was possible to practice natural history was itself transformed. The work of Biagioli and Findlen has recently revealed how differently the scientific world appeared when viewed from within the patronage culture of early modern Europe.[21] Between the sixteenth- and seventeenth-century court life experienced by such as Galileo or Aldrovandi and the dying struggles of absolutist France was a considerable distance. Yet patronage still played a crucial role for French naturalists in the age of Enlightenment. My account will chart the transformation and destruction of that system of social operations. As one world of scientific practice collapsed, a new one was being invented in which the naturalists of the Muséum d'Histoire Naturelle and their institution would become the models of natural historical authority for other European practitioners.[22] Such a transformation was not merely operated by the recruitment of naturalists to the broader administrative reforms of late old-regime France, as discussed in Charles Gillispie's invaluable work.[23] It also depended upon the naturalists' ability to present themselves as experts in defining the natural, while simultaneously renegotiating their definitions before a succession of different audiences. Such expertise was in part recruited through considerable transformations in the social status of the naturalist during the period. Before 1793 there was no chair of zoology in existence anywhere in Europe; before the 1760s there were precious few chairs of

20 On technologies of knowledge see Steven Shapin, "Pump and Circumstance: Robert Boyle's Literary Technology," *Social Studies of Science* 14 (1984): 481–520; Lissa Roberts, "The Death of the Sensuous Chemist: The 'New' Chemistry and the Transformation of Sensuous Technology," *Studies in History and Philosophy of Science* 26 (1995): 503–530. On the eighteenth century as an age of classification, see Foucault, *Order of Things*, chapters 2, 3, and 5.

21 Paula Findlen, *Possessing Nature: Museums and Collecting in Early Modern Italy* (Berkeley: University of California Press, 1994); Mario Biagioli, *Galileo, Courtier: The Practice of Science in the Culture of Absolutism* (Chicago: University of Chicago Press, 1993).

22 Nicolaas Rupke, *Richard Owen: Victorian Naturalist* (New Haven: Yale University Press, 1994), 24, 100–101; Sophie Forgan, "The Architecture of Display: Museums, Universities, and Objects in Nineteenth-Century Britain," *History of Science* 32 (1994): 139–162. This was not solely a function of the subsequent rise to prominence of Georges Cuvier, for as early as 1798, the naturalists at Göttingen sought to remodel their collections along the lines adopted at the Paris Muséum d'Histoire Naturelle.

23 Gillispie, *Science and Polity*.

natural history.[24] The Muséum was indeed, in this respect, an innovation, and so were the fully salaried natural history posts which its staff were able to obtain from the revolutionary government.

Naturalists' perceptions of the meaning of the Revolution were in part the same as those of their colleagues at other scientific institutions in late-eighteenth-century Paris, such as the Académie Royale des Sciences. A common concern with national and personal improvement underlay the political engagement of many savants during the Revolution: to write of Bailly, Condorcet, Lacepède, Grégoire, and Fourcroy is to name but a few who passed from scientific organizations to government after 1789.[25] But at the same time, the particular history of the Jardin naturalists is unusual, not merely among naturalists in France as a whole, who suffered considerably during the revolutionary decades, but among savants, most of whom did not fare nearly as well as the Jardin's naturalists. The undoubtedly high credit in which the latter stood through the early years of the Republic did not save them and their institution from the lean years of collapsing currency, wartime shortage, and urban violence, no matter what the rhetoric of their governmental supporters. Nonetheless, the new Muséum would be taken as a model of social harmony and of national utility by many contemporaries. Thus the ability of the Muséum to capture a particularly *Republican* nature was emphasized by the deputy Antoine-Clair Thibaudeau before the Convention Nationale in year III/1795: "Call all men to consider the great and magnificent spectacle of the power of nature, the variety of her productions, and the harmony of her phenomena. She is the source of good laws, of useful arts, of the sweetest pleasures and of happiness . . . The Muséum d'Histoire Naturelle is perhaps the only establishment which has remained intact in the midst of the storms of the Revolution; the destructive hand of Vandals, which has broken so many precious monuments of the arts, has respected the temple of nature."[26] If the Republican Muséum succeeded in representing the sacralization of nature, economy, and utility, at other times, and for other

24 René Taton, ed., *Enseignement et diffusion des sciences en France au XVIIIe siècle* (reprint, Paris: Hermann, 1986); Laurence Brockliss, *French Higher Education in the Seventeenth and Eighteenth Century: A Cultural History* (Oxford: Clarendon, 1987); Gillispie, *Science and Polity.*

25 Nicole Dhombres, *Les savants en révolution, 1789–1799* (Paris: Cité des Sciences et de l'Industrie, 1989), chapter 2. Indeed, some of these individuals, such as Jean-Sylvain Bailly, an astronomer who became the mayor of Paris, and Marie-Jean-Antoine-Nicolas de Caritat, marquis de Condorcet, the permanent secretary of the Académie Royale des Sciences but also a prominent spokesman on public-instruction reform, are probably better known as political figures.

26 Antoine-Clair Thibaudeau, *Rapport fait au nom des comités d'instruction publique et des finances, sur le Muséum national d'histoire naturelle . . . à la séance du 21 frimaire, l'an 3* (Paris: Imprimerie Nationale, year III/1795), 18.

commentators, the same establishment would embody quite different qualities which still allowed it to be seen as an institution devoted to the study of nature.[27] Likewise, the objects of French natural history would change their meaning substantially as natural history moved from being the pursuit of a leisured elite to being an integral part of the centralizing policies of revolutionary regimes.

Like books, natural history collections, whether in cabinets or in botanical gardens, were not merely collections of material objects. They possessed a coherent internal logic which related their contents to each other and to the whole. They were thus spaces which could be "read" as symbolic systems, and guides and systematic texts were published in order to direct visitors to decode appropriately.[28] "Reading" the Jardin was no straightforward act. It might serve as a site for symbolic representations on several levels: revealing royal power over distant lands and hands, displaying the increasing commitment of the Crown administration to medical and utilitarian reform, or demonstrating the central and natural power of the Republic. But it was also a place at which the physical, the material was always to be reckoned with. Gardens everywhere may be designed, but they must also be worked on in order to create and maintain that design. The choice of historical focus adopted by some recent cultural historians implies that all human activities can be reduced to texts, language, and symbolism, thus to political discourse.[29] But if historians are forced to resort to highly encoded objects

27 Hans-Christian Harten and Elke Harten, *Die Versöhnung mit der Natur: Gärten, Freiheitsbäume, republikanische Wälder, heilige Berge, und Tugendparks in der französischen Revolution* (Reinbek: Rowohlt, 1989), 57–63.

28 Denis Cotgrove and Stephen Daniels, eds., *The Iconography of Landscape: Essays on the Symbolic Representation, Design, and Use of Past Environments* (Cambridge: Cambridge University Press, 1988), 1; Paul Holdengräber, "'A Visible History of Art': The Forms and Preoccupations of the Early Museum," *Studies in Eighteenth-Century Culture* 17 (1987): 107–117; Eilean Hooper-Greenhill, *Museums and the Shaping of Knowledge* (London: Routledge, 1992), chapter 1; Bennett, *Birth of the Museum,* chapter 1; Jean Baudrillard, "System of Collecting," in *Cultures of Collecting,* ed. Elsner and Cardinal, 7–24; id., *The System of Objects,* trans. James Benedict (London: Verso, 1996). For models of reading, see Susan R. Suleiman and Inge Crosman, eds., *The Reader in the Text: Essays on Audience and Interpretation* (Princeton: Princeton University Press, 1980); Stanley Fish, *Is There a Text in This Class? The Authority of Interpretative Communities* (Cambridge: Harvard University Press, 1980); Robert C. Holub, *Reception Theory: A Critical Introduction* (London: Methuen, 1983); Chartier, "Texts, Printings, Readings"; Adrian Johns, *The Nature of the Book: Print and Knowledge in the Making* (Chicago: University of Chicago Press, 1998).

29 In particular Keith Michael Baker, introduction to *Inventing the French Revolution: Essays on French Political Culture in the Eighteenth Century* (Cambridge: Cambridge University Press, 1990); see also essays in Hunt, *New Cultural History.* Similar criticisms of the new cultural history, for the case of the Versailles garden, are in Chandra Mukerji, *Territorial Ambitions and the Gardens of Versailles* (Cambridge: Cambridge University Press, 1997), 36. A more differentiated model of culture is offered in Chartier, *Cultural History,* which emphasizes the contested nature of texts and discourses.

to pursue their historical inquiries, they need not abandon the possibility of speaking of anything besides such encodings. In so doing it is very easy to suppress those individuals and objects which do not generate texts, or which are excluded from particular discursive domains. The "history of meanings" sought by cultural historians sometimes risks suppressing the labor involved in making and then policing these meanings. Moreover, it ignores what must be crucial to a study of natural history conceived as practice: the labor involved in fashioning the recognizable object of scientific study from the raw stuff of everyday life. The making of gardens, buildings, dissections, collections, classifications, and texts, the transportation of objects from distant places to the Jardin, the creation and capture of posts, the recruitment of audiences, and the legitimation of knowledge-claims were all integral to natural history.

The acts of representing and of encoding natural objects thus formed part of a single continuous fabric of practice that composed what the Jardin/Muséum was for its staff, its rulers, and its visitors. Picturing natural historical practice as a network enables a consideration not only of the naturalists themselves, but also of the many different participants—day laborers, gardeners, customs officers, diplomats, curators, guards, students, family members, ministers, fashionable visitors, and the king—whose acts constituted natural historical knowledge in the late-eighteenth-century Jardin. Collections and greenhouses, herbaria and journals, become as important in this view of natural history as theoretical developments or changes in linguistic practices. Like eighteenth-century natural historical practices, this book is eclectic. It is not a political, social, or intellectual history of natural history, nor a history of ideas in society, nor a history of nature in politics. It borrows from recent developments in the social study of knowledge, but is far from subscribing all-inclusively to the language and concerns of these. Rather, it co-opts a variety of strategies and methods in fashioning instruments with which to explore natural history. Clearly, it should be seen as a partial treatment of its subject, not least because the avenues of inquiry opened by my approach are too many to be addressed in one short book. Certain specific issues and problems are touched upon only minimally, particularly chemistry and the theory of the earth, prominent aspects of the Jardin's natural historical enterprise, but too substantial to be treated in addition to my other themes.

I shall begin by discussing the ways in which the business of doing natural history went on at the Jardin between the publication of the *Histoire naturelle* in 1749 and Buffon's death in 1788. This account is set in the wider context of savants' attempts to negotiate social standing and good character through linking their expertise over the natural world to a new program for moral and rational government, while coping with the mechanics of the patronage system.[30] In chapter 1, then, I set the scene for the dismantling of patronage by discussing the problems of moral management which allowed the Jardin to function as a scientific establishment of the absolutist monarchy. In chapter 2 I study the correspondence and seed-exchange networks of the Jardin's head gardener, André Thouin. By working to increase the flow of specimens and money into the Jardin du Roi, Thouin rendered the Jardin and himself central to an economy of social and natural historical credit. Good practices in natural history resembled the precepts for good management laid down by writers on commerce and finance; the growth of opportunities for naturalists to serve as state consultants during this period meant that natural history increasingly became a science of the natural economy. How best to manage and control that economy was a primary concern for naturalists, whose social status was validated by their managerial role as consultants to an improving state. The form and direction of improvement, however, reflected the interests of those who controlled the meaning of the natural.

The cultural meanings of animals and plants and the changes that their bodies might experience are the subject of chapter 3. Here I consider the links between the concerns of naturalists employed at the Jardin in the 1780s and the project for natural history developed by Buffon and Louis-Jean-Marie Daubenton from the 1740s. By importing new accounts of the operations of the economy of nature produced by natural philosophers during the mid- to late-eighteenth century, French naturalists re-created natural history as a radical account of the relationship between humans and the natural world. In chapter 4 I examine the role played during the Revolution by the Jardin's naturalists in the breakdown of a patronage society. The transformation of organization, spokesmanship, and social status was closely linked with the problems of distribution and legitimation of power which were problematized in the new political world. Different strategies of association would be explored by the naturalists as they struggled to find an appropriate way of presenting themselves as "new men" before a succession of ruling groups.

These issues are taken up in chapter 5. Natural history at the Jardin du Roi differed from the experimental sciences, according to its bota-

30 Schaffer and Shapin, *Leviathan and the Airpump,* 332.

nists and zoologists, in offering a direct and unmediated sensory access to Nature. A delicate task faced naturalists in their attempts to construct a natural order. They needed to be able to remove evidence of their labor from the knowledge that they produced, to minimize the artifice in their accounts of Nature. Here I seek to draw an analogy between those strategies of self-effacement and the problems of inventing a public order which republican reforms of public instruction between 1793 and 1795 sought to address. Together with the fine arts, natural history would become an instrument for generating a self-educating, self-policing nation. The Muséum d'Histoire Naturelle would become a privileged center for the portrayal of a republican future to French citizens.

CHAPTER ONE

The Place of *Histoire naturelle* at the Jardin du Roi

> One day early in the eighteen hundred and sixties, I, being then a small boy, was with my nurse, buying something in the shop of a petty news-agent, bookseller, and stationer in Camden Street, Dublin, when there entered an elderly man, weighty and solemn, who advanced to the counter, and said pompously, "Have you the works of the celebrated Buffoon?"
>
> My own works were at that time unwritten, or it is possible that the shop assistant might have misunderstood me so far as to produce a copy of *Man and Superman.* As it was, she knew quite well what he wanted; for this was before the Education Act of 1870 had produced shop assistants who know how to read and know nothing else. The celebrated Buffoon was not a humorist, but the famous naturalist Buffon. Every literate child at that time knew Buffon's Natural History as well as Esop's Fables. And no living child had heard the name that has since obliterated Buffon's in the popular consciousness: the name of Darwin.
>
> George Bernard Shaw, *Back to Methuselah: A Metabiological Pentateuch*

The name of Georges-Louis Leclerc de Buffon is scarcely known today. It is hard to believe that it was a name which to many defined what "natural history" was for over a century, from 1749, when Buffon published the first three volumes of his best-seller, the *Histoire naturelle, générale et particulière, avec la description du Cabinet du Roi,* until the 1860s, when, if Shaw is to be believed, the name of Charles Darwin superseded that of Buffon in fame. From the time when Buffon took up his post at the Jardin in 1739 until his death in 1788, the intendant worked to further his natural historical project. That project was manifest in two major material objects: the *Histoire naturelle* and the Jardin du Roi.

The completion of the *Histoire naturelle* and the improvement of the Jardin were continued after Buffon's death by his protégés. The Jardin's conversion from a small medical garden, which served largely to teach student physicians and apothecaries, into the world's foremost center for natural history between 1800 and 1830, had commenced much earlier, but during Buffon's fifty years of intendancy, the establishment was wholly transformed.[1] On all occasions the intendant worked to reduce or even eliminate the medical function of the Jardin and to portray the establishment as a center of scientific activity, as Charles-François de Cisternay Dufay, his predecessor, had done. Although medical teaching continued—the Jardin served the students of the *école de médecine,* some minutes' walk down the road, and the trainee apothecaries—there was a significant change taking place in the establishment's function. Many of the Jardin's naturalists were trained physicians, but among the most prominent were several who had no medical training at all. Louis-Claude Bourdelin was the first professor of chemistry, in 1743, not to receive the added duty of pharmacy in his title. In 1745 Louis-Jean-Marie Daubenton took up the post which had previously held the title of "guard of the drugs cabinet." However, in Daubenton's brevet—the official document that served as a bond of loyalty, a job description, and a certificate of good morality for officeholders of the Crown—he was described as "guard and demonstrator of the cabinet of natural history and of the royal garden." This in itself was highly significant: it was only in the mid–eighteenth century that the Crown would explicitly acknowledge the practice of natural history as being the principal function of a royal establishment. Buffon also obtained the formation of several new posts: Daubenton's cousin Edmé-Louis was employed as the "guard and subdemonstrator of the king's natural history cabinet" in 1766, and Barthélémi Faujas de Saint-Fond, Buffon's long-term correspondent, as an "adjunct to the guard of the cabinet of natural history" in 1787.[2] The post of official correspondent to the King's Garden, created in 1773, replaced the vaguer title of "king's naturalist" that Buffon had previously negotiated to have conferred upon individuals undertaking a voyage which might yield useful specimens for the collections.[3]

Defining natural history as a scientific enterprise in its own right, independent of medicine, meant inventing a new expertise over the natural world which could be demonstrated to peers, patrons, and the

1 Rio Cecily Howard, "Guy de La Brosse: The Founder of the Jardin des Plantes in Paris" (Ph.D. dissertation, Cornell University, 1974).

2 Laissus, "Le Jardin du Roi." Brockliss, *French Higher Education,* and Laurence Brockliss and Colin Jones, *The Medical World of Early Modern France* (Oxford: Clarendon, 1997), suggest that a close relationship continued to exist between the medical and natural historical constituencies.

3 AN, AJ/15/510, piece 364.2.

wider scientific audience. Such expertise did not exist a priori for naturalists in the mid-eighteenth century. It was made possible by the existence of a highly complex patronage society which coexisted, and to a great extent overlapped, with a growing commerce of knowledge. Naturalists generally participated in both of these. Some, such as Bernard-Germain-Étienne de La-Ville-sur-Illon, comte de Lacepède, who assisted Buffon in the completion of the *Histoire naturelle* after 1784, were independently wealthy. But even Lacepède valued the public acknowledgment of savant status and the toehold in the royal hierarchy that came with a post at a royal establishment. Those who were not part of the royal patronage system celebrated their independence in print, but usually, unless they had other patrons, eked out a precarious existence selling natural history as entertainment for a polite public: prints, books, and specimens.[4] Buffon converted his financial capital and the social and scientific credit he gained from the *Histoire naturelle* into posts for naturalists within the royal hierarchy. These resembled other royal posts; they were lifelong, and their value resided more in the contacts they gave the holder with higher-placed patrons than in the income attached to them, which was often not enough to support a gentlemanly lifestyle. Thus the Jardin was a part of the patronage structures of eighteenth-century French society.

Patronage to some extent determined contemporary evaluations of Buffon and his natural historical project. However, two other complex social systems for evaluating writings about the natural world flourished during this period. The first was the world of print: since the *Histoire naturelle* was adjudged a literary masterpiece, it became the possession of all readers, and they freely used their right to comment upon Buffon's character as exemplifying the moral, scientific, and stylistic claims they found within the work.[5] Many of these assessments also reflected Buffon's high visibility in polite society; as a philosophe, he attended numerous polite salons in the 1750s and 1760s, and his presence was

4 E. C. Spary, "The Nut and the Orange: Natural History, Natural Religion, and Republicanism in Late Eighteenth-Century France," unpublished paper; id., "'Nature' of Enlightenment."

5 A thorough study of readings of the *Histoire naturelle* is sorely lacking, but see Wolf Lepenies, *Das Ende der Naturgeschichte: Wandel kultureller Selbstverständlichkeiten in den Wissenschaften des 18. und 19. Jahrhunderts* (Munich: Hanser Verlag, 1976); Paul-Marie Grinevald, "Les éditions de l'*Histoire naturelle,*" in *Buffon 88: Actes du colloque international pour le bicentenaire de la mort de Buffon (Paris, Montbard, Dijon, 14–22 juin 1988),* ed. Jean-Claude Beaune, Serge Benoît, Jean Gayon, and Denis Woronoff (Paris: Vrin, 1992), 631–637; Franck Bourdier, "Principaux aspects de la vie et de l'oeuvre de Buffon," in *Buffon,* ed. Roger Heim (Paris: Muséum National d'Histoire Naturelle, 1952), 15–86.

the object of both complimentary and critical comments, the common currency of salon life. As contemporary accounts reveal, he was considered one of the "great men" of French society, routinely named alongside Diderot, d'Alembert, Rousseau, Helvétius, and others.[6] But unlike many other philosophes, Buffon's activities stretched well outside the realms of literary life, thanks to his involvement with the Jardin du Roi and his growing credit with the Crown. It is testimony to the largely underexplored integration of scientific activity within polite society in mid-eighteenth-century France that sectarian sniping by other philosophes such as Marmontel or Grimm could subsequently hit home in Buffon's reputation as a scientific practitioner. Many assessments of his natural history as unscientific or artificial stemmed from the pens of individuals who are not today viewed primarily as naturalists, such as Voltaire, Condorcet, and the Genevan philosopher Charles Bonnet.[7] However, it is also important to stress that, broadly speaking, the scientific reputation of Buffon, the Jardin, and the *Histoire naturelle* were not successfully or very publicly challenged until the French Revolution. The interruption of the patronage system upon which the patterns of employment at the Jardin and the assessments of moral character of naturalists depended opened the way for new sorts of readings, and ultimately the rejection, of the *Histoire naturelle* as a piece of scientific writing and of Buffon as a scientific practitioner. The Jardin du Roi, on the other hand, never lost its credibility as a site for the making of true natural knowledge. I shall present here only a brief outline of the *Histoire naturelle*'s reception insofar as it contributed to contemporary evaluations of the Jardin du Roi as a site for natural historical practice.

6 Jean-François Marmontel, *Mémoires: Édition critique établie par John Renwick,* 2 vols. (Clermont-Ferrand: G. de Bussac, 1972), 119, 139; Maurice Tourneux, ed., *Correspondance littéraire, philosophique et critique par Grimm, Diderot, Raynal, Meister, etc. revue sur les textes originaux, comprenant, outre ce qui a été publié à diverses époques, les fragments supprimés en 1813 par la censure; les parties inédites conservées à la Bibliothèque ducale de Gotha et à l'Arsenal à Paris,* 16 vols. (Paris: Garnier frères, 1877–1882), 2:303.

7 Charles Bonnet, "Observations sur quelques auteurs d'histoire naturelle," in *Correspondance littéraire,* ed. Tourneux, 4:163–171; Marmontel, *Mémoires,* 224–225; Marie-Jean-Antoine-Nicolas Caritat, marquis de Condorcet, "Éloge de M. le Comte de Buffon," *HARS* 1788/1790, 50–84 (also separately published in 1790); Lepenies, *Das Ende der Naturgeschichte.* John Lyon and Phillip R. Sloan, *From Natural History to the History of Nature: Readings from Buffon and His Critics* (Notre Dame: University of Notre Dame Press, 1981), give translations of numerous reviews of the *Histoire naturelle,* but, for example, the magistrate, censor, and minister Chrétien-Guillaume Lamoignon de Malesherbes's extended attack upon the philosophical foundations of the work did not appear in print until 1798. On salon life see Dena Goodman, *The Republic of Letters: A Cultural History of the French Enlightenment* (Ithaca, NY: Cornell University Press, 1994); Alan Charles Kors, *D'Holbach's Coterie: An Enlightenment in Paris* (Princeton: Princeton University Press, 1976), 12, 95–96.

Natural history at the Jardin in the mid-eighteenth century also represented Buffon's attempt to lay claim to the sorts of natural knowledge that were being produced at the Académie des Sciences, where he had recently failed to be elected to the most powerful post, that of permanent secretary.[8] Here his most formidable enemy, Réaumur, had ruled supreme over natural history since the 1710s, and had constructed a large network of supporting naturalists around Europe, including Charles Bonnet and Abraham Trembley. So the Jardin's version of natural history was not the only one in existence even in Parisian scientific institutions. Yet during the half century of Buffon's intendancy, his name would come to be almost synonymous with natural history; and it was Buffon's vision of nature which was drawn upon by Jean-Jacques Rousseau and referred to by countless revolutionary legislators, and upon which the foundations of a new natural history in France were laid. The purpose of this chapter is to consider Buffon, the *Histoire naturelle,* and the Jardin as the three instruments for forging a natural historical program which will be discussed in detail in subsequent chapters, and to demonstrate the extent to which natural history's eighteenth-century preeminence in general, and the Jardin naturalists' preeminence in particular, depended on the relations of Buffon with his protégés and patrons.

DEFINING THE NATURALIST

The fame of the naturalists at the Muséum d'Histoire Naturelle in the early nineteenth century depended to a considerable degree upon the vast resources in terms of specimens, travelers, and social and scientific credibility that had been amassed by their predecessors at the Jardin du Roi in the eighteenth century.[9] Naturalists set about recruiting social and natural historical credit by fashioning natural history as a distinct program of inquiry, appealing to a number of different audiences. They offered the monarchy a policed setting in which royal glory could be displayed. For improving ministers they posed as scientific experts with the skills to enrich the nation, while before polite society they presented themselves as authoritative connoisseurs of the natural. Finally, they aligned themselves with other savants in claiming to share the desire to validate scientific practice as a socially significant enterprise in its own right.

8 Hahn, *Anatomy of a Scientific Institution.*

9 Limoges, "Development of the Muséum d'Histoire Naturelle."

Even by the mid-1780s it was rare for an individual to make a living from natural history. Professorial chairs in natural history began to appear in France only during the 1770s, such as that founded at the Benedictine college of Sorèze in 1778, or that occupied by Louis-Jean-Marie Daubenton at the Collège Royal from 1784.[10] Out of all the nearly 1,800 correspondents with whom André Thouin, the Jardin's head gardener from 1764 onward, maintained an exchange, only a tiny fraction were referred to in his files as "chimiste," "naturaliste," or even "botaniste." Natural history in late-eighteenth-century France, then, was an activity whose practitioners depended on other sources for their social status and income. This was as true of the naturalists employed at the Jardin as it was of those who engaged in natural history outside a recognizable "institutional" context. The demonstrator of botany, Antoine-Laurent de Jussieu, inherited a substantial sum of money from his uncle, the physician and former Jardin naturalist Bernard de Jussieu, and practiced medicine on his own account, as did his colleagues Antoine-François de Fourcroy, Louis-Guillaume Le Monnier, and Daubenton.[11] The growing audience for natural history made it possible to profit from lecturing, curating, collecting, and teaching. Public lecturers like Fourcroy and Jacques-Christophle Valmont de Bomare were able to add to their income through natural historical and chemical lecture courses starting in 1780 and 1756, respectively.[12] Print was also a popular means of self-presentation for naturalists, and Buffon's *Histoire naturelle* created new opportunities for a writerly industry supplying a growing audience for natural history works, such as Pierre-Joseph Macquer and Henri-Gabriel Duchesne's *Manuel du Naturaliste* of 1770. Naturalists with a salaried post, often in the house of a nobleman, were expected to act as tutors, or to collect on behalf of the employer and arrange the collection fittingly.[13] Others managed botanical gardens and public cabinets throughout Europe.[14] Many who engaged in natural historical practice possessed independent means; but most of those have

10 Robert Lemoine, "L'enseignement scientifique dans les collèges bénédictins," in *Enseignement et diffusion,* ed. Taton, 101–123.

11 Condorcet, "Éloge de M. de Jussieu," *HARS* 1777/1780, 94–117; *NBU,* s.vv. "De Jussieu," "Daubenton," "Le Monnier"; Louise Audelin, "Les Jussieu: Une dynastie de botanistes au XVIIIe siècle, 1680–1789" (dissertation, École des Chartes, Paris, 1987); for Fourcroy see David M. Vess, *Medical Revolution in France, 1789–1796* (Gainesville: University Presses of Florida, 1975).

12 *DSB,* s.vv. "Fourcroy," "Valmont de Bomare."

13 Cuvier, for example, began his working life as a tutor to the son of a noble Norman family. See Outram, *Cuvier,* chapter 2.

14 Thouin's "Etât" refers to François Bonami in Nantes, Antonio José Cavanilles in Madrid, William Forsyth at Chelsea.

since been ejected from the realms of "scientific" natural history by institutional naturalists' attempts to shift the boundaries of natural historical expertise.[15]

Finally, and increasingly during the reign of Louis XVI, naturalists began to be recruited by the Crown as consultants. This process reveals the extent to which their claims of hegemony over the natural world were becoming accepted in government circles: decades of appointing ministers as honorary academicians had paid off. But the act of serving as a consultant consolidated that expertise, especially within the public domain. Throughout the second half of the eighteenth century, the state liaised with savant bodies—the Académie Royale des Sciences, the agricultural societies, the Jardin du Roi, and the Observatoire. Savants were often employed as administrators and consultants, as for example in the royal manufactures of Sèvres and Gobelins, the Ponts et Chaussées, the countless inspectorates of fisheries, *eaux et forêts,* manufactures, and hospitals, of the old regime. Successful savants held simultaneous posts in different royal organizations or commissions. Macquer, the Jardin's professor of chemistry from 1777 to 1784, was consultant chemist to the Sèvres royal porcelain manufacture after 1766.[16] From 1766, Daubenton was commissioned by the ministerial dynasty of the Trudaines to study the best ways of improving French sheep.[17] During the 1780s, he served on two of the most important royal commissions at the Académie Royale des Sciences, examining the plan to reform the Hôtel-Dieu by building four new hospitals on the outskirts of Paris, and the plan to distance the slaughterhouses from the center of Paris.[18] The perpetual secretary of the Paris Société Royale de Médecine, Félix Vicq d'Azyr, was recruited in Crown attempts to halt the spread of disastrous epidemics among livestock. André Thouin was one of a committee hurriedly assembled from the members of the

15 Such as the *président* Laurent du Chazelles at Metz, or Paul Demidoff in Moscow (ibid.).

16 *DSB,* s.v. "Macquer."

17 Louis-Jean-Marie Daubenton, "Mémoire sur l'amélioration des bêtes à laine," *MARS* 1777/1780, 79–87.

18 Joseph-Marie-François de Lassone, Louis-Jean-Marie Daubenton, Jacques-René Tenon, Jean-Sylvain Bailly, Antoine-Laurent de Lavoisier, Pierre-Simon Laplace, Charles-Augustin Coulomb, and Jean D'Arcet, "Rapport des Commissaires chargés, par l'Académie, de l'examen du Projet d'un nouvel Hôtel-Dieu," *MARS* 1785/1788, 2–110; id., "Rapport des commissaires chargés par l'Académie de l'examen du projet d'un nouvel Hôtel-Dieu," *HARS* 1786/1788, 1–12; id., "Deuxième Rapport des commissaires chargés par l'Académie de l'examen des projets relatifs à l'état des quatre hôpitaux," *MARS* 1786/1788, 13–41; id., "Rapport des mémoires et projets pour éloigner les Tueries de l'intérieur de Paris," *MARS* 1787/1789, 19–43. Daubenton's involvement with this powerful "core group" of academicians, which also included the *agronome* Mathieu Tillet, is poorly studied, but may bear upon his importance as a patron late in life.

Société Royale d'Agriculture to prepare advice on dealing with a hailstorm which devastated crops in July 1788.[19] Buffon's earliest administrative post had been as manager of the tree nursery set up with Crown funding at his own estate near Montbard, along with the *agronome* Henri-Louis Duhamel du Monceau, who ran one in Denainville on his brother's estate.[20]

Natural historical practitioners, even those now revered in the historiography as the "great naturalists" of the eighteenth century, depended upon the links of a patronage society, but also upon a wealthy, leisured elite which could afford the cost of amassing collections, and of employing naturalists to curate them. New natural historical knowledge was produced in the gardens and cabinets of those who possessed the most splendid natural resources. Since the origins of museums and cabinets, these owners had been the very rich, nobles and princes; however, by the eighteenth century, natural history collections were coming to be the property of gentlemen and the well-to-do.[21] When natural history became fashionable in 1730s France, its popularity had a great deal to do with the fact that it was a representational science. The visible manifestations of scientific practice displayed in cabinets and gardens could confer status, both moral and economic, upon the owner. Collections could demonstrate the owner's power to move objects at a distance, to control the rare and the unusual as perfectly as possible. Thus problems of audience and display are central to our understanding of the shaping of natural historical knowledge in the eighteenth century. The basis for royal support for the Jardin lay as much in its value as a showcase and spectacle as in its role in producing useful knowledge through teaching or research. In June 1746, Buffon applied to Jean-Frédéric Phélypeaux, comte de Maurepas, the minister of the king's household, for the transferal of twelve cabinets and a table encrusted with precious stones from the Salle des Gardes at the Académie Royale des Sciences "to adorn the king's natural history cabinet." These were lavish items: one was described as a "large cabinet in ebony framed with

19 Caroline C. Hannaway, "The Société Royale de Médecine and Epidemics in the *Ancien Régime*," *Bulletin for the History of Medicine* 46 (1972): 257–273; André Thouin, Charles-Germain Bourgeois, and Antoine-Augustin Parmentier, *Avis aux Cultivateurs dont les Récoltes ont été ravagées par la Grêle du 13 Juillet 1788. Rédigé par la Société Royale d'Agriculture, et publié par ordre du Roi* (Paris: Imprimerie Royale, 1788).

20 *Correspondance,* 1:22. Buffon's nursery was founded in 1736, enlarged in 1741, and suppressed in 1777.

21 For a selection of studies exploring this transition, see Findlen, *Possessing Nature;* Hooper-Greenhill, *Museums and the Shaping of Knowledge;* Oliver Impey and Arthur MacGregor, eds., *The Origins of Museums: The Cabinet of Curiosities in Sixteenth- and Seventeenth-Century Europe* (Oxford: Clarendon, 1985); Krszysztof Pomian, *Collectionneurs, amateurs, et curieux: Paris, Venise, XVIe–XVIIIe siècle* (Paris: Gallimard, 1987).

tin in three parts, the first decorated with six large Corinthian columns of lapis with bases and capitals of gilded copper, in the middle, a door with its principal façade decorated with two Corinthian columns in marble of Sicily with bases and capitals of gilded copper."[22]

Both Buffon and his patrons viewed such items as appropriate showcases for the royal natural history collection; the sumptuous nature of the king's Cabinet was echoed in those belonging to members of the royal family and wealthy nobles. The same concerns appeared in the kinds of specimens that were displayed by Buffon and Daubenton in the Cabinet. Published guides dwelt upon the more showy pieces in the royal collection for the delight of the connoisseur: "The mineral kingdom adorns the second room with that which is most rich and most brilliant in nature. One can admire here precious morsels, such as ores of gold, silver, copper, tin, lead, and iron."[23] Judging by such guides, the "public" which entered the Jardin and the Cabinet on Tuesday and Thursday afternoons was eager to see pomp and splendor. Indeed, such concerns shaped the expert observer's response to the Cabinet.[24] Visibility was, consequently, a motivating factor in the reforms at the Jardin—or, at least, an adequate excuse for Buffon to make requests which resulted in such royal pronouncements as that of 1766, where, because the "most complete suite [of specimens] in Europe cannot be put into view [*mis en evidence*], nor exposed to the examination of foreign and domestic savants . . . , his Majesty prefers to take, to augment the same cabinet, the lodging which the Sieur de Buffon occupies at the same royal garden."[25] The palaces, cabinets, and gardens of royal life were sites of display, whose value resided in their visibility. The *contrôleur des bâtiments du roi,* Charles-Claude Flahaut, comte de La Billarderie d'Angiviller, was in charge of the Jardin and other establishments of the sciences, but from the mid-1770s he was also charged with opening up the royal art collections as displays of moral virtue and monarchical supremacy to a broad public.[26] The collection served to display the patron's power, so that naturalists needed to maximize

22 AN, AJ/15/512, piece 492: "Etat des Cabinets et Tables incrustez de différentes pierres précieuses . . . ," April 25, 1748.

23 Dezallier d'Argenville, *La Conchyliologie,* 201.

24 See E. C. Spary, "Forging Nature at the Republican Muséum," unpublished paper.

25 AN, AJ/15/507, piece 161, September 5, 1766.

26 James A. Leith, *The Idea of Art as Propaganda in France, 1750–1799: A Study in the History of Ideas* (Toronto: University of Toronto Press, 1965), 77–79; Andrew McClellan, *Inventing the Louvre: Art, Politics, and the Origins of the Modern Museum in Eighteenth-Century Paris* (Cambridge: Cambridge University Press, 1994), chapter 2. On the apogee of monarchical display in absolutist France, see Peter Burke, *The Fabrication of Louis XIV* (New Haven: Yale University Press, 1992).

its striking visual qualities. But it simultaneously served to provide naturalists with witnesses for their claims about the natural world. As Denis Diderot, probably using Daubenton's notes, would write in the *Encyclopédie* article "Histoire naturelle": "One has found the means of shortening and flattening the surface of the earth in favor of the naturalists; one has assembled individuals of every species of animals and plants, and examples of minerals in cabinets of natural history. Here one sees productions from all the countries in the world, and so to speak an abridgment of the whole world."[27] Buffon's careful construction of a kingly collection was also a glittering pleasure palace for polite society. Only the wealthy could afford his *Histoire naturelle;* only they would have the education and experience required to employ the aesthetic standards of collecting, to be able to place a value upon a seal, an engraving, or a shell, or even to know just how rare a natural production or savage artifact was. Thus the *experience* of wealth was necessary to make relative evaluations of financial and aesthetic merit. The ability to comment critically upon a collection was a matter of education.[28] Although, apparently, little record of the Jardin's visiting public remains, the demonstrator of chemistry, Antoine-Louis Brongniart, kept a register of those who attended his lecture courses between 1777 and 1791, several of which were held at the Jardin. This reveals that the courses were frequented by the elite: nobles and the clergy were heavily, sometimes predominantly, represented.[29] By the mid-eighteenth century, natural history collections were no longer the exclusive preserve of rulers. The Jardin du Roi was still the King's Garden, as it had been since its foundation. But the literate, urban well-to-do from all estates formed a growing audience for scientific display in cabinets and experiments, and for scientific works which combined education with entertainment, works which were a little risqué because their authors verged on political and religious heterodoxy, and which yet offered their read-

27 Denis Diderot and Jean Le Rond d'Alembert, eds., *Encyclopédie, ou Dictionnaire raisonné des sciences, des arts et des métiers, par une société de gens de lettres* (Paris: Briasson, 1765), 8:225–230. Certain claims in this article seem very close to statements made by Daubenton in his later course of comparative anatomy at the Collège de France, although the style seems to owe more to Diderot.

28 Pomian, "Marchands, connaisseurs, curieux à Paris au XVIIIe siècle," in *Collectionneurs, amateurs, et curieux,* 163–194.

29 AN, AJ/15/509, piece 274. In his "Second cours de Phisique février 1778," Brongniart recorded eighteen students, including four marquises, two barons, two chevaliers, one comte, one lord, and two abbés. He earned 840 livres, equivalent to two-thirds of his annual salary at the Jardin as demonstrator of chemistry. For scientific audiences in France, see Jean Torlais, "La physique expérimentale," and Yves Laissus, "Les cabinets d'histoire naturelle," both in *Enseignement et diffusion,* ed. Taton, 342–384, 619–645.

ers a new politics of active citizenship through the exploitation of natural resources. Indeed, most savants were themselves part of that same polite culture.

THE *HISTOIRE NATURELLE*

Buffon and Daubenton used their joint history of quadrupeds, the first part of the *Histoire naturelle,* to take advantage of that developing audience for enlightenment. The first three volumes were published in 1749, in a quarto format, richly illustrated with engraved and sometimes hand-painted plates, including those illustrating Daubenton's anatomies, which were often luridly colored.[30] The work was ambitious, as much in style and comprehensiveness as in its professed epistemological and disciplinary goals. Its initial discourse was written by Buffon, and the original version had been presented at a meeting of the Académie Royale des Sciences some years before. This programmatic statement has generally been read as a text written by a naturalist for other specialists.[31] But it was also Buffon's first address to his wider audience, and he started with the *uninitiated* reader.

Rather like the first man, the newcomer to natural history would open his or her eyes upon a confusing variety of natural productions. True knowledge of Nature and natural laws would emerge only gradually, through repeated observation which at the same time would develop remarkable qualities within the individual, such as "force of genius," "courage," "taste"; the opposing abilities to grasp everything "at a glance" and to attend to minutiae with painstaking labor. Like his contemporaries, Buffon laid stress upon both the utility of natural historical knowledge for the worldly man, and its charm and emotive appeal. Among those useful qualities was natural history's didactic value. The study of natural history benefited the young, Buffon argued, because it reduced their self-love and brought home the extent of their ignorance. But, above all, "even a light study of Natural History will elevate their ideas, and give them knowledge of an infinity of things

30 On the publication and illustration of the first part of the *Histoire naturelle,* see Alain M. Bassy, "L'oeuvre de Buffon à l'Imprimerie Royale, 1749–1789," in *L'art du livre à l'Imprimerie Nationale* (Paris: Imprimerie Nationale, 1973), 171–189; Grinevald, "Les éditions de l'*Histoire naturelle.*"

31 Phillip R. Sloan, "The Buffon-Linnaeus Controversy," *Isis* 67 (1976): 356–375; Lyon and Sloan, introduction to *From Natural History to the History of Nature;* Jacques Roger, *Buffon: Un philosophe au Jardin du Roi* (Paris: Fayard, 1989); Georges-Louis Leclerc de Buffon, *Oeuvres philosophiques de Buffon,* ed. Jean Piveteau (Paris: Presses Universitaires de France, 1954), vii–xxxviii.

which the ordinary man is ignorant of, and which can often be found useful in life."[32]

Here was a powerful dual appeal to the audience which Buffon was attempting to recruit to the *Histoire naturelle.* In a society in which a substantial number of commoners were using wealth to buy or marry their way into nobility and out of the undifferentiated third estate, Buffon presented natural history as a route to social superiority, a form of distinction from the "ordinary man."[33] As the rise of many philosophes from obscurity to the glittering circles of salon life reveals, in the mid-century knowledge, self-cultivation, and worldly acumen were coming to be valued as marks of social superiority almost as much as rank and connections.[34] Buffon was not alone in ascribing to natural history the power to develop appropriate qualities in the young, while demonstrating the possession of exceptional qualities in its practitioners. One predecessor for the *Histoire naturelle* was the abbé Noël-Antoine Pluche's *Le Spectacle de la nature,* published from 1732, which offered a distinctly natural theological version of natural history.[35] Edmé-François Gersaint, a merchant of antiquities and paintings, discovered a whole new market in natural history specimens in the early 1730s, which he ascribed to the popularity of Pluche's work.[36] Although Buffon probably sought to capitalize upon that success, the *Histoire naturelle* proved, if anything, even more popular. In a survey of private libraries during the latter half of the eighteenth century, Mornet revealed the *Histoire naturelle* to be the third most commonly owned work, despite its length and undoubted expense.[37] The first edition sold out in a matter of

32 "Discours," in Buffon, *Oeuvres,* 8; see also "Des sens en général," in *Histoire naturelle,* vol. 3, chapter 8 (1749). Such arguments for undertaking study were comparatively new in England, according to Steven Shapin, "A Scholar and a Gentleman: The Problematic Identity of the Scientific Practitioner in Early Modern England," *History of Science* 29 (1991): 279–327, esp. 295–296.

33 David D. Bien, "Manufacturing Nobles: The Chancellerie in France to 1789," *Journal of Modern History* 61 (1989): 445–486; Guy Chaussinand-Nogaret, *The French Nobility in the Eighteenth Century: From Feudalism to Enlightenment,* trans. William Doyle (Cambridge: Cambridge University Press, 1985).

34 On philosophes and merit see Peter Gay, *The Enlightenment: An Interpretation,* 2 vols. (London: Weidenfeld and Nicolson, 1967); also comments concerning the exclusivity of salon circles in Robert Darnton, *The Literary Underground of the Old Régime* (Cambridge: Harvard University Press, 1982); id., *Mesmerism and the End of the Enlightenment in France* (Cambridge: Harvard University Press, 1968); Goodman, *Republic of Letters,* 5.

35 Noël-Antoine Pluche, *Le Spectacle de la Nature, ou entretien sur les particularités de l'histoire naturelle, qui ont paru les plus propres à rendre les jeunes gens curieux, et à leur former l'esprit,* 2d ed., 8 vols. (Paris: Veuve Estienne, 1732–1751).

36 Laissus, "Les cabinets d'histoire naturelle."

37 Daniel Mornet, "Les enseignements des bibliothèques privées, 1750–1780," *Revue d'histoire littéraire de la France* 17 (1910): 449–496.

weeks, and the publisher, Panckoucke, labored to bring out another in time to satisfy demand.[38]

Buffon himself simultaneously represented and appealed to upwardly mobile, managing, moral, educated individuals. But the popularity of the *Histoire naturelle* brought other consequences for its principal author. A considerable hagiographic literature grew up around Buffon, even during his own lifetime. He was described in glowing terms in works such as Nicolas Ponce's *Les illustres Français,* in the writings of his successors, and in periodical publications.[39] The appearance of engravings or portraits of him and his family in the 1760s to 1780s was recorded in the press; of the portrait by the well-known salon painter, François-Hubert Drouais, the *Mercure* noted in 1761 that "[t]he Public has long been demanding the Portrait of M. *de Buffon* from the printers of his *Histoire naturelle.*"[40]

Such engravings were collected by Buffon's acolytes. The *Histoire naturelle* itself was a touchstone for statements about animals and man, cited in British satirical cartoons attacking man-midwifery ("A Man-*Mid*-Wife, or a newly discover'd animal, not Known in Buffon's time") and used as an authority by Louis-Sébastien Mercier in the *Tableau de Paris.*[41] It saved the comte de Tilly from utter, unfashionable boredom at the house of his provincial uncle, being almost the only book possessed by that worthy.[42] Lacepède read it in sublime solitude as a youth, looking over the Garonne plain, and Georges Cuvier colored in the pictures.[43] Readers had the same rapturous, personal response to the *Histoire naturelle* as Robert Darnton describes for Rousseau's work.[44] At Buffon's death, in a letter to André Thouin, one of Buffon's chief protégés, the cultivator Gérard wrote: "Knowing the regard in which

38 Bassy, "L'oeuvre de Buffon," 172–173.

39 See also Jean-Louis Ferry de Saint-Constant, *Génie de M. de Buffon, par M.**** (Paris: Panckoucke, 1778); Bernard-Germain-Étienne de La-Ville-sur-Illon, comte de Lacepède, "Éloge du Comte de Buffon," in *Histoire naturelle des serpens* (Paris: Hôtel de Thou, 1789), 1–8.

40 *Mercure de France,* October 1761, 163. Other Buffon portraits were mentioned in the *Mercure* for April and October 1775, July 1777, and throughout the 1780s, especially after his death in 1788. Daubenton, by contrast, was mentioned only once during this period, when the bust by Lecomte was exhibited at the Salon of 1783. See Franck Bourdier, "Buffon d'après ses portraits," in *Buffon,* ed. Heim, 167–180.

41 S. W. Fores's cartoon "Man-mid-wifery exposed," 1793, is reproduced in Adrian Wilson, *The Making of Man-Midwifery: Childbirth in England, 1660–1770* (London: UCL Press, 1995), 4; see also [Louis-Sébastien Mercier], *Tableau de Paris, critiqué par un solitaire du pied des Alpes,* 3 vols. (Nyon: Natthey, 1783).

42 Chaussinand-Nogaret, *French Nobility,* 77–78.

43 Roger Hahn, "L'autobiographie de Lacepède retrouvée," *Dix-huitième siècle* 7 (1975): 49–85; Outram, *Cuvier,* 19.

44 Robert Darnton, *The Great Cat Massacre and Other Episodes in French Cultural History* (Harmondsworth, Middlesex: Penguin, 1985), chapter 6.

he held you I judge your sensibility by that which I have experienced, although I only knew M. de Buffon through his works, he belonged to the whole universe whose masterpiece he was."[45]

Such approval was not unmixed. Some, like Friedrich Melchior Grimm, charged the *Histoire naturelle* with being a work of imagination.[46] That particular criticism has received much attention, since it served as an important weapon with which natural history was dismantled during the nineteenth century. But it is far from clear that such criticisms always met their mark during Buffon's lifetime; there were almost certainly more members of the French elite who agreed with Ferry de Saint-Constant and Lacepède that Nature, site of the sublime, *ought* to be described in poetic terms.[47] Style was an important part of the *Histoire naturelle*'s function as a patronage tribute to the king, and it was no coincidence that Buffon was elected to the Académie Française in 1753.[48]

For the *Histoire naturelle* was not just a work of natural history, nor an appeal to a literate elite concerned with self-cultivation. It was also—indeed primarily—a product of the patronage system, containing as it did the description of the king's Cabinet. It was prefaced by a lengthy dedication to the king, from Buffon and Daubenton. In its rich illustrations and binding, the *Histoire naturelle* demonstrated royal

45 BCMHN, MS 1975, letter 898: Gérard to André Thouin, May 8, 1788.

46 *Correspondance,* 1:109.

47 Ferry de Saint-Constant, *Génie,* xii; Louis-François Métra regarded literary quality as an essential element in forming natural knowledge, "puisqu'aucune observation ne peut lever nos doutes dans cette recherche [knowing the first laws of nature]" (*Correspondance secrete, politique et littéraire ou mémoires pour servir à l'histoire des Cours, des Sociétés et de la Littérature en France, depuis la mort de Louis XV,* 18 vols. [London: John Adamson, 1788], 9:155). My argument runs counter to a historiographical tradition which has sought to rehabilitate Buffon by denying his literary importance; see, e.g., Louis Roule, *L'histoire de la nature vivante d'après l'oeuvre des grands naturalistes français,* 6 vols. (Paris: Ernest Flammarion, 1924–1932), vol. 1, *Buffon et la description de la nature,* 9, 242–243 (1924); likewise, Roger, *Buffon,* 14, claimed that "[t]oo often Buffon has only been viewed as a writer, and I have tried to react, no doubt excessively, against that tendency." An otherwise excellent treatment of Buffon as author nonetheless sides with those who, in the 1780s, were trying to ensure that the natural sciences were nonliterary subjects (Lepenies, *Das Ende der Naturgeschichte*). Such a view is problematized by the work of Wilda A. Anderson, *Between the Library and the Laboratory: The Language of Chemistry in Eighteenth-Century France* (Baltimore: Johns Hopkins University Press, 1984); Mary Terrall, "Salon, Academy, and Boudoir: Generation and Desire in Maupertuis's Science of Life," *Isis* 87 (1996): 217–229.

48 The "Discours prononcé à l'Académie Française par M. de Buffon, le jour de sa réception," commonly known as the "Discours sur le style," is still used by French schools as the subject of *dictées* (reproduced in Buffon, *Oeuvres,* 500–509). Not all praised Buffon's style: Malesherbes complained of his use of "the disjointed phrases and the cropped style [which] are presently the fashionable style" (*Observations de C.-G. Lamoignon de Malesherbes sur l'histoire naturelle générale et particulière de Buffon et Daubenton,* 2 vols., ed. Paul Abeille [Paris: Pougens, year VI/1798], 1:190).

Figure 1 The Drouais portrait of Georges-Louis Leclerc de Buffon, which readers of *Histoire naturelle* collected. Engraving by C. Baron, after F.-H. Drouais, 1761. From *Histoire naturelle, Supplément* (1774), vol. 1. By permission of the Syndics of Cambridge University Library.

power to move objects at the ends of the earth. Likewise, the Jardin itself served to display royal power to polite viewers and thus, in one respect, conferred upon them a political role by demanding their assent to visions of monarchical glory.[49] The success of the *Histoire naturelle* and the Jardin as locations for representing natural history depended upon Buffon's ability to excel in two spheres—first, in the appeal to the self-perceptions of a fashionable readership in the fast-growing print culture of mid-eighteenth-century France, and second, in the negotiation of patronage relations within Crown hierarchies. In the last fifty years of the old regime, the monarchical hierarchy remained the focus of the aspirations of many well-to-do families. Insofar as the *Histoire naturelle* had a message for Buffon's natural historical rivals, it was one derived from Buffon's explicit commitments to the imported English philosophy. The admirer of Newton was also a supporter of John Locke's attacks upon essentialist accounts of nature.[50] Unlike contemporary English natural philosophers, though, Buffon decried the type of natural theological explanations advanced by Réaumur and his

49 Dominique Poulot has likewise suggested that d'Angiviller's salon programs of the 1770s implied a public which already assented to monarchical power ("Le Louvre imaginaire: Essai sur le statut du musée en France, des Lumières à la République," *Historical Reflections* 17 (1991): 184).

50 Sloan, "Buffon-Linnaeus Controversy."

circle.[51] Attempts to use living nature to prove God's creative powers and perfect goodness in having ordered the world as it was were, he claimed, invalid.

Because of his rejection of outright appeal to God as natural cause, Buffon was viewed by some contemporaries as attempting to advance a covert pantheism under a royal *privilège.*[52] Buffon's skeptical stance was both an attack against the competing natural theology of Réaumur, his institutional rival, and a philosophe's attempt to position the new moral authority of nature within the framework of a patronage society. But the *Histoire naturelle* appealed to both camps; read with approval as natural theology by some, it was equally approved or denounced as an atheistic work by others.[53] The early volumes were published at a time when interest in the writings of the philosophes was very intense, but while Crown censors were still clamping down upon any work that threatened the maintenance of the moral and political status quo.[54] Among those with whom Buffon commonly associated were many who had suffered at the hands of the censors. Besides Diderot and François-

51 In a letter to a Florentine natural philosopher concerning the debates about generation that preoccupied naturalists after 1740, Buffon attacked Haller and Bonnet for "imagining that [the system of preexisting germs] is connected with religion" (*Correspondance,* 1:327–331, letter 259, November 8, 1776). See also Aram Vartanian, "Trembley's Polyp, La Mettrie, and Eighteenth-Century French Materialism," *Journal for the History of Ideas* 11 (1950): 259–286; Shirley A. Roe and Renato G. Mazzolini, eds., *Science against the Unbelievers: The Correspondence of Bonnet and Needham, 1760–1780* (Oxford: Voltaire Foundation, 1986). Compare English uses of Newton to support natural theology, especially Marina Benjamin, "Elbow Room: Women Writers on Science, 1790–1840," in *Science and Sensibility: Gender and Scientific Enquiry, 1780–1845,* ed. Marina Benjamin (Oxford: Basil Blackwell, 1991), 27–59; Larry Stewart, *The Rise of Public Science: Rhetoric, Technology, and Natural Philosophy in Newtonian Britain, 1660–1750* (Cambridge: Cambridge University Press, 1992), part 1; Neal C. Gillespie, "Natural History, Natural Theology, and Social Order: John Ray and the 'Newtonian Ideology,'" *Journal of the History of Biology* 20 (1987): 1–49.

52 Lyon and Sloan, *From Natural History to the History of Nature,* 235–252; John Pappas, "Buffon vu par Berthier, Feller, et les *Nouvelles ecclésiastiques,*" *Studies on Voltaire and the Eighteenth Century* 216 (1983): 26–28; Thomas Royou, *Le monde de verre reduite en poudre, ou Analyse et réfutation des Epoques de la nature de M. le comte de Buffon* (Paris: n.p., 1780); for Scotland, Paul B. Wood, "The Natural History of Man in the Scottish Enlightenment," *History of Science* 27 (1989): 89–123. On the Enlightenment as secular enterprise, see Gay, *Enlightenment,* vol. 2, *The Rise of Modern Paganism.*

53 The best-known account of Buffon as materialist, which remained in manuscript for many years, was [Marie-Jean Hérault de Séchelles], *La visite à Buffon, ou Voyage à Montbard* (Paris: Solvet, 1801). What Hérault de Séchelles published at the time of his visit in 1786 was a more innocuous piece, approved by Buffon himself: "Parallèle de Jean-Jacques Rousseau et le comte de Buffon," *Journal Encyclopédique* 3 (1786): 329–330. Compare the natural theological sentiments of the comtesse de Beauharnais, "Aux incrédules. Épitre envoyée à M. le Comte de Buffon," *Journal de Paris,* November 7, 1778; [Ane (possibly Charles-Joseph Panckoucke)], *De l'homme, et de la reproduction des differens individus. Ouvrage qui peut servir d'introduction & de défense à l'Histoire naturelle des animaux par M. de Buffon* (Paris: n.p., 1761).

54 Darnton, *Literary Underground.*

Marie Arouet de Voltaire, there was the salon circle of Paul Thiry, baron d'Holbach, himself a secret materialist, which was frequented by Claude-Adrien Helvétius, author of a scurrilously Hobbesian *De l'Esprit,* and the naturalist Charles-Georges Leroy, known for his materialist account of animal habits in the *Lettres sur les animaux,* published in the first half of the 1760s. Madame d'Holbach was a close friend of Madame de Buffon. Buffon was thus close to many of the materialists and atheists on whom the lieutenant general of the Paris police force kept dossiers.[55] But he never got into such deep waters. As the king's intendant, hoping to publish a description of the king's Cabinet, with the king's approval, he was not in a position to publicize radical philosophical speculations. Volume 4 of the *Histoire naturelle* contained a public recantation—mocked by some as being a spoof—of the heretical statements that the Sorbonne's theologians had uncovered in the first three volumes.[56] Moreover, he criticized his peers for overly unsubtle approaches to radical issues. The young *parlementaire* Marie-Jean Hérault de Séchelles anonymously circulated an account of a visit to Montbard in which Buffon was reported as having said that "[o]ne should never clash in the forefront, as Voltaire, Diderot, Helvétius did."[57] Buffon dropped out of the social circle of d'Holbach and Helvétius when they became too radical and was consequently attacked by the writer Jean-François Marmontel.[58]

Modern commentators have emphasized the more controversial aspects of the *Histoire naturelle,* as for example the theory of the earth, in which Buffon speculated about a much extended timescale for the earth's history, thus conflicting with Scriptural accounts.[59] And indeed, such claims drew attacks from theologians. But what of the aspects of the *Histoire naturelle* which it was possible to relate to the personal

55 On Buffon's social circles see *Correspondance,* passim; Roger, *Buffon;* Louis Bobé, ed., *Mémoires de Charles Claude Flahaut, comte de La Billarderie d'Angiviller: Notes sur les mémoires de Marmontel publiés d'après le manuscrit* (Copenhagen: Levin and Munksgaard, 1933); Elizabeth Anderson, introduction to Charles-Georges Leroy, *Lettres sur les animaux* (Oxford: Voltaire Foundation, 1994), 51; on his friendship with Lenoir, AN, AJ/15/514, piece 639; on police surveillance of philosophical writers, see Darnton, *Great Cat Massacre,* chapter 4; id., *Literary Underground;* Kathleen Wellman, *La Mettrie: Medicine, Philosophy, and Enlightenment* (Durham: Duke University Press, 1992).

56 Lyon and Sloan, *From Natural History to the History of Nature,* 283–293.

57 Hérault de Séchelles, *La visite à Buffon.* Buffon often expressed his desire to avoid public controversy, e.g., in *Correspondance,* 1:66–70, letter to abbé Le Blanc, Montbard, March 21, 1750, concerning Montesquieu.

58 Bobé, *Mémoires,* 52; Marmontel, *Mémoires,* 224–225.

59 Lyon and Sloan, introduction to *From Natural History to the History of Nature;* Hans-Jörg Rheinberger, "Buffon: Zeit, Veränderung, und Geschichte," *History and Philosophy of the Life Sciences* 12 (1990): 203–223.

experience of the educated well-to-do in cabinets and gardens? The bulk of the first part of the *Histoire naturelle* was composed of descriptions of animals and varieties of man. Buffon's readers were not being asked to admire, like those of Réaumur, the intricate fashioning of minute body parts as evidence of divine skill. Instead, as we will see in chapter 3, they were being presented with continual glimpses of the links between nature and society. In much of his writing, Buffon imported the moral world and social concerns which he shared with many of his readers into the natural world, and there converted them into prescriptive claims about human appropriation of nature which were faithfully represented in the natural history collections of the Jardin du Roi, with their emphasis on the human value of specimens. But, paradoxically, he was also silent on the providential design of nature for human use. Buffon's nature was not the pristine source of virtue envisaged by Rousseau as being corrupted by human contact. Instead it was constantly degenerating, falling into gradual decay and ultimate death. In the *Histoire naturelle,* Buffon offset this negative trend with man's ingenuity and ability to combat degeneration through the advancements that the sciences, the arts, and society could bring.[60] Labor lay at the heart of Buffon's political perceptions, both in the Jardin and on his estate, where he replaced traditional charity with projects to put his poor tenants to work; in his later years he allied himself with the paternalistic style of the Necker ministry, becoming an intimate friend of Suzanne Curchod de Necker, wife to the Swiss banker Jacques Necker who became *contrôleur-général des finances* in 1776.[61]

In portraying scientific enterprise as a means of contributing to national improvement, Buffon was expressing the concerns of many savants in late-eighteenth-century France. But in representing themselves as improving experts, savants needed to create a new value for mental labor, one in which the superiority of mind over hand was emphasized.[62] Such self-portrayals fitted well with the development of middling concerns about work as the legitimation for political existence. Laboriousness offered a scale of evaluation which sometimes mapped onto the patronage hierarchy. When Jean-André Thouin, the head gardener, died, Daubenton wrote to his seventeen-year-old son André

60 "De la Nature. Première Vue," in *Histoire naturelle* (1764), 12:viii; see also Claude Blanckaert, "Buffon and the Natural History of Man: Writing History and the 'Foundational Myth' of Anthropology," *History of the Human Sciences* 6 (1993): 30–31.

61 *Correspondance,* passim.

62 Antoine-Laurent de Lavoisier and Armand-J. Séguin, "Premier mémoire sur la respiration des animaux," *MARS* 1789/1793, 566–581; Louis-Jean-Marie Daubenton, "Mémoire sur les laines de France, comparées aux laines étrangères," *MARS* 1779/1782, 1–2.

Thouin: "You may find some consolation in the good use that you have made of your time, I hope that you will reap the fruit for I have learned that M. de Buffon is well disposed toward you; he likes to protect people who apply themselves to their duties."[63] Daubenton went on to offer his patronage to the young man, in the form of a letter of support on Thouin's behalf to Buffon; André succeeded to the post in 1764. Similarly, in eulogies or *éloges,* which served as expressions of the characteristics of ideal savants, the virtues of laboriousness and zeal were repeatedly stressed.[64] Such concerns persisted in revolutionary discourse as well.[65] But the moral value of labor also underpinned savants' criticisms of the patronage society in which they existed, where social advancement was achieved and validated in quite a different way. The appeal to laboriousness could be a way of distancing savant activities from the leisured pursuit of natural history recommended in polite manuals. Denunciations of the "corruption" of sinecures, favoritism, and nepotism were polemics for a different moral code, and not, as many historians have implied, unbiased reflections upon the state of society.[66] In part, such criticisms were the outcome of the success of a relatively new account of intellectual merit as the proper basis for good government. In polite society, social change was mediated through patronage relationships, and this form of government was still largely in place until the Revolution.[67]

63 Yvonne Letouzey, *Le Jardin des Plantes à la croisée des chemins avec André Thouin, 1747–1824* (Paris: Muséum National d'Histoire Naturelle, 1989), 49: Daubenton to Thouin, February 3, 1764.

64 Charles Paul, *Science and Immortality: The Éloges of the Paris Academy of Sciences* (Berkeley: University of California Press, 1980); Dorinda Outram, "The Language of Natural Power: The Éloges of Georges Cuvier and the Public Language of Nineteenth-Century Science," *History of Science* 16 (1978): 153–178.

65 But see the conclusions of Jay M. Smith, *The Culture of Merit: Nobility, Royal Service, and the Making of Absolute Monarchy in France, 1600–1789* (Ann Arbor: University of Michigan Press, 1996).

66 See, e.g., Albert Soboul, *La France à la veille de la Révolution,* vol. 1, *Economie et société* (Paris: Société d'Enseignement Supérieure, 1966); George Rudé, *The French Revolution* (London: Weidenfeld and Nicolson, 1988). David D. Bien, "Offices, Corps, and a System of State Credit: The Uses of Privilege under the Ancien Régime," in *The French Revolution and the Creation of Modern Political Culture,* vol. 1, *The Political Culture of the Old Regime,* ed. Keith Michael Baker (Oxford: Pergamon, 1987), 89–114, and Robert Darnton, *The Business of Enlightenment: A Publishing History of the Encyclopédie, 1775–1800* (Cambridge: Harvard University Press, Belknap Press, 1979), 72, both map the distinction between good and bad officeholders onto criteria of activity and efficiency. Such concerns, however, themselves probably originated as a particular political critique of court elites. See Norbert Elias, *The Court Society,* trans. Edmund Jephcott (Oxford: Blackwell, 1983); Goodman, *Republic of Letters.*

67 Most work on patronage or clientage centers on the sixteenth and seventeenth centuries: e.g., Peter Burke, *The Italian Renaissance: Culture and Society in Italy* (Cambridge: Polity Press, 1987); id., *Fabrication of Louis XIV;* Roland Mousnier, *The Institutions of France under the Absolute Monarchy, 1598–1789,* trans. Brian Pearce and Arthur Goldhammer, 2 vols. (Chicago: University

Patronage Power Plays

Eighteenth-century educated individuals needed to compete for patronage in order to gain advancement. Patronage stretched downward throughout society in a network of complex ramifications which controlled the getting of finances and posts. While making swift social mobility possible for some, patronage could also cause enormous frustration and bitterness among individuals who were less successful in playing the patronage game.[68] The Revolution would place particular demands upon savants to generate new accounts of the nature of the scientific life, in response to changing political circumstances which destroyed the validity of patronage as a political system.[69] Such pressures problematize the use of accounts of personal relationships, careers, and feelings, such as autobiographies and *éloges*, as historical sources.[70] These texts reflected either the careful negotiation of associations within a patronage society, or the subsequent need to present oneself as independent of such immoral ties. Good character was mediated along the lines of patronage relationships, and proposals for the reform of government by Enlightenment writers, including savants, simultaneously entailed new programs of moral conduct.

Not all savants wished to abandon patronage as a necessary consequence of Enlightenment; some, like Buffon, were particularly adept at welding the new morality of Enlightenment and the framework of patronage in order to pursue the sciences. Among the major changes at the Jardin was the doubling of the establishment's surface area in a series of reforms between 1778 and 1788. As Falls shows, Buffon profited by several hundred thousand livres from the land transactions undertaken during the enlargement, chiefly by buying the sites adjacent

---

of Chicago Press, 1979–1984); Sharon Kettering, *Patrons, Brokers, and Clients in Seventeenth-Century France* (New York: Oxford University Press, 1986); Biagioli, *Galileo, Courtier;* Findlen, *Possessing Nature.*

68 See, e.g., Hahn, *Anatomy of a Scientific Institution,* chapter 6, on Marat's revolutionary attacks against the Académie des Sciences, which had earlier rejected his natural philosophical work. My account, in which scientific activity within institutions was the outcome of patronage negotiations rather than "professionalization," is closer to Outram, *Cuvier,* than to earlier literature such as Roger Hahn, "Scientific Research as an Occupation in Eighteenth-Century Paris," *Minerva* 13 (1975): 501–513; id., "Scientific Careers in Eighteenth-Century France," and Maurice P. Crosland, "The Development of a Professional Career in Science in France," both in *The Emergence of Science in Western Europe,* ed. Maurice P. Crosland (London: Macmillan, 1975), 139–160, 127–138.

69 Hahn, "L'autobiographie de Lacepède retrouvée"; Dorinda Outram, "The Ordeal of Vocation: The Paris Academy of Sciences and the Terror, 1793–1795," *History of Science* 21 (1983): 251–273.

70 As in Outram, *Cuvier.*

to the Jardin with his own capital, and then receiving handsome returns from the Crown—the selling price plus a bonus on top. Falls plainly considers Buffon's actions in this matter to have been of a somewhat shady nature.[71] But the picture of Buffon as corrupt manager and embezzler deserves to be reevaluated in the light of his position in the patronage networks. Every transaction and request for money regarding the Jardin had to be approved by the ministers of the king's household and of his finances.[72] Buffon's honesty vis-à-vis the king and his ministers cannot have been in question. The sums of money that were being made available to Buffon were of a different kind. One might say that they were returns on an investment. Buffon's financial practice at the Jardin followed a well-established pattern for the redistribution of credit in Old Regime society: he gave financial credit and received social credit.[73] In pledging what was legally his own property, or his own lodgings, to the king, he was also playing by the rules of a society in which personal and royal property were not always linguistically distinct. Even in little ways, Buffon ran the Jardin like a speculation, riding the wave of ministerial support for the sciences and arts in Louis XV's and Louis XVI's reigns; it provided his lands in Montbard with vegetables and fruit trees from the 1760s onward, and his forge in Montbard supplied the iron railings that were erected around the new boundaries of the Jardin after its enlargement.[74]

Effectively, the patronage system allowed the delegation of power from the king to able administrators, who always acted in the name of the

71 William P. Falls, "Buffon et l'agrandissement du Jardin du Roi à Paris," *Archives du Muséum d'Histoire Naturelle,* series 6, 10 (1933): 131–200. See also Léon Bertin, "Buffon, homme d'affaires," in *Buffon,* ed. Heim, 87–104; Beaune et al., *Buffon 88,* section 1, esp. Françoise Fortunet, Philippe Jobert, and Denis Woronoff, "Buffon en affaires," 13–28. Contemporary debates concerning the morality of financial probity and efficiency are superbly explored in J. F. Bosher, *French Finances, 1770–1795: From Business to Bureaucracy* (Cambridge: Cambridge University Press, 1970).

72 Anna Raitières, "Lettres à Buffon dans les registres de l'ancien régime, 1739–1788," *Histoire et nature* 17–18 (1980–1981): 85–148. Attacks on Buffon's honesty originated with an untitled, printed petition of 1790, addressed to the king and signed by Verdier, Delaune, and Picquenard, three dispossessed tenants of the former Abbaye de Saint-Victor lands engulfed by the Jardin's 1780s enlargement project. But this document also suggests that, according to the Old Regime's ministerial and police authorities, Buffon's actions were not illegal, or even dishonest.

73 Bien, "Offices, Corps, and a System of State Credit." Bosher, *French Finances,* chapter 5, argues that the Crown depended upon a system of loans. Daniel Roche, *The People of Paris: An Essay in Popular Culture in the Eighteenth Century,* trans. Marie Evans and Gwynne Lewis (Leamington Spa: Berg, 1987), 82–83, shows that even small wage earners were investing in royal *rentes* prior to the Revolution.

74 AN, AJ/15/149: "Mr. Thouin. Dépenses pour le Jardin, 1760 à 1793." In 1765 Thouin recorded the cost of sending three dozen tubers, 300 stands of asparagus, fig trees, and vegetable seeds to Montbard.

king, but had a considerable degree of freedom—of *licence*—in how they interpreted the king's desires within their own sphere of authority. The advantages of such a system worked both ways: the king could have influence in many places at once, and the licensee could call on a greater source of authority than himself to force his peers to acquiesce in his views, and those below him to carry out his wishes. The network of royal administrators and inspectors controlled and filtered the demands of royal subjects, and managed the royal business in a rambling hierarchy of power delegation which emanated from the king. Buffon's actions can be compared with the typical practice of administering government concerns as private speculations—if such a task was carried out to the satisfaction of the state, speculative administrators stood to reap large rewards. In return for the king's favor, Buffon would invest his own capital in the Jardin du Roi and carry out reforms and repairs, police staff and specimens . . . in short, he would farm out the Jardin.

At the Jardin, royal and administrative power were to some extent interpermeable. There was no such thing as a "state-employed scientist": by the terms of their brevets, Buffon and his subordinates were not accountable to the faceless bureaucracy of a state for their work, but to the king; the Jardin was, in this respect, literally the King's Garden.[75] The king's will concerning this garden was mediated through ministers who were themselves personally answerable to the king. The individuals who worked in and oversaw the Jardin du Roi were thus connected in an intricate hierarchy, subordinate to the monarch. They were effectively linguistically disempowered, since all events at the Jardin purported to occur as a consequence of the exercise of the royal will. But this did not prevent enterprising individuals from succeeding with ambitious projects which required royal support on many levels: financial, legal, personal, or institutional, for example. Action on behalf of the king enabled action on one's own behalf, through a series of complex maneuvers which Outram has termed "micronegotiations."[76]

75 The term *bureaucratie* was coined in this period but referred, pejoratively, to ministerial secrecy rather than to the neutral administrative state. See Jean Dubois, ed., *Dictionnaire de la langue française* (Paris: Larousse, 1987); Baker, *Inventing the French Revolution,* 160–162. Alternative accounts of patronage at early modern scientific institutions are in Alice Stroup, *A Company of Scientists: Botany, Patronage, and Community at the Seventeenth-Century Parisian Royal Academy of Sciences* (Berkeley: University of California Press, 1990); David J. Sturdy, *Science and Social Status: The Members of the Académie des Sciences, 1666–1750* (Woodbridge, Suffolk: Boydell, 1995); David S. Lux, *Patronage and Royal Power in Seventeenth-Century France: The Académie de Physique in Caen* (Ithaca, NY: Cornell University Press, 1989); Bruce T. Moran, ed., *Patronage and Institutions: Science, Technology, and Medicine at the European Court, 1500–1750* (Woodbridge, Suffolk: Boydell, 1991).

76 Outram, *Cuvier,* 3, and cf. her extended discussion of Cuvier's patronage relations, passim. Her approach, however, confers considerable autonomy upon Cuvier as historical agent.

In one respect, prerevolutionary "politics" resided in the structure of such micronegotiations, which involved all individuals of the French elite in a language of duty, honor, and obligation.

Because of the nature of royal power, all suggestions for improvements to the institution, or legal changes such as the creation of new posts, had to emanate from the king. Accordingly, such suggestions followed a tortuous linguistic route from Buffon or his subordinates, to the ministers, and thence to the king. This was the case for every attempt that Buffon made to produce change at the Jardin: in the case of one abbé Galloys, appointed naturalist to the Jardin in 1763, for example, the request had to go via the minister of the king's household, the duc de La Vrillière, to the king, and then there would be a confirmatory letter to Buffon from his direct patron, the minister, as well as the official permission from the king.[77] One literally needed to "make demands to be made" to the ministers; and the recruitment of royal patronage rested upon a complex language of paternalism which depended upon the king's approval of his subjects.[78]

Honorific language was the vehicle for the kind of business which Buffon transacted within the Jardin and with the Crown. This was no legally binding contract, in which he was promised a certain financial return for his administrative and improving activities at the establishment. Instead, he poured his capital into the institution, as a way of proving his devotion to the Crown; and, in return, "His Majesty desiring more and more to make known the esteem and the particular goodwill with which he honors . . . Sieur Buffon, so celebrated in the Republic of letters and who by his care and the extent of his knowledge has conducted the Cabinet of His Majesty to the degree of superiority where it finds itself today" would periodically give marks of that benevolence, on this occasion as a pension of 6,000 livres a year, 4,000 of which were to revert to Buffon's wife and son on his death.[79] Money and social status were not conferred by the Crown in return for specific services, but in return for a more general "service" of loyalty and submission to the king. Power was delegated in the same way, so that Buffon was officially licensed to administer the Jardin in the king's name by virtue of his personal qualities—birth, morality, zeal, and devotion to

77 Raitières, "Lettres à Buffon"; AN, AJ/15/509.

78 On the changing uses of eighteenth-century patriarchalism, see, e.g., Jay Fliegelman, *Prodigals and Pilgrims: The American Revolution against Patriarchal Authority, 1750–1800* (Cambridge: Cambridge University Press, 1982), chapters 1 and 2; Jeffrey Merrick: "Patriarchalism and Constitutionalism in Eighteenth-Century Parlementary Discourse," *Studies in Eighteenth-Century Culture* 20 (1990): 317–330; Lynn Hunt, *The Family Romance of the French Revolution* (London: Routledge, 1992).

79 AN, AJ/15/507, piece 162.

serving the Crown. More, the remarkable success of the *Histoire naturelle* made it into a valuable patronage gift, and Buffon into an especially desirable protégé; he could represent French—and royal—excellence within the international Republic of Letters.[80]

By virtue of the license which Buffon had obtained from the king, he had considerable control over events and people within the Jardin, whether he chose to intervene personally or whether—as became increasingly common in the last years of his life—he delegated again to his most trusted protégés, André Thouin, the head gardener, or Louis-Jean-Marie Daubenton, his collaborator on the *Histoire naturelle.* However, he himself was not without answerability. On the occasion of the appointment of André Thouin's father, Jean-André, as head gardener to the Jardin in 1745, Maurepas, the then *ministre de la maison du roi,* wrote to Buffon: "You may Monsieur name to the post of Gardener of the Jardin Royal which you inform me is vacant I am persuaded that you will only make a good choice you know that this post requires a careful and faithful man and I strongly trust to the attention that you will give to [the choice]."[81] Buffon was under an obligation to the Crown in choosing which candidate to present to the king as the successor to a vacant post. He was required to present the individual as able, Catholic, faithful, and moral. Incoming officeholders swore an oath of fealty before taking up their posts, which was recorded in their *brevet.* This system continued until 1792, when Étienne Geoffroy Saint-Hilaire's *brevet* for the post of "adjoint à la garde du Cabinet d'Histoire naturelle" contained the same claims as his predecessors', the only concession to the Revolution being the reference to the Jardin as national rather than royal. Such matters might seem like irrelevant remnants of an outdated system of loyalties which collapsed in the French Revolution. However, this system of expressions of love, honor, faith, and loyalty long bound the upper echelons of Old Regime society. Royal *pensions* and *brevets* were awarded as a recompense for good behavior, and the "candidates" could be nominated by individuals already in the king's favor.

Although Buffon appears, as Raitières suggests, to succeed in obtaining "all he wants" from the ministers, nevertheless much work went into that success.[82] Throughout his intendancy, whenever posts became vacant or letters of succession were available, and in obtaining the cre-

80 On the man of letters as international yet patriotic, see Goodman, *Republic of Letters,* chapter 1.

81 Raitières, "Lettres à Buffon," 91–92.

82 Ibid., 85.

ation of new posts, Buffon used every opportunity to advance his own protégés. This is demonstrated from the very start of his career, by the choices of Daubenton as guard to the Cabinet and of Jean-André Thouin as head gardener. Both men came from Montbard in Burgundy, the town where Buffon's château stands. Buffon requested Daubenton to come to Paris as his collaborator on the Jardin and *Histoire naturelle* projects in 1740, obtaining an official post for him five years later. Their families continued to be closely linked. Buffon's son married Daubenton's grand-niece Betzy. Daubenton's brother Pierre served as the family *avocat* for the Buffons, Daubentons, and Thouins for many decades; his son Georges-Louis was Buffon's godchild and Philibert Guéneau de Montbeillard, another *Histoire naturelle* collaborator, was the uncle of Georges-Louis's wife.[83]

Jean-André Thouin was a gardener whom the intendant shipped over to Paris, where he settled in a house within the Jardin grounds, married, and produced a family of four sons and two daughters. The eldest son, André, predictably succeeded his father as head gardener; the second, Jacques, went into financial administration, and became the duc d'Orléans' accountant; the third, Gabriel, became a garden designer; and the fourth, Jean, worked as second gardener at the Jardin, under André. The daughters also benefited: the elder, Marie-Jeanne, married a tutor to Buffon's son, and the younger, Louise, became a servant to the third of the royal children.[84] Many of these posts were respectable advances within the royal hierarchy, demonstrating the power of a patron such as Buffon over the future of an entire family. But the intendant's success in patronage negotiations was never the success of autonomy. Negotiations over power, posts, and people were "played" like games, with their own language and rules. Buffon was far from being the only player in this game; ministers, savants, and relatives of interested individuals all participated in the appointment process.

There was a choice of candidates for the succession to Le Monnier, the Jardin's professor of botany until his retirement in 1786.[85] Le Monnier, a wealthy physician, was appointed to the service of the royal armies in 1770. He then adopted an apparently common, although rarely officially recorded, course of taking a young understudy or *suppléant* to lecture on his behalf at the Jardin. He chose an obvious candidate, the

83 *Correspondance,* 1:110, 215–216.

84 Letouzey, *Jardin des Plantes,* 27–33.

85 Joseph Laissus, "La succession de Le Monnier au Jardin du Roi: Antoine-Laurent de Jussieu et René-Louiche Desfontaines," *Comptes-rendus du 91e congrès national des sociétés savantes, 1966, Section des sciences* 1 (1967): 137–152.

twenty-three-year-old Antoine-Laurent de Jussieu. Antoine-Laurent was the son of Christophle de Jussieu, the eldest of four brothers originating from the second marriage of a successful Lyons apothecary. Christophle remained in Lyons, his son's birthplace. The second brother, Antoine, had held the post of professor of botany at the Jardin until his death in 1758, when it passed to Le Monnier. The third brother, Bernard, had joined Antoine, first as guard to the natural history cabinet, and then, after his displacement by Daubenton, as demonstrator of plants—a less prestigious post than the professorship, with a lower salary. The fourth brother, Joseph, had been ship's surgeon on the La Condamine and Bouguier voyage to Peru to measure the arc of the meridian. Stranded in South America when Crown funding ran out, Joseph spent fifty years there, returning to die at the age of seventy-one, having forgotten his native tongue. Their nephew Antoine-Laurent came from Lyons to join his remaining uncle, Bernard, in 1764.[86]

It was a common practice in French families to "invest" in the eldest son's financial and social status, since his success drew with it the promise of benefit for the rest of the family. Individuals holding a post, particularly those working for the Crown, were both able and expected to use their resulting connections to advance their family members, friends, countrymen, and protégés. *Éloges,* which were standard forms of tribute from protégés, usually stressed the paternal characteristics of the patron. Family relationships were woven into this rhetoric of honor and loyalty that cemented society in a wider sense. Buffon expressed the patriarchal society simply but revealingly: "An Empire, a Monarch, a family, a father, here are the two extremes of society: these extremes are also the limits of Nature."[87] As can be seen from the complexity of interconnections within the Montbard families at the Jardin, these "social limits" were representative of the actual pattern of recruitment from the provinces for many individuals. The award of posts in the Jardin always depended upon personal relationships between individuals: candidates and their patrons, patrons with ministers, and ministers with the king. Such relationships were often familial, geographical, or tutelary.

Antoine-Laurent's family had, then, a lengthy and fruitful relationship with the Jardin du Roi which predated the arrival of the Burgundian faction, headed by Buffon. He was an obvious candidate for the letters of succession for Le Monnier's post—not only did he have pow-

86 *DSB,* s.vv. "Antoine de Jussieu," "Bernard de Jussieu," "Antoine-Laurent de Jussieu," "Joseph de Jussieu"; Audelin, "Les Jussieu."

87 Buffon, "Les animaux carnassiers," in *Histoire naturelle* (1758), 7:3–38.

erful support within the Jardin itself, he had also lectured in Le Monnier's stead from 1770 until 1778, in the periods when the professor's medical duties prevented him from doing so. *Suppléance* was an act that was very often tantamount to obtaining letters of succession in any case. The de Jussieu family was making a bid to reclaim the post which it had lost to an outsider when Antoine died and there was no male relative in a position to leap into the breach. In 1774 de Jussieu was granted permission by Buffon to reorganize the botany school at the Jardin to display his uncle Bernard's natural method of classification; and in 1778 Buffon approved the succession of Antoine-Laurent to Bernard's chair of demonstrator.[88] In spite of these marks of approval, however, Buffon did not look kindly upon the young man's plan to continue lecturing in Le Monnier's stead and to obtain letters of succession to the professorship. As intendant, Buffon had the final say in the negotiation over the succession, despite all Antoine-Laurent's claims of experience and family ties and despite Le Monnier's support. De Jussieu saw the intendant's interest turning toward the gardener of the Jardin des Apothicaires, Jean Descemet, as Le Monnier's successor.

Meanwhile, however, an election at the Académie to fill the post of *botaniste-adjoint* vacated by Brisson seemed to offer new hopes. Although Descemet received the highest number of votes, de Jussieu thought that he detected Buffon's favor toward the candidate lessening, in favor of Jean-Baptiste-Pierre-Antoine de Monet, chevalier de Lamarck, the second in line; and he turned to his patrons Le Monnier and d'Angiviller, both well known at Versailles, to massage this relationship along: Lamarck was "a man of merit and worth infinitely more than [Descemet]." Buffon was induced to use his influence in favor of Lamarck, who obtained the adjunction to the Académie by royal appointment. De Jussieu further encouraged Le Monnier to consider as successor one of his own protégés and friends, René-Louiche Desfontaines, an impoverished Breton studying for his medical *licence*. The emotional relationship of patronage is perfectly expressed in Desfontaines's letter to a friend, the naturalist Savary: "I am very much loved by M. Le Monnier . . . he is an excellent man who has pure and honest morals in the middle of the court, an upright, sensible and beneficent heart; with all these qualities, a curious thing, he wields the greatest credit . . . I am still loved by M. de Jussieu."[89] This passage also reveals how character and public office became mixed up in assessments of the out-

88 Antoine-Laurent de Jussieu, "Exposition d'un nouvel ordre de plantes adoptés dans les démonstrations du Jardin Royal," *MARS* 1774/1778, 175–197.

89 Desfontaines to Savary, September 10, 1779, quoted in Laissus, "La succession de Le Monnier," 148.

come of patronage acts: protégés like Desfontaines helped to make the reputation of the patron. Once Desfontaines had secured his *licence*, Le Monnier was able to obtain a post on a voyage to Barbary for him, which gave him the experience necessary to make a bid for the succession in 1786. This was successful, and it was he who obtained the professorship in 1786, when Le Monnier retired. The processes whereby this dispute over preferment was settled reveal the scope and limitations of maneuvers around official posts. In a patronage society advancement was often a matter of greatest claim. Had Antoine de Jussieu remained alive until his nephew was old enough to succeed to the post, the professorship would have gone to Antoine-Laurent—but, by losing the post, the family also lost its power to determine who should fill it. Since Le Monnier had no natural successors, the question of the succession then spun off into a more open-ended scramble among comparable claimants. Antoine-Laurent succeeded his uncle Bernard as demonstrator, unofficially in 1771, and officially in 1778 at Bernard's death.

A similar problem beset Buffon himself. In 1771 he fell gravely ill, and rapidly negotiated from his sickbed in an attempt to secure the succession of the intendancy to his son, Louis-Marie Leclerc de Buffon. His patrons, however, were unresponsive, arguing that Buffonet (as he was known) was too young, at the age of seven, to receive letters of succession. Buffon's plan backfired: the letters of succession were granted, but to the comte d'Angiviller, and Buffon recovered to live until his son was certainly old enough to have had the succession granted. Nominally the reversion of the post was to pass to Buffonet when he reached the age of twenty-five, but this could not be expressed formally within the text of appointment for d'Angiviller's profession of faith—his "title" to the post. In a telling letter from the *commis* L'Echevin to Buffon in April 1771, the former noted that only d'Angiviller's knowledge of the gravity of Buffon's state had determined him to accept the post in order to transmit it to the son, "otherwise it was lost for him, and one of the factions which wished for [its] union to the post of *premier médecin* would certainly have carried it off . . . M. de Buffon is too just and too honest to require that M. d'Angiviller should only give the appearance of passing into that post and that he should only have occupied himself with it as tutor to his [i.e., Buffon's] son."[90] But the post was nevertheless lost, for when Buffon died in 1788, the intendancy passed to d'Angiviller's elder brother, Auguste-Charles-César de Flahaut, marquis de La Billarderie. The only consolation was that

90 *Correspondance,* 1:202. L'Echevin was one of the *premier commis* of La Vrillière.

d'Angiviller became the *contrôleur-général des bâtiments du roi* in 1778. In this post he had responsibility for the advancement of all things connected with the Jardin, and proved an extremely important patron, especially in supporting the enlargement of the establishment after 1778.

Similarly, Buffon was not always successful in promoting other individuals to posts at the Jardin. On one occasion which was (besides Buffonet) perhaps the biggest defeat of his administrative career, Buffon's plans to install a protégé at the Jardin were defeated by his own patrons. Félix Vicq d'Azyr, a rapidly rising and reform-minded young physician, stood in for the Jardin du Roi's professor of anatomy, Antoine Petit, from March 1775, during the latter's illness. However, Antoine Portal, another young anatomist, had just managed to secure the letters of succession as soon as Petit had announced the possibility of his retirement. Portal was able to achieve this because of his support by the *ministre de la maison du roi* at the time, the duc de La Vrillière. From the surviving letters to Buffon in the old-regime registers, it is clear that he supported Petit in wishing for the reversion of the professorial chair to Vicq d'Azyr instead of Portal. Vicq d'Azyr was Daubenton's student in comparative anatomy and his relation, having married his niece Zoë. Moreover, in his capacity as a physician, Petit had been harshly criticized by Portal in the latter's widely read *Histoire de l'anatomie et de la chirurgie.*[91] Familial ties and corporate conflicts both operated in Vicq d'Azyr's favor. However, Buffon's ultimate answerability, and subordinacy, to the ministers meant that he was unable to push Vicq d'Azyr through. On March 15, 1775, La Vrillière reminded him of this: "Monsieur Portal who has worked in this area for a long time and whose talents you know as well as I do desires to be able to succeed [Petit] and I cannot prevent myself from making known to you the interest which I take in his affairs." On March 31, two weeks later: "Monsieur Petit, Monsieur, having caused a demand to be made to me that Sieur Vicq d'Azir might replace him in his anatomy lectures at the Jardin du Roi until his health should be reestablished I cannot see any problem but always understood that this permission will not give him any right to a post which as you know is promised to another."[92]

91 *Correspondance,* 1:110; Antoine Portal, *Histoire de l'anatomie et de la chirurgie, contenant l'origine & les progrès de ces Sciences,* 6 vols. (Paris: P. Fr. Didot le jeune, 1770–1773), 1:xiij, 5: 389–405. The importance of this attack in determining Petit's opposition to Portal reveals the extent to which the Jardin du Roi continued to function as a property of competing medical groups.

92 AN, AJ/15/509, pieces 238, 234: La Vrillière to Buffon, March 15 and 31, 1775.

Buffon had to capitulate: it was not possible for him to go against the wishes of his patrons, even when they conflicted with his own, as La Vrillière's veiled warning showed.

Chrétien-Guillaume Lamoignon de Malesherbes was appointed to the ministry of the *maison du roi* in mid-1775. Since new patrons might have new interests, Buffon wrote to Malesherbes in December on the Vicq d'Azyr–Portal problem. The content of the letter is not available, but Malesherbes's reply shows that Buffon had suggested some kind of compromise to satisfy both Portal, the official successor, and Petit, the holder of the chair, who supported Vicq d'Azyr and wanted him as a stand-in for his anatomy lectures. Malesherbes suggested, too, that Vicq d'Azyr should be given the succession to Portal.[93] However, Malesherbes's term of office was too brief to allow him to carry out the proposals, and nothing resulted from the correspondence.

Buffon did not give up. When Petit eventually decided to retire, in 1777, the minister of the king's household was a close friend, the former intendant of Burgundy, Antoine-Jean Amelot de Chaillou. With such auspicious patronage conditions, Buffon could not resist one more try to have Portal replaced by Vicq d'Azyr, now the permanent secretary of the Société Royale de Médecine. He apparently prepared an official presentation document for Portal (as a candidate to be proposed to the king to fill the vacant place) but "the first condition that you impose upon Sieur Portal is that of giving the adjunction to his chair at the Collège Royal to Sieur Vic d'Azir [*sic*]." Amelot warned: "It is not possible for me to acquiesce in the arrangement that you propose . . . This adjunction would be absolutely contradictory with the last letters patent issued for the administration of the Collège Royal in 1772. Besides, I will not dissimulate to you that I have different views for the first chair which falls vacant in this Collège."[94]

Subsequent communications from Amelot on Portal and the chair of anatomy and surgery at the Jardin contained no mention of Vicq d'Azyr. Buffon had had to back down. Evidently, it was useful for aspiring savants to acquire more than one patron, and particularly to enlist patrons as high up the patronage scale as possible. Despite Buffon's repeated efforts, almost every biographical source on Vicq d'Azyr, starting with an *éloge* in the revolutionary journal *La Décade philosophique,* portrays Buffon as having opposed Vicq d'Azyr's claims to the Petit chair, a high-handed intendant who preferred favor to merit.[95] Perhaps

93 AN, AJ/15/509, piece 239: Malesherbes to Buffon, December 7, 1775.

94 AN, AJ/15/509, piece 241: Amelot to Buffon, May 24, 1777.

95 Joseph-Jerôme Lefrançois de Lalande, "Éloge de Vicq d'Azir," *La Décade philosophique* 3 (1794): 513–521, 4 (1795): 1–10.

Buffon did not inform Vicq d'Azyr of the reasons for his rejection, or possibly the intendant merely provided a convenient target for criticism of the old regime's scientific institutions. Certainly Portal had far more cause to complain of heavy-handed treatment by Buffon.

Conclusion

The Vicq d'Azyr affair shows that it was possible to misuse the patronage system, and that such misuses were often a way of testing or extending the limits of personal power by one patron.[96] Acts of patronage formed an ongoing process of social negotiation which served to establish the power of patron and protégé simultaneously, as well as the relative power of their bond when compared to other patronage relationships. Thus Vicq d'Azyr's failure did not merely determine his power but also had consequences for the power of others involved in the dispute as well: Petit, Daubenton, Portal, Buffon, Amelot, Malesherbes, and La Vrillière. Patronage relationships linked many individuals in such processes, and the outcome of each negotiation affected future actions by the same individuals; having obtained the post, Portal then had power over his own succession. Patronage was thus a historicizing and historical process in which social (and natural historical) status was continually constructed and reconstructed. Successful advancement, then, was frequently a matter of recruiting the highest-level patron, the one with greatest power to decree the fate of a post, in a given hierarchy. In doing so, and in general during the course of patronage relationships, a great range of resources was deployed. From Buffon's letters to Thouin, it is clear that he himself was continually thinking of ways to flatter and amuse his patrons as part of his attempt to sustain the relationship, as for example with presents and the use of honorific language. In 1780 Thouin's register of accounts listed "a present of ornamental shrubs and plants made by Mr. the Count to M. Amelot."[97] This was at the time of Amelot's ministry. Similarly, Buffon awarded patrons and important correspondents other gifts such as miniature portraits of himself, or gold medals depicting his bust.[98] As his own fame grew, such objects themselves entered the realms of collection and display.

The case of Vicq d'Azyr also offers a cautionary tale regarding histor-

96 Compare the account of Tudor bribery in Steven G. Ellis, *Tudor Ireland: Crown, Community, and the Conflict of Cultures, 1470–1603* (London: Longman, 1985); see also Biagioli, *Galileo, Courtier*.

97 AN, AJ/15/149.

98 *Correspondance*, 2:292–294, Buffon to Thouin, Montbard, August 9, 1785.

ical narrations of scientific character in a patronage society. Buffon was responsible for ensuring that his subordinates possessed the appropriate moral qualities; in the eyes of his own patrons, their moral quality reflected upon him. Likewise, the intendant's moral character was fashioned through patronage texts such as eulogies and letters. The relationship of love and trust that was expressed in exchanges between the fatherly patron and the filial protégé was a form of linguistic government to which all members of the French eighteenth-century elite were subject in making their way through society. However, the characteristic mode of discourse of the patronage society was subsequently interpreted as a straightforward historical account of the temperament of an individual, suppressing the enormous political freight that such characterizations bore. Buffon's activities as a patron are thus often represented by biographers as deeply revealing of his character.[99] Attacks against his personal and scientific credentials, however, often emanated from those in competing patronage networks. Thus Buffon's secession from the company of d'Holbach and Helvétius, and his new allegiances to Mme. Necker and her salon, produced a revengeful literature from individuals such as Condorcet, a protégé of Necker's political opponent Turgot, and Marmontel, a protégé of d'Holbach. The fundamental problem underlying psychobiographical histories of eighteenth-century individuals is clearly evident here. The patronage system was a way of doing business which could not be separated from the construction of character; only after the Revolution was the literature arising from disputes between patronage groups and sociable networks reinterpreted as neutral evaluations of character.

If men of learning in the eighteenth century wished to achieve social advancement, they could do so only by engaging in the exchanges, language, and system of values which characterized the patronage system. In an insightful paper, Robert Darnton once suggested that politics, *politesse,* and policing were three intimately linked problems. In other words, when politics was revised, so too must be the structures of courtesy and politeness which framed old-regime interactions. And so too must be the basis of police. If the king's moral role was the control or policing of events in the public domain, that of his chief of police was the maintenance of moral order in the public realm. Thus the *lieutenant*

99 See Roger Heim, "Preface à Buffon," in *Buffon,* ed. Heim; also Gillispie, *Science and Polity,* esp. 150–151. Goodman, *Republic of Letters,* and Anne Goldgar, *Impolite Learning: Conduct and Community in the Republic of Letters, 1680–1750* (New Haven: Yale University Press, 1995), offer excellent treatments of the patronage networks of men of letters and of the subsequent pressures on philosophes to rewrite the nature of their relationships.

de police Jean-Charles-Pierre Lenoir, Buffon's friend, was in charge of the policing of texts and authors, of property and public places, as well as of the enforcement of enlightened bodily and moral values upon the poor.[100] Buffon and his staff at the Jardin du Roi sought to develop a self-presentation which justified their positions as, in Schaffer's term, "moral managers" of others.[101] Besides the local problem of establishing order within their establishment, the Jardin's managers would endeavor to extend order beyond its boundaries during the last decades of the eighteenth century. The natural history collection was the site at which disciplining people and disciplining specimens became similar problems. While naturalists ordered their specimens, they were simultaneously ordering society, predicating what kinds of natural knowledge were to be trusted and how to ensure the trustworthiness of participants in the natural historical enterprise. Simultaneously, they were engaged in accumulating social credit: good character, political authority, and moral standing. Natural history was based upon a material economy of objects which had to be controlled by a social economy of morals. This economy will be the concern of chapter 2.

100 Darnton, *Literary Underground;* id., "Le lieutenant de police J.-P. Lenoir, la guerre des farines, et l'approvisionnement de Paris à la veille de la Révolution," *Revue d'histoire moderne et contemporaine* 16 (1969): 611–624; Thomas McStay Adams, *Bureaucrats and Beggars: French Social Policy in the Age of the Enlightenment* (New York: Oxford University Press, 1990); George Rudé, *The Crowd in the French Revolution* (Westport, CT: Greenwood Press, 1986); Ludmilla J. Jordanova, "Policing Public Health in France, 1780–1815," in *Public Health,* ed. Teizo Ogawa (Tokyo: Saikon Publishing, 1981), 12–32; George Rosen, *From Medical Police to Social Medicine: Essays on the History of Health Care* (New York: Science History Publications, 1974); the medical program was outlined in *HSRM* 1776/1779. The approach of Dorinda Outram, *The Body and the French Revolution: Sex, Class, and Political Culture* (New Haven: Yale University Press, 1989), has plainly informed my own.

101 Simon Schaffer, "Measuring Virtue: Eudiometry, Enlightenment, and Pneumatic Medicine," in *The Medical Enlightenment of the Eighteenth Century,* ed. Andrew Cunningham and Roger French (Cambridge: Cambridge University Press, 1990), 281–318.

CHAPTER TWO

Acting at a Distance: André Thouin and the Function of Botanical Networks

> Because one cannot see everything oneself, and because one can do so still less, the more extensive and more distant one's commerce, it is necessary to have intelligent, attentive correspondents, whose capacity is recognized; otherwise false advice would engage [one] in ruinous enterprises. It is no less necessary to assure oneself of the exactitude and of the fidelity of all those to whom one confides the guard and the sale of one's wheat; and it is necessary to have men accustomed to transport it, and on whom one can equally count; it is through the concourse of a multitude of agents, always in movement, that the circulation of wheat takes place.
>
> Étienne Bonnot de Condillac, *Le Commerce et le Gouvernement, considerés relativement l'un à l'autre*

André Thouin, the Jardin's head gardener from 1764 to 1793, illustrates the problems and opportunities of old-regime patronage. Born at the Jardin in 1747, he was the eldest son of the head gardener, Jean-André Thouin. By moving to Paris, the family's social status was considerably altered. The children were sent out to rural wet nurses, a common practice among the urban elite. When his father died, leaving a widow and six children, André was only seventeen.[1] Increasingly after André succeeded to the post of head gardener, and particularly with the development of a fashion for an idealized simple life, as advocated by Jean-Jacques Rousseau (who visited and corresponded with André), the Thouin household became a model of Rousseauist simplicity and virtue for several members of the elite.[2] Something of the nature of French society of the late eighteenth century is revealed by the fact that the son of a man who worked with his hands, through patronage, could associate with Malesherbes, a landowner and minister from an old

1 Letouzey, *Le Jardin des Plantes,* 26–27.

2 Ibid., 27; *DSB,* s.v. "Thouin"; Outram, *Cuvier,* 171.

parlementaire family.[3] Toward the end of the old regime, the wealthy and powerful were prepared to allow some individuals of comparatively humble origins into their society.

Buffon was undoubtedly Thouin's most important patron. In Thouin's youth Buffon financed his *collège* education, enabling him to serve as an efficient administrator and accountant, but also effectively conferring upon him the skills to participate in polite society. Particularly from 1778 onward, Thouin became the intendant's indispensable second-in-command in the negotiations over the Jardin's enlargement.[4] Meanwhile, he was also establishing a network of correspondents within and outside France, a typical practice of naturalists. In such networks different groups were enrolled to play different but relatively stable roles in the transmission of correspondence and plants. I shall show how the successful maintenance of a correspondence network enabled Thouin to control an economy of specimens centering around the soil of the Jardin du Roi.

Exchanges of correspondence between eighteenth-century naturalists have hitherto attracted little attention from historians of the sciences.[5] At best they have served as a means of dating changes in the theories advanced by individual naturalists, or as a way of determining friendships between "great" figures of eighteenth-century natural history. Yet such textual exchanges occupied a large proportion of naturalists' lives. During 1786, when his network reached its largest extent, Thouin was in correspondence with more than 400 individuals. Such a vast correspondence was not unique, however; other managers of large

3 Malesherbes was Thouin's most assiduous correspondent; the "Etât" records 467 letters from him between 1778 and 1791.

4 Falls, "Buffon et l'agrandissement du Jardin du Roi."

5 See, however, Anne Secord, "Corresponding Interests: Artisans and Gentlemen in Natural History Exchange Networks," *British Journal for the History of Science* 27 (1994): 383–408. For other, briefer analyses of scientific correspondence, see Dorinda Outram, introduction to *The Letters of Georges Cuvier: A Summary Calendar of Manuscript and Printed Materials Preserved in Europe, the United States of America, and Australasia,* ed. Dorinda Outram (Chalfont St. Giles: British Society for the History of Science, 1980), 1–11; J. L. Pearl, "The Role of Personal Correspondence in the Exchange of Scientific Information in Early Modern France," *Renaissance and Reformation* 8 (1984): 106–113. More general accounts of the social function of correspondence are in Janet Gurkin Altman, "Teaching the 'People' to Write: The Formation of a Popular Civic Identity in the French Letter Manual," *Studies in Eighteenth-Century Culture* 22 (1992): 147–180; id., "Political Ideology in the Letter Manual (France, England, New England)," *Studies in Eighteenth-Century Culture* 18 (1988): 105–122; Alain Pagès, "La communication circulaire," in *Ecrire, publier, lire: Les correspondances,* ed. Jean-Louis Bonnat and Mireille Bossis (Nantes: Université de Nantes, 1983), 343–361; Goodman, *Republic of Letters,* chapter 4. The function of correspondence was far more extensive than I am able to suggest here. For additional perspectives, see the collected papers in *Men/Women of Letters,* special issue of *Yale French Studies* 71 (1986).

gardens were in the same situation. Joseph Banks, the naturalist and president of the Royal Society, wrote over fifty letters a day in his Soho townhouse.[6] Because of Thouin's current obscurity, letters written by him are not often extant or catalogued. But his orderliness meant that details of many correspondents found their way into his records, notably into his vast manuscript "Etât de la Correspondance de M. A. Thouin," written in 1791.[7] Additionally, large numbers of his correspondents' letters to him survive. These allow the documentation of salient features of that correspondence which suggest that letters played a far more important role for the whole of the natural historical community throughout Europe and the rest of the world than has previously been argued. The same resources can also be used to suggest how individuals at the center of networks attempted to achieve positions of power by making themselves indispensable to the interests of the groups on whom they depended. One way of analyzing such botanical networks is in terms of the model proposed by Bruno Latour, who suggests that the successful accomplishment of eighteenth-century voyages of exploration enabled the exotic natural world to be constructed as a set of inscriptions stored at European centers of calculation. I shall argue that this perspective needs to be linked both with Simon Schaffer's account of disciplining the vision of scientific observers, and with the language of patronage discussed in chapter 1.[8]

Thouin's centrality as administrator and accountant of the Jardin was reflected in his rising social status. Throughout his career as head gardener, Thouin developed increasing expertise in managing the many-sided economy of the Jardin du Roi, an economy which was simultaneously social, financial, and natural historical. These managerial skills enabled him to increase the Jardin's botanical income considerably by the time of Buffon's death. The Jardin's economy was not primarily an

6 Patrick O'Brian, *Joseph Banks: A Life* (London: Collins Harvill, 1987), opp. 209.

7 BCMHN, MS 314.

8 Simon Schaffer, "Astronomers Mark Time: Discipline and the Personal Equation," *Science in Context* 2 (1986): 115–145; Latour, "Visualization and Cognition"; id., *Science in Action,* chapter 6. My use of Latour's model differs from that of David Philip Miller, "Joseph Banks, Empire, and 'Centers of Calculation' in Late Hanoverian London," in *Visions of Empire: Voyages, Botany, and Representations of Nature,* ed. David Philip Miller and Peter Hanns Reill (Cambridge: Cambridge University Press, 1996), 21–37; and Dirk Stemerding, *Plants, Animals, and Formulae: Natural History in the Light of Latour's* Science in Action *and Foucault's* The Order of Things (Enschede: Universiteit Twente, 1991). On botanical gardens as centers see Lucile H. Brockway, *Science and Colonial Expansion: The Role of the British Royal Botanic Gardens* (New York: Academic Press, 1979), 7; Staffan Müller-Wille, *Botanik und weltweiter Handel: Zur Begründung eines Natürlichen Systems der Pflanzen durch Carl von Linné, 1707–1778* (Berlin: Verlag für Wissenschaft und Bildung, 1999), chapter 7.

economy of money. In eighteenth-century texts *économie* possessed a variety of meanings, all sharing some common ground, and in using the term to describe the function of the Jardin, I wish to avoid its current connotations of a global and inescapable financial structure, as in "the economy."[9] "Économie politique" was being invented during this period as the science in which the laws of the circulation of wealth—in the form of grain, money, and other goods—were studied with the aim of enriching the state. Often this was to be achieved through the harnessing of private interests to the wider public good. Political economists were thus simultaneously occupied with the true laws of society, with the natural laws of human behavior, and with the morality of human motivation.[10]

The related science of rural economy involved the application of the laws of good management to agricultural practice. It thus involved savants in attempts to establish a particular authority over peasants. The natural economy was the balance book of the world, in which the continuity of the natural order was assured; it was generally observed in the global circulation of active principles such as electricity, and in the constant regeneration of species. The study of its laws was, again, the preserve of savants. Likewise, the circulation of principles within the individual body, as well as its powers to increase and replicate itself through generation and regeneration, were known to naturalists as the animal or vegetable economy. Thus the term *économie,* as used by the naturalists of the Jardin du Roi, implied a set of claims about society, nature, and the body as circulating systems subject to rational comprehension and intervention.[11] The Jardin du Roi was certainly an economy

9 Bernard Balan, *L'ordre et le temps: L'anatomie comparée et l'histoire des vivants au XIXe siècle* (Paris: Vrin, 1979), chapter 2; Foucault, *Order of Things,* chapter 6; William Reddy, *Money and Liberty in Modern Europe: A Critique of Historical Understanding* (New York: Cambridge University Press, 1987); Steven Laurence Kaplan, *Bread, Politics, and Political Economy in the Reign of Louis XV,* 2 vols. (The Hague: Nijhoff, 1976); Myles Jackson, "Natural and Artificial Budgets: Accounting for Goethe's Economy of Nature," *Science in Context* 7 (1994): 409–431. These approaches contrast with economic histories which explain debates over state management purely in terms of twentieth-century categories, e.g., James C. Riley, *The Seven Years War and the Old Regime in France: The Economic and Financial Toll* (Princeton: Princeton University Press, 1986).

10 See esp. Catherine Larrère, *L'invention de l'économie au XVIIIe siècle: Du droit naturel à la physiocratie* (Paris: Presses Universitaires de France, 1992).

11 M. Norton Wise, "Work and Waste: Political Economy and Natural Philosophy in Nineteenth Century Britain," part 1, *History of Science* 27 (1989): 263–301; Camille Limoges,"Économie de la nature et idéologie juridique chez Linné," *Actes du XIIIe congrès international d'histoire des sciences 1971* 9 (1974): 25–30; Jean-Marc Drouin, "Linné et l'économie de la nature," in *Sciences, techniques, et encyclopédies,* ed. D. Hue (Paris: Association Diderot, 1991), 147–158; E. C. Spary, "Political, Natural, and Bodily Economies," in *Cultures of Natural History,* ed. Jardine, Secord,

in this sense, for it was the center for a circulation of seeds and letters, where problems of social and natural historical order were to be solved by the good management of people and specimens. In effect, Thouin would import his managerial role at the Jardin into his natural historical practices.

Everything in the Garden's Lovely

In 1780s France, gardens were often portrayed as settings for erotic pleasure. The fashion for the rococo had established a vast industry of pictures portraying lovers before, after, and even during coitus in gardens.[12] In redesigning the Jardin du Roi from 1778 onward, André Thouin had been careful to preserve the romantic aspect of the garden, with its labyrinths, its dovecote, and its secluded bowers.[13] Inevitably, however, this led to problems. A young man who suspected his wife of having an affair followed her to the Jardin one morning, where he found her, indeed with her *galant.* Infuriated, the cuckolded husband promptly attacked the lover, and a fight broke out. Some of the Jardin guards hurried up to the scene of the outrage, and promptly arrested both participants. They were marched off to Thouin, and after some deliberation it was decided that the husband should be locked up in the case containing the skeleton of an elephant dissected by Daubenton after it had died at the Versailles menagerie. As Thouin commented in his fortnightly letter to Buffon, "While this man was sequestrated, the public crowded to the windows and one could say that the elephant, as extraordinary as it must be for the inhabitants of Paris, seemed less so than that surly husband, so little informed of the ways of his country."[14] Such events were not unique. Although few miscreants were actually kept on display, Thouin's records reveal that the problem of

---

and Spary, 178–196. As Müller-Wille, *Botanik und weltweiter Handel,* 315–317, points out for the case of Linnaeus, natural historical economies did not precisely coincide with any particular contemporary political economical theory; nonetheless, both in France and in Sweden, political economists had close relations with naturalists and agricultural reformers. See Sven-Eric Liedman, "Utilitarianism and the Economy," in *Science in Sweden: The Royal Swedish Academy of Sciences, 1739–1989,* ed. Tore Frängsmyr (Canton, MA: Science History Publications, 1989), 23–44; Gillispie, *Science and Polity.*

12 R. G. Saisselin, "The French Garden in the Eighteenth Century: From Belle Nature to the Landscape of Time," *Journal of Garden History* 5 (1985): 284–297; Thomas Eugene Crow, *Painters and Public Life in Eighteenth-Century Paris* (New Haven: Yale University Press, 1985).

13 Letouzey, *Le Jardin des Plantes,* 63.

14 From a letter of 1786, reproduced in William P. Falls, "Buffon et les premières bêtes du Jardin du Roi: Histoire ou légende?" *Isis* 30 (1939): 494.

managing the Jardin's public was an ongoing one. In 1788 Thouin proposed to Buffon's successor, the marquis de La Billarderie, a set of regulations which would give Thouin official power over the actions of the subgardeners in his charge. Besides the fixing of the posts and wages of the subgardeners, Thouin also stressed laws of behavior and moral conduct.

> Article 12.—We expressly forbid . . . the *maîtres-garçons* as well as the other *garçons* and extraordinary workers and day laborers, from giving [or] selling seeds, flowers, plants, or any of the objects cultivated in the Jardin . . . under whatever pretext . . .
>
> Article 13.—Similarly [we] forbid, as much for the decency due to this establishment as for good morals, the *garçons-jardiniers* and other employees from allowing suspect women into the Jardin, from walking with them, from introducing them into their rooms . . . [We] forbid the Porters from letting them in under the pretext of family relationships.[15]

Thouin's proposed reforms were made more urgent by the proliferation of subordinate gardening posts that the massive reforms of the establishment between 1778 and 1788 had necessitated. Where Jean-André, his father, had started out with "one boy and four women" to carry out all the work on the establishment, André had to negotiate with a full-time locksmith, plumber, painter, two grass cutters, two pavers, a sculptor, a glazier, a carpenter, a roof-mender, and a barber, and managed a host of day laborers, the first subgardener or *maître-garçon* mentioned above (his younger brother Jean), a greenhouse gardener, a nurseryman, a gardener for the formal beds, a guard for the stores, and a supplementary gardener.[16] As the establishment grew in importance, then, the administrative problems and responsibilities grew along with it. In the Cabinet the picture was much the same: from a first cleaner appointed in 1745, the number rose to three by the early 1790s.[17] Reflecting their status as royal servants, the cleaners had to wear livery.

The anecdote about the irate husband and the skeletal elephant sharing a cage, however, serves to remind us that the problem of manage-

15 Letouzey, *Le Jardin des Plantes,* 255–256.

16 Ibid., 26, 254; AN, AJ/15/505: Jean-André's work agreement drawn up by Buffon, 1745. By 1793 seven subgardeners (besides Jean Thouin) were listed in the Jardin's outgoings: Jean Moreau, Antoine Poiteau, Louis Hurelle, Gonot, Macée, André Sinet, and the widow of a former gardener, Pallé. There were four "gardes-bosquets," four porters, two cleaners, and two subnaturalists employed to prepare specimens, Fatory (retired) and Valenciennes (AN, F/17/1221).

17 AN, AJ/15/510, piece 353: "Arrêt du conseil pour l'etablissement d'un domestique au Jardin du Roi," January 10, 1747; Letouzey, *Le Jardin des Plantes,* 300, on the three cleaners, Feuillet, Villeduc, and François.

ment was twofold: it was not just people who needed to be controlled, but also specimens. The elephant's skeleton could not be allowed to decay, disintegrate, be consumed by insects or rodents, or be stolen. Louis-Jean-Marie Daubenton and his cousin Edmé-Louis were "guards" in this sense also. The loss of valuable specimens would have indicated the keepers' failure to fulfill their moral responsibility as employees of the king. The same thing applied if the flowers were picked, or the Cabinet visitors were drunk, disorderly, dishonest, or improperly dressed. Thouin, the keeper of the garden, and Daubenton, the keeper of the cabinet, confronted a set of similar problems relating to the maintenance of good order within the Jardin du Roi; that preservation necessitated the development of new techniques for the social management of natural bodies.

Naturalists were thus involved in the preservation of order not only in the natural world, but also in the social world; they were under an obligation to maintain the moral order among their subordinate employees and among the Jardin's visitors, as well as establishing and preserving order among other lower forms of life. Throughout Buffon's intendancy a small army of lower-level administrative staff were employed at the Jardin, often for many years: ushers, police, porters, and curators. The two most important of these were Lucas, a relative of the Thouin family, the Cabinet's usher, who received a brevet for long service on December 13, 1788, and Guillotte, the *inspecteur commandant,* who received his brevet under similar circumstances on January 1, 1786.[18] Both lived at the Jardin. Guillotte's duties were to "assure good order": to guard the doors and interior of the Jardin, the botany school and amphitheater during lectures in botany, anatomy, and chemistry, and the doors and rooms of the Cabinet in public opening hours; also to confiscate forged keys to the schools and greenhouses.[19]

To carry out his task, Guillotte employed a host of lesser guards, described as "Cavaliers" or "Gendarmes" in the list of payments made to the Jardin between 1764 and 1790. Guillotte himself was a former cavalry officer from the company of the provost of the Isle de France, who succeeded his father and brother in the post, which had been held by the family for over fifty years.[20] The importance of having such systems of control was amply demonstrated in 1749 and 1750, for example, when disturbances and vandalism took place at the Jardin.[21] But

18 AN, AJ/15/510, pieces 356, 354, 355. Lucas was also the *huissier* (usher or chief caretaker) for the Académie des Sciences.

19 Letouzey, *Le Jardin des Plantes,* 288: Thouin to Comité des Finances, circa 1789.

20 AN: AJ/15/505, piece 128: "Ordonnances 1764–1790."

21 AN: AJ/15/507, piece 159: "Expéditions."

the spaces of the Jardin were also carefully policed. In 1770 and 1771, describing his visit to Paris, Linnaeus's protégé Carl Thunberg noted of the Cabinet that "[i]n every room there is a sentinel who only allows well-presented people to enter."[22] Since naturalists possessed power filtered down from the king, via Buffon, they had authority not only over the ordering of specimens but also, simultaneously, over the ordering of the public. They were the ones who determined whether specimens and people should be allowed to traverse the spatial boundaries of the Jardin.

The issuing of keys to different parts of the establishment was tightly monitored; the Jardin employed a permanent locksmith, Mille. After the botany school was replanted according to Bernard de Jussieu's classificatory method in 1774, Thouin made a list of the key holders in 1776. They included "M^{r}. de Jussieux the uncle," "M^{r}. Laurent de Jussieux," "M^{r}. Antoine de Jussieux," "D'aubenton the Doctor," "D'aubenton the Younger," "M^{r}. Métra," "M^{r}. Cels," "M^{r}. Aublet," "M^{r}. le Mis [marquis] de la Billiardrie," "M^{r}. le Comte d'Angiviller," "M^{r}. Rouelle," "M^{r}. Guiotte," "Thoüin," "Jacque," "Garçons Jardiniers," and "M^{r}. Du Doyeu."[23] In about 1779, when the list was revised, Thouin listed the deaths of Rouelle and Antoine-Laurent's uncle, and added the names of Descemet, the gardener of the Jardin des Apothicaires; Lamarck, Buffon's new protégé; and de Boiseaujeu, a former musketeer who traveled with Thouin, Lamarck, and André Michaux to the Auvergne in 1779.[24] Other versions of the list included the names of Brongniart, Rouelle's successor; the comte de Saint-Germain (possibly the minister); Achille-Guillaume Le Bègue de Presle, censor and physician,

22 Carl Petter Thunberg, *Voyages de C. P. Thunberg au Japon, par le Cap de Bonne-Espérance, les isles de la Sonde, etc.,* trans. J. Langlès and Jean-Baptiste-Pierre-Antoine de Monet Lamarck, 2 vols. (Paris: Dandré, year IV/1796), 1:39. Compare Thouin, letter to Buffon, undated: "malgré l'affluence du monde qui se porte au jardin les Dimanches et festes et les jours du Cabinet comme c'est toutes personnes raisonnables et Elitiés qu'on y laisse entrer, il ne si commet aucun desordre tout si passe avec clemence et traquillitée [*sic*]" (roughly, "despite the flow of people coming to the garden on Sundays and holidays and on the days of the Cabinet as it is all reasonable and Elite people that are allowed in, no disorder occurs everything happens in harmony and tranquility") (AN, AJ/15/514, piece 626, f°. 2).

23 In order, Bernard de Jussieu; probably two keys for Antoine-Laurent de Jussieu; Louis-Jean-Marie Daubenton; Edmé-Louis Daubenton; Métra, Thouin's correspondent between 1774 and 1786 (BCMHN, MS 314); Jacques-Martin Cels, *receveur* at the barrière Saint-Jacques and owner of a famous garden; Jean-Baptiste-Christophe-Fusée Aublet, voyager naturalist; the marquis de La Billarderie; his younger brother, *contrôleur-général des bâtiments du roi* from 1778; Hilaire-Marin de Rouelle, demonstrator of chemistry at the Jardin from 1768; Guillotte; André Thouin; Jacques, the first subgardener before Thouin's brother Jean. Du Doyeu I have been unable to identify so far.

24 Letouzey, *Le Jardin des Plantes,* 51.

working on a book on rural economy; and Papilloz.[25] The keys were given out by Buffon; one could view them as badges of the holder's authority to act and move around in the Jardin's space—physical emblems of the ways in which power was diffused through Old Regime society.[26] Le Monnier and Desfontaines are not listed, although they almost certainly possessed keys.

Honesty was a prime requisite for Thouin and Daubenton, for over time they handled sums and objects of increasing value, besides their moral responsibility for the employees under their supervision. Jean-André and André Thouin kept the Jardin's accounts in a large ledger covering the period from 1760 to 1793, at which time the Jardin became the Muséum d'Histoire Naturelle.[27] Even allowing for the inflation in the second half of the century, the Jardin's income from the state increased substantially after Louis XVI came to power. The largest sums were disbursed for the purchase of new buildings to be annexed to the Jardin. Thouin's activities in helping Buffon bring about the enlargement of the Jardin during the same period are well documented.[28] This mirrors an upsurge in building speculation which led to the transformation of many previously unfashionable areas of Paris between 1760 and 1790 by the construction of new symbols of wealth and learning, such as the many mansions of the Faubourgs Saint-Germain and Saint-Honoré, the new École de Chirurgie, and the Royal Mint.[29]

The natural historical resources upon which naturalists could draw increased similarly. During the 1760s most of Thouin's outgoings consisted of fares to nearby gardens and herborizing sites to collect for the Jardin and especially for the annual demonstrations by Bernard de Jussieu.[30] In addition, there was the cost of equipment: wire, sand, repairs, weeding, digging, brooms, trellis, and so on. By 1776 the labor costs had risen substantially, as the establishment began to be reformed. A multiplicity of jobs was being carried out all over the Jardin: clearing alleys, staking trees, cultivating land, sweeping the greenhouses. But above all, a new kind of cost was beginning to appear with increasing

25 Again, I am unable to identify this last individual.

26 AN, AJ/15/514, piece 678: "Liste des Personnes a qui Monsieur le Comte de Buffon a accordé la Clef de la Nouvelle Ecolle en 1776."

27 AN, AJ/15/149: "Mr. Thouin. Dépenses pour le Jardin, 1760 à 1793."

28 Falls, "Buffon et l'agrandissement du Jardin du Roi."

29 Roche, *People of Paris,* 12, 14, 32.

30 AN, AJ/15/149. In 1765 Thouin went to Trianon twice, to Saint Prix in May with Antoine-Laurent de Jussieu, on the annual four herborizations for the botany course in June and July, to Mme. de Marsan's garden, and to Saint-Germain in July and August—all for "plantes utiles au jardin."

frequency in the accounts. Thouin recorded the payment of carriage for consignments of plants, trees, or seeds sent to him by correspondents, from elsewhere in France, from other European gardens, or from all over the world. Between 1774 and 1786 the number of seeds being sown at the Jardin, year by year, increased accordingly, from 1,096 to 2,200.[31] The number of his correspondents, similarly, increased from 147 in 1774 to 403 in 1786, as shown in figure 2.[32] The amount spent on labor increased steadily over the period, until in 1781 Thouin began to prepare a separate memoir concerning outgoings on the workforce, which was sent fortnightly to Buffon.[33] During the 1780s the responsibility for handling the Jardin's financial affairs came to rest solely in Thouin's hands; it remained there until the foundation of the Muséum d'Histoire Naturelle in June 1793, whereupon Thouin prepared his last budget for the minister of the interior with the words "Here ends the accountability of André Thouin toward the Intendants of the Jardin des Plantes."[34]

Thouin was accountable to the intendants in many senses. During the 1780s he wrote a series of fortnightly letters to Buffon while the intendant was on his Burgundian estates. Many of these letters no longer survive; however, Buffon's responses give some impression of what Thouin thought it necessary to account for.[35] Clearly, the letters that Buffon received were full of minute details of Thouin's expenditure on the Jardin, after the style of the ledger. By 1790 Thouin had adopted the practice of breaking down this expenditure into a number of categories established during the period of the Jardin's most rapid growth: labor, purchases, accessories, extraordinary expenses, maintenance of buildings, maintenance of the Cabinet, expenses relating to the cultivation of the Jardin, wages for the gardeners, wages of the Cabinet's employees, and salaries of the professors and demonstrators. Piles of receipts from the 1780s and 1790s survive in the archives, attached together with string and neatly initialed by Thouin or Daubenton.[36] As the Jardin grew into a large-scale enterprise, employing ever more personnel and absorbing larger sums of money, Thouin perfected a sys-

31 BCMHN, MS 1384: "Catalogue des Graines qui ont été semé le 14 Avril 1770 au Jardin du Roy . . . "; BCMHN, MS 1321: "Catalogue des graines semées au Jardin du Roi en 1785. 1786. & 1787. 1788."

32 "Etât."

33 AN, AJ/15/149.

34 Letouzey, *Le Jardin des Plantes,* 310.

35 *Correspondance,* 2:297–304, Buffon to Thouin, August 31, September 8, October 3, and October 17, 1785, all from Montbard; BCMHN, MS 882.

36 AN, F/17/1221.

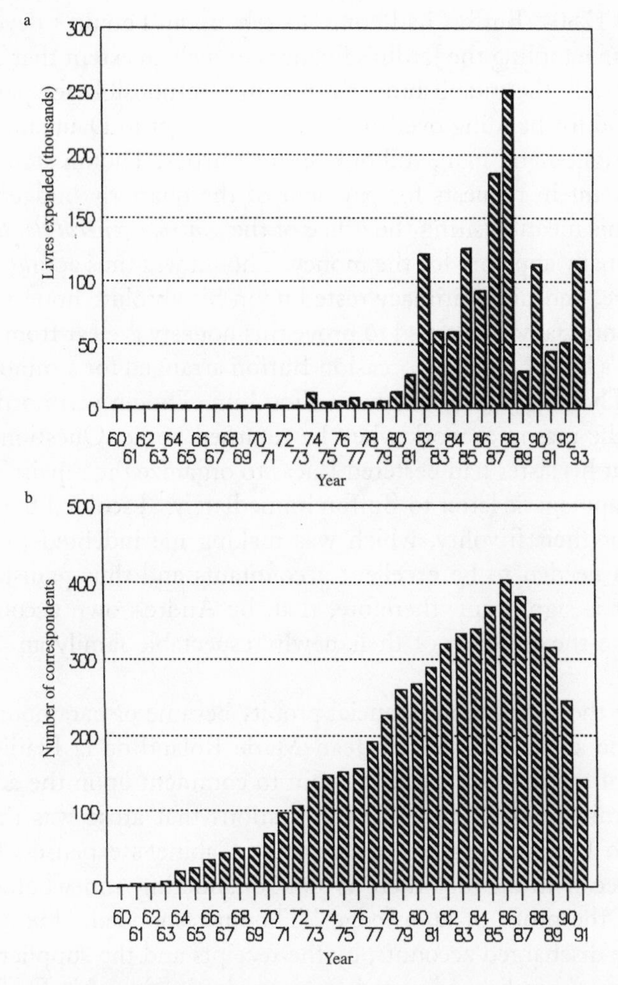

Figure 2 The rise and fall of the Jardin's fortunes: *a,* expenditure on the Jardin from 1760 to 1793; *b,* André Thouin's total correspondence between 1760 and 1793. Based on AN, AJ/15/149; BCMHN, MS 314.

tem of accounting to deal with an increasing volume of outgoings, salary demands, and bills. His records of the influx of specimens to the Jardin's collections reveal the same growing complexity and resultant streamlining. By the mid-1780s, Thouin had developed a system of numbered drawers containing different categories of seeds, such as ornamental or useful, and would select from the most suitable drawers for the purposes of his correspondents.[37]

37 Revealed by the notes Thouin made on his correspondents' letters: e.g., BCMHN, MS 1975 (2), letter 988: Ange Gualandris to Thouin, from Mantua, March 17, 1784.

By the 1780s, Buffon had come to rely upon Thouin's developing expertise in handling the Jardin's finances to such an extent that Thouin was effectively the establishment's treasurer, responsible for paying all its staff and for handing over the Cabinet's budget to Daubenton.[38] As Buffon's right-hand man, still only in his thirties, Thouin stood in for the intendant in requests for payment of the quarterly budget of the Jardin. This meant visiting the office of the *contrôle général des finances,* and personally applying for the money. Thouin was thus acting in Buffon's name, and his legitimacy rested upon his absolute honesty. That he was continually concerned to prove this honesty is clear from several of Buffon's letters. On one occasion Buffon arranged for a minor alteration to Thouin's house, without telling him. Thouin was horrified to discover the mason at work when he returned home. Questioning revealed that his sister had pestered Lucas to organize the repair; Thouin wrote an apologetic letter to Buffon immediately: "I scolded them both sharply for their frivolity, which was making me indebted to you."[39] Managers needed to be excellent accountants and thus painstakingly honest. It is significant, therefore, that, by André's own account, his father gave the children of their newly respectable family an "honest education."

During the Revolution financial probity became of paramount concern to the authorities. When Jean-Marie Roland de la Platière, the minister of the interior, asked Thouin to comment upon the accounts of the Jardin in 1792, one of the questions that arose was the item relating to Daubenton's payments of the Cabinet's expenses. Roland noted: "See if in all that there is not squandering and whether it is turned to the profit of knowledge." Thouin responded, "He . . . will see by the discharged accounts, by the receipts and the suppliers' bills, that not a *sol* has been diverted from its destination. M. Daubenton who makes these payments is scrupulously exact in this respect, as in all things."[40] The many pensions and gratifications conferred by the Crown underwent ministerial assessment in the early 1790s, and the Jardin was no exception. Thus Denormandie, the director of the liquidation of the public debt, wrote to the intendant, Bernardin de Saint-Pierre, in 1793: "It is necessary for my work, Citizen, that you should make known to me which are the objects that Citizen Faujas St. Fond has deposited in the National Cabinet; among others he announces

38 Thouin continued in this role after the 1793 transformation of the Jardin into the Muséum d'Histoire Naturelle, when he was elected the establishment's treasurer.

39 Letouzey, *Le Jardin des Plantes,* 81.

40 Ibid., 300–301.

comparatively precious volcanic products. What is their quantity and estimated value."[41] Such investigations reveal the extent to which Old Regime administrative standards of honesty and accountability were now being incorporated into new structures of government. Accountability was not just financial: the Jardin's staff spent much time before and after the reform of the establishment assessing the nation's wealth in terms of natural historical specimens. But because of the increased emphasis on financial honesty and the changing moral status of indebtedness during the 1780s, Thouin's own status as economist at the Jardin grew in significance.

CORRESPONDENCE NETWORKS

Thouin is not commonly recognized as an outstanding naturalist of the eighteenth century, despite his respected position in the late-eighteenth- and early-nineteenth-century French scientific world. Only a recent biography has brought to light his many surviving manuscript materials.[42] From these Thouin appears as the center of a network of botanical correspondence which stretched around the world, encompassing all the major botanical institutions in Europe, and involved a vast range of individuals from ambassadors, ministers, and foreign rulers to peasants and subgardeners. Moreover, these materials indicate that Thouin's activities were probably representative of developments at all major botanical centers in Europe and the colonies in the later eighteenth century.[43]

Many of those involved in Thouin's correspondence network have not succeeded in entering the canon of "scientific" botanists, so that they are either unknown or, at best, are identified as nobles who were seized by the renowned eighteenth-century "botanizing fad." Such distinctions are hard to sustain in a botanical culture in which private and

41 AN, AJ/15/512, piece 504: Denormandie, Directoire général provisoire des liquidations, to Jacques-Henri Bernardin de Saint-Pierre, May 8, 1793. Faujas had been awarded a *pension* of 6,000 livres per annum in April 1790 (*AP*, 13:181). Denormandie also inquired concerning Pierrart and Nicolas Gabriel Leclerc, who had received *pensions* in consequence of their contributions to the Cabinet (AN, AJ/15/512, pieces 500, 501).

42 Letouzey, *Le Jardin des Plantes.* David Elliston Allen, "Arcana ex Multitudine: Prosopography as a Research Technique," *Archives of Natural History* 17 (1990): 349–359, emphasizes the value of prosopographical studies of scientific practice.

43 The renowned Swedish botanist Carolus Linnaeus, one of Thouin's heroes, possessed one of the most devoted networks of travelers in this period. See Lisbet Koerner, "Purposes of Linnaean Travel: A Preliminary Research Report," in *Visions of Empire,* ed. Miller and Reill, 117–152; and compare David Mackay, "Agents of Empire: The Banksian Collectors and Evaluation of New Lands," in *Visions of Empire,* ed. Miller and Reill, 38–47. There was considerable overlap between Banks's, Linnaeus's, and Thouin's traveling correspondents.

royal systems of plant introduction and exchange were interdependent. With the growing popularity of agriculture from the 1760s onward, landowners were active in establishing experimental farms and research gardens on their properties. Such programs were fueled by writers who pleaded for research into the useful properties of exotic naturalia.[44] These authors' deft interrelation of political economical, natural historical, and agricultural matters stimulated ministerial interest in the introduction of new plants into French agriculture, resulting in the privileging of botanical networks of plant transfer. Botanical networks in the eighteenth century cannot, therefore, be differentiated from agricultural and economic concerns. Such concerns formed the link between many of the disparate groups represented in Thouin's list of correspondents, and were often the means by which European naturalists validated their collecting activities to patrons.

The exchange of letters between individual botanists largely followed an established pattern. In the 1770s Thouin's incoming correspondence indicates that he himself often wrote an initial letter approaching a potential correspondent, and he would usually accompany it with a selection of seeds taken from the harvest at the Jardin.[45] Such a gift served to place the recipient under an obligation which could only be thrown off by replying and returning the favor. Usually, such favors were returned in kind, but the abbé André-Pierre Ledru provided Thouin with quantities of fowl at Christmas instead: "One favor deserves another."[46] Since it was unsuitable for a botanist employed at a royal garden to earn money from the sale of what, after all, were the king's seeds, other means had to be used to organize individuals at a distance in order to obtain seeds of new species of plants. The language of patronage was the vehicle for disciplining individuals and controlling the influx of new specimens to the Jardin and other similar centers.

In commencing correspondence with other botanists in the 1760s and 1770s, Thouin primarily addressed individuals at other French or European institutions, with established botanical gardens. Examples are Charles-Nicolas Desmoueux, the head of the Caen botanical gar-

44 André Bourde, *Agronomie et agronomes en France au XVIIIe siècle*, 3 vols. (Paris: SEVPEN, 1967); Gillispie, *Science and Polity*, chapter 5.

45 E.g., BCMHN, MS 1973, letter 424: Desmoueux to Thouin, undated, from Caen.

46 BCMHN, MS 1978, letter 1667: Ledru to Thouin, December 11, 1788, Le Mans. In original, "Une politesse se paye par une autre." As Müller-Wille, *Botanik und weltweiter Handel*, 317, shows, Linnaeus's English supporter Peter Collinson was driven to write to the Swedish botanist in protest against his failure to respond to botanical gifts in kind.

den, with whom Thouin corresponded from 1776 onward, and Franz Anton Ranffls in Salzburg, with whom Thouin corresponded after 1779. In addressing prospective correspondents, naturalists seem to have adopted the language used by a young or inexperienced individual soliciting an established member of the savant community for patronage.[47] Often the aim of Thouin's solicitations of other centers was the contraction of a long-term arrangement for the exchange of plants and seeds. The correspondence could continue on this basis of mutual obligation, with individuals bound together by the ties of patronage. In 1779 the Dutch botanist Nahuys wrote from Utrecht in response to Thouin's letter and consignment of seeds: "If our Garden, although much inferior to yours should happen to contain plants whose seeds I might send you, you will always find me ready to render you reciprocal services, insofar as I am able."[48] As the number of Thouin's correspondents rose, he could send out a given number of species to all his European correspondents and expect to receive a certain number back from each. A proportion of each consignment received would usually consist of species new to the Jardin du Roi. Several years of correspondence and cultivation could thus substantially increase the number of species grown at a center. As an individual succeeded in controlling more plant resources, he or she moved upward in the patronage scale: hence the significance of Thouin's appeal to the new intendant, de La Billarderie, in 1788: "At this moment [the Jardin] contains approximately 6,000 different species and more than 60,000 individuals. It is without doubt the most numerous collection in Europe."[49] Two decades of continual negotiations with other botanical centers had considerably added to the riches of the collection.

Thouin's success as a networker meant that he came to be solicited in his turn by individuals wishing to develop their own botanical centers through correspondence.[50] With his peers—managers of gardens recognized throughout Europe—the system of mutual obligation which

47 E.g., BCMHN, MS 1976, letter 1165: draft, Thouin to Jacquin, March 25, 1776.

48 BCMHN, MS 1981, letter 2074: Nahuys to Thouin, August 27, 1779, Utrecht. On gift-giving see Marcel Mauss, *The Gift: The Form and Reason for Exchange in Archaic Societies,* trans. W. D. Halls (London: Routledge, 1990); Nicholas Thomas, *Entangled Objects* (Cambridge: Harvard University Press, 1991); Alain Guery, "Le roi dépensier: Le don, la contrainte, et l'origine du système financier de la monarchie française d'ancien régime," *Annales: Économies, sociétés, civilisations* 39 (1984): 1241–1269; Biagioli, *Galileo, Courier,* 41–54.

49 AN, AJ/15/502: André Thouin, project for the regulation of the Jardin, 1788.

50 E.g., BCMHN, MS 1975, letter 954: Granier to Thouin, January 12, 1783, Nîmes; BCMHN, MS 1982, letter 2323: Poiret to Thouin, arrived February 25, 1788.

bound these individuals was expressed by the mutual use of the language of honor and obligation, which is to say that of protégé to patron, in their letters. For example, Nahuys addressed Thouin as follows in a letter in 1783, four years after the commencement of their correspondence "contract": "I am infinitely obliged by this new present, it is an unequivocal mark of your benevolence and of the sincere friendship which you have for me."[51] The confirmation by the patron of his continuing goodwill toward the protégé was an essential part of the maintenance of patronage relationships, as was the protégé's acknowledgment of the friendship gesture.[52] Thus in 1784 Buffon wrote: "I pray you, my dear Monsieur Thouin, . . . to believe me in all those sentiments which you could desire and with the most sincere attachment your friend rather than your superior."[53] In the Nahuys letter the writer's thanks to Thouin for his patronage cannot be distinguished from his thanks for the botanical specimens. Nahuys used the language of mutual obligation that made failure to respond to one's correspondents a matter of dishonor. Thouin's Toulouse correspondent, Philippe Picot, baron de La Peyrouse, likewise promised that "all that we possess is at your service."[54] The cultivators of plants were also cultivating personal character: in 1779 La Peyrouse commented that Thouin's letter "gave me so much pleasure . . . I especially love those people who join as many talents as you [possess], to a sincere modesty; and I judge myself very happy to be in relation with you Monsieur; Believe that I will withhold nothing in cultivating it."[55] Letters like these also provided

51 BCMHN, MS 1981, letter 2080: Nahuys to Thouin, January 23, 1782, Utrecht. On the long history of denoting social distinction in correspondence and its transformations, see Roger Chartier, "Secrétaires for the People? Model Letters of the Ancien Régime: Between Court Literature and Popular Chapbooks," in Roger Chartier, Alain Boureau, and Cécile Dauphin, *Correspondence: Models of Letter-Writing from the Middle Ages to the Nineteenth Century,* trans. Christopher Woodall (Cambridge: Polity Press, 1997), 59–111; Altman, "Teaching the 'People' to Write"; id., "Political Ideology in the Letter Manual"; Maurice Daumas, "Manuels épistolaires et identité sociale, XVIe–XVIIIe siècles," *Revue d'histoire moderne et contemporaine* 40 (1993): 529–556.

52 Biagioli, *Galileo, Courtier,* chapter 1. See also Goodman, *Republic of Letters,* esp. 113–118, although my examples conflict with her portrayal of polite society as an "egalitarian" sphere distinct from the Crown hierarchy.

53 Buffon to Thouin, June 11, 1784, quoted in Letouzey, *Le Jardin des Plantes,* 92. See also Thouin, draft letter to Buffon, 1783, written to congratulate his patron on his convalescence: "Sans Médecin sans remèdes la Nature Seule a opéré votre Guérison, il n'appartenait qu'à elle de Sauver son plus précieux Chef d'oeuvre" (Without a Physician without remedies Nature Alone has operated your Cure, it was her prerogative to Save her most precious Masterpiece), "Lettres inédites de Buffon," in *Buffon,* ed. Heim, 221.

54 BCMHN, MS 1977, letter 1369: La Peyrouse to Thouin, November 8, 1785, Toulouse.

55 BCMHN, MS 1977, letter 1350: La Peyrouse to Thouin, October 28, 1779, Toulouse.

the basis for the management of protégés: Petit of Nevers asked for a place as *garçon-jardinier* at the Jardin in the 1790s; Denesle, director of the Poitiers botanical garden, asked Thouin to find a place for a protégé gardener in late 1788.[56]

Each of the letters exchanged between the owners or managers of established gardens, whether these were municipal botanical gardens endowed by the king or the local academy of sciences, or agronomic gardens run by private landowners, would be accompanied by a consignment of seeds or plants. Thouin labeled each letter received with the date of his reply and the number of species sent. Throughout the 1780s, it became increasingly common for Thouin's botanical peer group—those with whom he had been in correspondence for many years—to send a simple greeting accompanied by a list of desiderata for the forthcoming growing season, or even a copy of the sender's catalogue of his garden, from which the missing species could be determined by the recipient.[57] It was thus only necessary to utilize the language of patronage where such correspondence relationships were not yet stable. However, when new correspondence relationships were being formed, when a considerable length of time had elapsed between consignments, or when there were large differences between the resources of the two, the language of patronage was implemented. The individual's botanical wealth, therefore, determined his social status in the patronage/correspondence network. The professor of botany Jean-Auguste L'Estiboudois, in Lille, acknowledging receipt of Thouin's catalogue of the Jardin in 1778, replied that "according to the time and the occasion, I shall send you other plants suitable to your State."[58] Here the term used, "État," was both one's condition and one's social status. The letters written by Thouin's correspondents abound with references to the "wealth," the "economy" of the sender's and recipient's garden, as much as they do with references to the success of the cultivation of rare specimens.[59] Thus the imagery of the garden as economy was explicit in botanical correspondence, as is revealed by a letter written by Jean Hermann, the director of the Strasbourg botanical garden, to a different correspondent at the Jardin during the Revolutionary

56 BCMHN, MS 1981, letters 2247–2248: Petit to Thouin, undated, Nevers; BCMHN, MS 1973, letter 444: Denesle to Thouin, undated.

57 E.g., BCMHN, MS 1981, letter 2178: Ortega to Thouin, April 9, 1981, Madrid; BCMHN, MS 1974, letter 672: Durazzo to Thouin, January 24, 1790, Genoa.

58 BCMHN, MS 1978, letter 1718: L'Estiboudois to Thouin, April 12, 1778, Lille.

59 E.g., BCMHN, MS 1977, letter 1360: La Peyrouse to Thouin, January 7, 1784, Toulouse; BCMHN, MS 1975, letter 959: Granier to Thouin, November 1791, Nîmes; BCMHN, MS 1982, letter 2510: Roussel to Thouin, June 17, 1791, Caen.

years: "Alas, I am properly disgraced where M. Thouin is concerned, by not having anything [i.e., anything new] to offer him. I have been obliged to restrain myself singularly, and suspend several correspondences, given the poverty of our garden. The interior economy is maintained uniquely by the contributions of the pupils."[60] Hermann's letter reveals the most important problem for an individual wishing to engage in botanical correspondence. Without new resources to offer correspondents, the botanical-cum-social status of a naturalist would drop rapidly. The range and quality of new specimens coming into the center were vitally important as currency for the "interior economy" of the center. Thus many different potential sources for obtaining plants were recruited by networking botanists in the 1780s. Wealthy private landowners could purchase specimens from the many specialist plant merchants who flourished in the second half of the century.[61] The horticultural interests of private individuals meant that considerable sums might change hands for rare or unusual specimens, as had been the case for shells in the 1750s.[62] In the 1780s Thouin also purchased some plants from merchants and kept a number of catalogues, including those of the plant merchant Young in Pennsylvania and the Schuylkill botanical garden in Philadelphia, run by William and John Bartram.[63] But many botanists in public gardens did not have the funds for such purchases, so correspondence with private landowners was of great value to them.[64] This was true of the Jardin in the early 1770s. At this stage in his involvement with the botanical network, it was comparatively rare for Thouin to receive seed consignments from other botanists.[65] Most of the specimens that passed into the Jardin came from gardens in and around Paris: those belonging to members of the royal family, such as the comte d'Artois's garden at Bagatelle, the Trianon garden and others, and those belonging to courtiers, who could afford

60 BCMHN, MS 1975(2), letter 1092a: Hermann to unknown, after 1789.

61 Thouin sent La Peyrouse's list of desiderata to the traveler James Bruce to enable him to add prices from his catalogue of exotic plants in 1785 (BCMHN, MS 1977, letter 1369: La Peyrouse to Thouin, November 8, 1785, Toulouse).

62 S. Peter Dance, *A History of Shell Collecting* (Leiden: E. J. Brill, 1986), 53–54; see also Keith Thomas, *Man and the Natural World: Changing Attitudes in England, 1500–1800* (Harmondsworth, Middlesex: Penguin, 1984), 233, on English hyacinth crazes.

63 AN, AJ/15/511, pieces 468, 467.

64 Desmoueux, informing Thouin that he had paid for the carriage of a consignment of plants from Caen, claimed that "[a] botanist whose funds are in total 73 livres 10 sous has made all possible efforts to demonstrate his gratitude to you" (BCMHN, MS 1973, letter 426: January 8, 1777, Caen).

65 AN, AJ/15/149.

the inflated prices of merchants. An example was the comtesse de Marsan, a close friend, perhaps mistress, of Louis-Guillaume Le Monnier, who was the king's first physician, the professor of botany at the Jardin, and Thouin's teacher.

Increasingly, however, the Crown became involved in the process of obtaining and transporting natural historical specimens. The patrons of the Jardin had access to many diplomatic travelers, as well as to diplomatic routes for the transport of important objects. The Jardin du Roi, harboring the king's own natural history collection, also received specimens offered as patronage gifts, by individuals seeking to demonstrate their attachment to the Crown, or as diplomatic sweeteners. Usually these were presented to one of the ministers, who would then forward them to the Jardin. In his ledgers Thouin recorded consignments from Malesherbes, d'Angiviller, Gravier de Vergennes, and Joly de Fleury.[66] D'Angiviller, *contrôleur-général des bâtiments du roi* in the 1780s, was particularly active in employing his court position and postal privileges for the Jardin's benefit. From 1776 onward he sent packets of seeds and boxes of living plants obtained from diplomats and other official travelers to the Jardin, and in 1778 he also gave permission for Thouin to request all his correspondents to address their consignments of seeds and plants to the Jardin under an outer envelope bearing d'Angiviller's name.[67] By this means the consignments avoided the numerous taxes and duties that were levied at municipal customs barriers, as well as the time-consuming and damaging customs searches.[68] In addition, d'Angiviller's offer enabled individual correspondents to use the new *diligence* in sending specimens to the Jardin.[69] Prohibitive costs had hitherto prevented the use of these rapid forms of transport, which ensured that specimens suffered minimal deterioration in transit. Thouin and most of his long-term correspondents used this route; it is repeatedly mentioned in the letters of Desmoueux, Hermann, Nahuys,

66 BCMHN, MS 1321.

67 E.g., BCMHN, MS 1972, letter 219: Caqué to Thouin, March 20, 1790, Reims. The cost of transport could be remarkably high without such immunities. In the early 1770s Thouin had a short-lived correspondence with the privy counsellor to the court of Baden-Durlach. Transport costs were 51 livres, 6 sous for one consignment, and 107 livres, 13 sous for another (BCMHN, MS 1984, letters 2617, 2618: Schmidt von Rossen to Thouin, June 7, 1771, and undated, Frankfurt).

68 Apparently a not uncommon complaint of eighteenth-century naturalists, as is demonstrated by Hugh Torrens, "Under Royal Patronage: The Early Work of John Mawe (1766–1829) in Geology and the Background to His Travels in Brazil in 1807–1810," unpublished paper.

69 Roger Price, *The Economic Modernisation of France, 1730–1875* (London: Macmillan, 1975), chapter 1.

L'Estiboudois, and others.[70] Other European botanists had their own diplomatic channels for the movement of specimens: Winold Munniks in Groningen and Nahuys in Utrecht requested Thouin to send them to the Dutch ambassador in Paris.[71] Caspar Christopher Schmidel used the secretary to the margrave of Anspach; C. F. Rottboll, a professor of medicine in Copenhagen, used the French ambassador there, the marquis de la Hausse.[72] The Lund professor of natural history, Anders Jan Retzius, used the Swedish ambassador to France, Noringen.[73] Many made use of individuals traveling from one country or town to another where they had a correspondent, to send a single consignment.[74] The streamlining of the movement and selection of specimens was one of the most powerful concerns of the eighteenth-century naturalist.

But perhaps the most compelling concern of networkers was access to travelers, who could be displaced from the centers of botany in Europe and the colonies. Whether they were simply "Païsants botanophyles" (botanophile peasants), paid to collect, as recruited by Thouin's correspondent La Peyrouse in Toulouse, or travelers officially connected with public institutions, all centers derived their resources in terms of new species, ultimately, from these mobile elements of the network.[75] Both the degree of botanical training possessed by these collectors and their geographical range, as we shall see, determined the level of "income" they could generate for the botanical center to which they sent their specimens. In order to marshal new resources successfully, the central botanist depended crucially upon controlling the actions of distant individuals who had access to those resources. Even local travelers were desirable sources for specimens, and for many central naturalists they were the only means of getting new species. But the most important vehicles for specimens were individuals traveling outside Europe, by means of whom the naturalists could have access to the more exotic species. It was these which were most valuable to botanists in the system of plant bartering which underpinned botanical

70 E.g., BCMHN, MS 1979, letter 1894: Martin to Thouin, October 8, 1791, Toulon; BCMHN, MS 1981, letter 2078: Nahuys to Thouin, May 13, 1782.

71 BCMHN, MS 1981, letter 2074: Nahuys to Thouin, August 27, 1779, Utrecht; BCMHN, MS 1980, letter 2060: Munniks to Thouin, February 20, 1780, Groningen.

72 BCMHN, MS 1983, letter 2613: Schmidel to Thouin, October 18, 1783, Ansbach; BCMHN, MS 1982, letter 2504: Rottbolls to Thouin, October 28, 1781, Copenhagen.

73 BCMHN, MS 1982, letter 2427: Retzius to Thouin, October 5, 1787, Lund.

74 L'Estiboudois used the naturalist Palisot de Beauvois, traveling from Lille to Paris (BCMHN, MS 1978, letter 1717: L'Estiboudois to Thouin, April 9, 1778, Lille). The noble Genoan botanist Ippolito Durazzo used his cousin, the marquis Jacques Balbi (BCMHN, MS 1974, letter 673: Durazzo to Thouin, October 25, 1790, Genoa).

75 BCMHN, MS 1977, letter 1350: La Peyrouse to Thouin, October 28, 1779, Toulouse.

correspondences. Although the botanists obtained some new species from Europe after 1750, most unfamiliar species were arriving from other continents and from the colonies. During the period there was a large increase in traffic between the colonies and Europe, as well as an increase in the number of exploratory voyages made, both reflecting the increasing international importance of colonial trade. Well-known examples are the Cook voyages of the 1770s and the Lapérouse expedition of the 1780s.[76] In the former case Joseph Banks and Daniel Solander collected approximately 3,600 species of plants, of which about 1,400 were new to European botany, from what is now known as Botany Bay.[77] Networkers were very eager to harness the botanical outcome of such voyages. To exploit this, merchants paid high prices to sailors for specimens brought back from their travels, often waiting on the quayside with their competitors for the ships to dock.[78] Local traders on the Isle de France and in Canton made use of the distant European market for plants by preparing standard "kits" for travelers with little time to spare—portable herbaria containing dried specimens and seeds of local plants, which were sold at the ports along with other souvenirs.[79] Thus many travelers, diplomats, and merchants played an ephemeral role in the botanical correspondence network. Although these are hard to identify, they were an important element in moving specimens along the connecting lines carved out, as it were, by correspondence.[80]

Far better for the naturalists' attempts to direct the flow of specimens into their centers were travelers whose collecting practices could be shaped in advance. In epistolatory exchanges, such control was exerted by the patronage obligations that bound individuals. After informing Thouin that he was sending a consignment of seeds to the Jardin, his

76 Catherine Gaziello, *L'expédition de Lapérouse, 1785–1788: Replique française aux voyages de Cook* (Paris: CTHS, 1984); Danielle Faugue, "Il y a deux cents ans: L'expédition Lapérouse," *Revue d'histoire des sciences* 38 (1985): 149–160; David Mackay, *In the Wake of Cook: Exploration, Science, and Empire, 1780–1801* (London: Croom Helm, 1985).

77 Harold B. Carter, *Sir Joseph Banks, 1743–1820* (London: British Museum [Natural History], 1988), 95.

78 Dance, *Shell Collecting,* 69; David Elliston Allen, *The Naturalist in Britain: A Social History* (Harmondsworth, Middlesex: Penguin, 1978), 37; Nicholas Thomas, "Licensed Curiosity: Cook's Pacific Voyages," in *Cultures of Collecting,* ed. Elsner and Cardinal, 116–136.

79 BCMHN, MS 56; AN, F/10/201: "Liste d'Arbres et de Plantes les plus agréables ou utiles qui croissent dans le voisinage de Canton et qui sont inconnus dans les Jardins de l'Europe."

80 Other savants such as Gaspard Monge and Lazare Carnot were engaged in calculating the geometry of communications to improve the speed of correspondence in military campaigns at this time. Such military concerns formed the basis for early-nineteenth-century uses of the term *réseau,* or "network." See André Guillerme, "Network: Birth of a Category in Engineering Thought during the French Restoration," *History and Technology* 8 (1992): 151–166.

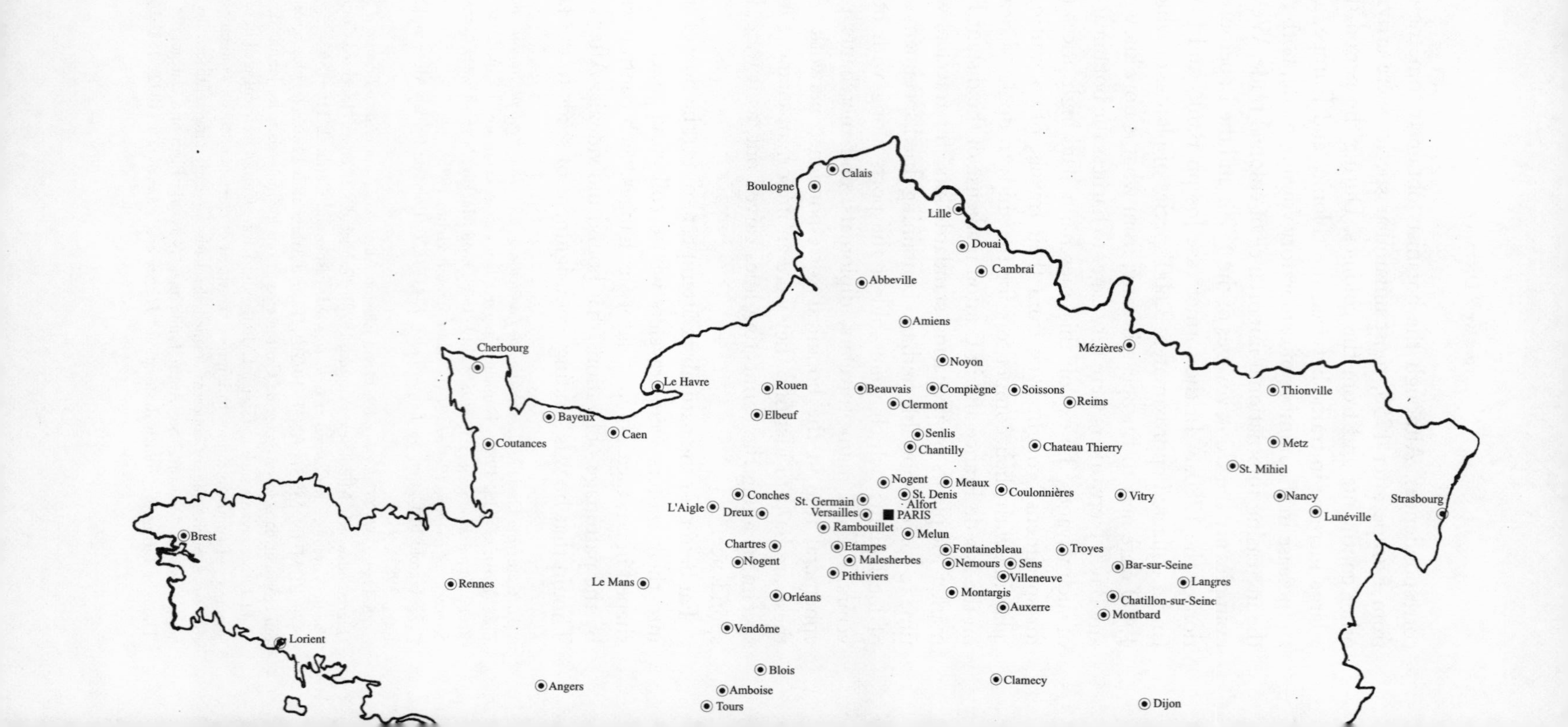

Calais
Boulogne
Lille
Douai
Cambrai
Abbeville
Amiens
Cherbourg
Mézières
Noyon
Le Havre
Rouen
Beauvais
Compiègne
Soissons
Thionville
Clermont
Reims
Bayeux
Elbeuf
Caen
Senlis
Coutances
Chantilly
Chateau Thierry
Metz
St. Mihiel
Nogent
Meaux
Conches
St. Germain
St. Denis
Coulonnières
Vitry
Nancy
L'Aigle
Alfort
Strasbourg
Dreux
Versailles
PARIS
Lunéville
Rambouillet
Melun
Brest
Chartres
Etampes
Fontainebleau
Troyes
Nogent
Malesherbes
Nemours
Sens
Bar-sur-Seine
Rennes
Le Mans
Pithiviers
Villeneuve
Langres
Orléans
Montargis
Chatillon-sur-Seine
Auxerre
Montbard
Vendôme
Lorient
Blois
Angers
Clamecy
Amboise
Dijon
Tours

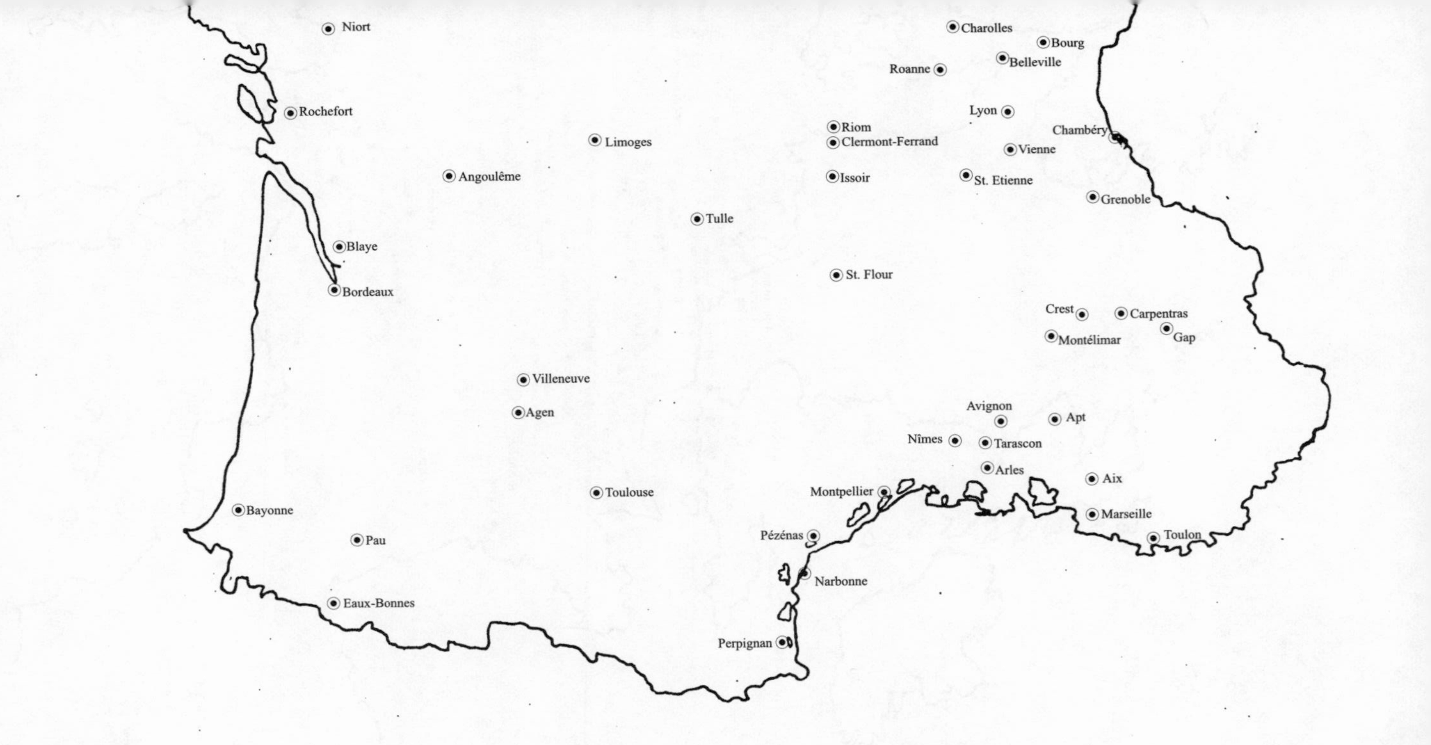

Figure 3 Map showing the distribution of Thouin's French correspondents, 1760–1791. Based on BCMHN, MS 314.

Figure 4 Map showing the distribution of Thouin's European correspondents, 1760–1791. Based on BCMHN, MS 314.

St Petersburg
GALICIA
Constantinople

Figure 5 Map showing the distribution of Thouin's correspondents worldwide, 1760–1791. Shaded areas denote approximate locations from which Thouin received seeds. Based on BCMHN, MS 314, 1321, 1384.

correspondent Emmanuel Baillon, from Montreuil-sur-Mer, continued: "May I flatter myself that you will ask me freely for all that may cause you pleasure here? I should be infinitely flattered to be of value to you in something."[81] When this letter was written, in 1781, Baillon was not yet officially connected with the Jardin, like most of the individuals who offered specimens. The title of correspondent to the Cabinet

81 BCMHN, MS 1971, letter 53: Baillon to Thouin, November 19, 1781.

and Jardin, obtained from the king by Buffon in 1773, was conferred on a number of individuals in the 1770s and 1780s, including Pierre Sonnerat (1773), Charles-Sigisbert-Nicolas Sonnini de Manoncourt (1775), André Michaux (1779), Lamarck (1781), and Baillon (1786).[82] It was a desirable title, for it carried with it royal recognition and could serve as a financial and social safety net, allowing the individual to appeal to the Crown for *pensions* and gratifications. The title might be

82 Laissus, "Le Jardin du Roi," 298; id., "Les voyageurs naturalistes du Jardin du Roi et du Muséum d'Histoire Naturelle: Essai de portrait-robot," *Revue d'histoire des sciences* 34 (1981): 259–317. On Sonnerat see Madeleine Ly-Tio-Fane, *Pierre Sonnerat, 1748–1814: An Account of His Life and Work* ([Reduit, Mauritius]: n.p., 1976).

a reward to a faithful protégé, as in Baillon's case, but it was also a means of controlling such individuals, who often received it when departing upon a journey where their collecting skills would benefit the Jardin. It conferred a special, legalized obligation toward the collections of the Jardin and Cabinet. Individuals at the centers of botanical networks rarely traveled themselves; usually, others formed the moving elements which fed the networks. Before the Revolution, Thouin never went farther from the Jardin than the Auvergne. But many of his correspondents, particularly those who did not possess institutional posts, traveled on diverse voyages, often as physician-naturalists or ships' surgeons. Naturalists were increasingly successful in getting their collecting interests represented on major national voyages of exploration.

Central networkers and travelers could not undergo the sort of "contracts" for mutual exchange that were possible between separate centers. But there were potential benefits from adopting the hazardous lifestyle of the voyager naturalist. If the voyage yielded useful discoveries for the Crown, such as new plants that might be exploited, the naturalist might be the object of public acclaim and financial reward on his return. The Banks visit to Australia possibly represents the traveling naturalist's dream come true. On his return, Banks succeeded in obtaining the presidency of the Royal Society and a baronetcy.[83] Besides making a name for himself, a traveling naturalist might also be able to sell his collections on his return. Several of the Jardin's correspondents, including Michel Adanson, Joseph Dombey, and Philibert Commerson, achieved fame in this way, and the Jardin reaped the benefit of their collecting as well. The difficulty of traveling, with its lack of space, variable weather conditions, and continual risk of shipwreck, piracy, or disease, made prolonged preservation of specimens a virtual impossibility. Many travelers sent their collections back to Europe piecemeal, retaining duplicates on shipboard. If they employed the established trade routes between Europe and the colonies, their consignments had a relatively good chance of arriving safely in European centers, where they would be cared for by the recipient until the collector's return. The spoils of voyages were often divided up among a number of "home" botanists in this way.[84]

83 Carter, *Sir Joseph Banks*, 176; John Gascoigne, *Joseph Banks and the English Enlightenment: Useful Knowledge and Polite Culture* (Cambridge: Cambridge University Press, 1994).

84 Catherine Lang, "Joseph Dombey (1742–1794), un botaniste au Pérou et au Chili: Présentation des sources," *Revue d'histoire moderne et contemporaine* 35 (1988): 262–274. That such divisions were also bitterly contested is demonstrated for the case of Dombey's herbarium—after its arrival in France—by Frans A. Stafleu, "L'Héritier de Brutelle: The Man and His Work," in Charles-Louis L'Héritier de Brutelle, *Sertum Anglicum 1788: Facsimile with Critical Studies and a Translation* (Pittsburgh: Hunt Botanical Library, 1963), xxii–xxiii.

Such acts also helped to reinforce patronage links. Joseph Dombey traveled in South America between 1766 and 1785. Chronically short of cash, he wrote to Micault d'Harvelay in 1777, offering to sell him a portion of the specimens that he expected to collect from the next leg of his journey. But Micault d'Harvelay quashed this hope with urbane courtesy: "No one could be more honored than I am by the offer that you have made me of curiosities in natural history from the country which you are about to traverse. I would not wish to deprive you of the fruit of your researches and your labors."[85] Dombey's suggestion of a vulgarly commercial exchange revealed his ignorance of the rules of a gift-exchange network, the system of polite indebtedness on which botanists' social interaction was grounded. Following his failure with Micault d'Harvelay, Dombey apparently wrote to Thouin for advice. The head gardener's reply made explicit the strategies of self-advancement for the traveling naturalist which usually went unspoken: "Don't miss any opportunity to send memoirs [and] observations to this country on all the objects which merit your attention. On the third or fourth consignment, request the title of foreign correspondent from the Académie. Sometime later, introduce into your memoirs the desire that you have to belong more closely to that respectable body, address to MM. d'Angiviller, de Buffon, de Malesherbes, de Maurepas, etc. objects of natural history suitable to their tastes."[86] Armed with this advice, Dombey finally succeeded in enlisting Buffon as a patron. The Jardin's intendant arranged for him to receive 2,000 livres during an illness in 1781. In 1784 Dombey wrote to Buffon from Rio de Janeiro: "You have, Monsieur, incontestable rights over all that I have and you may dispose of it as you see fit. I shall think myself adequately recompensed for my pains if I should be so lucky as to merit your approbation."[87] When Dombey ran into difficulties on his return to Europe in attempting to achieve financial recompense for his travels and collecting, Buffon helped to get him a pension. Thus Buffon and Dombey fruitfully deployed the language of patron and protégé to achieve a trade-off between the former's need for otherwise inaccessible specimens, and the latter's desire for financial security and success in a world where there was no such thing as a "professional scientist."

During the later eighteenth century, the rate of introduction of new

85 AN, AJ/15/511, piece 404. Joseph Micault d'Harvelay, a keeper of the Royal Treasury, moved in the highest financial circles and annually loaned millions of livres to the Crown. See Bosher, *French Finances,* 91, 96.

86 Quoted in Laissus, "La succession de Le Monnier," 152; see also Laissus, "Les voyageurs naturalistes du Jardin du Roi," 291–292.

87 AN, AJ/15/511, piece 408.

plants into Europe increased dramatically.[88] Successful harnessing of travelers and patrons contributed to this increase. In the 1780s and beyond, central networkers attempted to organize the training of voyagers on a larger scale, either by founding schools for this purpose, or by having their relationships with travelers legally acknowledged by the Crown. This prompts a closer examination of just what it was that travelers needed to be educated to do when on a voyage, and why so much value was placed by central naturalists in the correspondence network upon the recruitment of *trained* travelers.

A Little Light Reading

In May 1785 Captain Jean-François Galaup de Lapérouse visited the Jardin to take his leave of Buffon before embarking upon his ill-fated voyage around the world. In the account of the meeting written by Thouin in 1791 for the members of the Société d'Histoire Naturelle de Paris, Lapérouse requested Buffon for information on what to collect for the Jardin and Cabinet.[89] Buffon suggested suitable kinds of observations to be made on geographical issues, and made Lapérouse a present of "his works with his collection of birds in color" to aid him in recognizing desirable specimens for the Cabinet. He then directed the captain to Thouin, who suggested that the most advantageous way of obtaining specimens for the Jardin was to take a gardener on the voyage. When Lapérouse agreed, Thouin chose one of his brightest protégés, a young man named Jean Collignon, to care for plants on the voyage and to help in collecting and preserving plants. The botanist Lamartinière was his fellow traveler, so that Thouin could expect the most expert choice and care of the specimens that reached France from this voyage. Collignon was just one of a number of individuals whom Thouin had helped to obtain official posts on voyages and in gardens. To have access to travelers like Collignon was particularly valuable, because such protégés had been inducted into the language of botany.

Not all incoming specimens from foreign countries were of use to Thouin. On occasions he would receive packets of seeds from various sources, often connected with the Court, which he recorded in his catalogues as being "useless."[90] A combination of two circumstances

88 Thomas, *Man and the Natural World,* 224.

89 AN, AJ/15/565, "Instructions pour les voyageurs 1780s–1792": Thouin, "Mémoire historique rélatif à la partie d'agriculture de l'expédition de Mr. le chevalier de La Peyrouse."

90 AN, AJ/15/511, piece 450.

resulted in this judgment: first, if the seeds were jumbled up and unlabeled; second, if the seeds were in bad condition, for example, if they had not been well packaged or cared for and were thus rendered unviable. At a distance from the centers, there were periods during which seeds and plants escaped the surveillance of the central naturalist. When no natural historical expert could supervise the process of translation of specimens, the *état* of specimens was at risk. In 1787, for example, the entire crew of the *Sainte Charle* signed a document stating that they had found all the trees in two chests being transported on their ship to be dead.[91] Sometimes, however, central naturalists never discovered what had happened to consignments. To counteract such problems, Thouin, like other central naturalists, prepared numerous lists and memoirs to give to travelers, which would train them to know what plants to look for in different parts of the world, and how to label, package, and care for them.[92] Thus, requesting seeds from China, he asked for the Chinese characters giving the local name of each species to be written on separate packets, each containing seeds from a single species.[93] He followed a similar procedure when sending out lists of plants required to different parts of the world. Particularly if his correspondents were not educated in botany, he instructed them to separate kinds of seeds into different packets, to label each, and to list the different species according to their given labels in a catalogue.[94] The value of incoming specimens could thus depend both upon their state of preservation and upon the legibility of the accompanying inscription. It was less important to have a meaningful inscription than to have a distinctive one, so that Chinese characters and local and Linnaean names all served similar purposes. For example, when d'Angiviller sent Thouin specimens collected by the deceased abbé Noroña, he noted that the containing envelopes had torn and the seeds were mixed up in the bottom of the container. The majority would not grow; but d'Angiviller had preserved the envelopes carrying the descriptions of the source plants, and so their value was not lost, for Thouin could tell that there were many new species among the consignment.[95]

The importance of this textual accompaniment derived from the na-

91 BCMHN, MS 314.

92 AN, AJ/15/511, pieces 468, 469, 478, 486.

93 AN, AJ/15/511, piece 463.

94 For example, AN, F/10/210: "Mémoire abrégé sur les moyens les plus surs de Transporter par mer les Productions Végétales de l'Amérique septentrionale et autres Pays Temperés Analogues."

95 AN, AJ/15/511, piece 482.

ture of the correspondence network. Without the text or image that referred to plants, Thouin, unable to leave the center of his network, could not direct or control the movement of plants. When Thouin's correspondents requested species from him, they relied upon such inscriptions, increasingly the Linnaean binomial nomenclature, to communicate. Sometimes a full-length description was given, always in highly stylized form.[96] Thouin and his correspondents in the provinces, in Europe, throughout the world, formed a network which orchestrated the global movements of specimens. The influx of seeds and plants in the late eighteenth century depended upon texts: the letters sent to Thouin in Paris by Collignon on the Lapérouse expedition, William Aiton at Kew, or Jean-Nicolas Ceré on the Isle de France, the catalogues of provincial gardens used by Thouin as the basis of his consignments to these gardens, or the lists sent out by Thouin to travelers and colonists.[97]

Inscriptions were linked to the objects of natural history at all stages in their movements. The two-dimensional arrangement of the specimens in the Cabinet or Jardin meant that they could be read as inscriptions themselves by visitors literate in books and plants. It was commonplace to refer to "the immense book of nature," of which gardens and cabinets were "abridgments."[98] The systems published by naturalists enabled others to read that book. The classificatory debates sparked off by the arrival of new specimens in cabinets and gardens often revolved around how, and whether, one could represent real aspects of the nature of natural productions through writing.[99] The connection between knowing a natural production and converting it into inscriptions was very close. If Thouin received seeds which were not linked to recognizable names and descriptions, he would plant them in the Jardin in the following spring, and label the entry in his catalogue with an "x" and

96 E.g., BCMHN, MS 1973, letter 429: Desmoueux to Thouin, April 9, 1778, Caen.

97 BCMHN, MS 1973, letter 370: Denesle to Thouin, 1790, Poitiers: "Catalogue des arbres, arbustes et plantes qui manquent au jard. bot. de Poitiers," for example. On Ceré see Madeleine Ly-Tio-Fane, *The Triumph of J. N. Ceré and His Isle Bourbon Collaborators,* preface by Y. Perotin (Paris: Mouton, 1970).

98 "Rapport du citoyen Lakanal, Député de l'Arriège à la Convention, membre du Comité d'Instruction publique; lu le 10 juin 1793," in Hamy, "Les derniers jours du Jardin du Roi," 133.

99 Vicq d'Azyr summarized this view in proposing the reform of anatomical language to match that of botany: "[Linnaeus] understood that the basis of any construction of the spirit is the elementary science of words" (*Traité d'Anatomie et de Physiologie, avec des Planches coloriées Représentant au naturel les divers organes de l'Homme et des Animaux* [Paris: Didot l'aîné, 1786], 1: 47). See also Jean-Baptiste-Pierre-Antoine de Monet, chevalier de Lamarck, "Nomenclature," in *Dictionnaire de Botanique* (part of the *Encyclopédie méthodique*) (Paris: Panckoucke, 1783), 4:498–499; Louis-Jean-Marie Daubenton, "Introduction à l'histoire naturelle," in *Histoire naturelle des Animaux* (part of the *Encyclopédie méthodique*) (Paris: Panckoucke, 1782), ij–viij; Antoine-Laurent de Jussieu, *Genera plantarum secundum ordines naturales disposita* (Paris: Hérissant, 1789).

an "inconnu"—unknown.[100] The plant could not be known until it had been converted into text, both by describing and by naming. Until such time, it would not enter the correspondence network. Thus Georges-Louis-Marie Dumont, baron de Courset, at Boulogne-sur-Mer, sent Thouin a branch of an exotic shrub "which I do not know and whose name I pray you will tell me."[101] Most French naturalists publishing in the 1780s offered an opinion on the chances of finding the arrangement of natural productions which matched Nature's own. Systematists such as Linnaeus and Lamarck achieved fame in part because their books offered the wide botanical audience of the late eighteenth century a key (Lamarck's term) to the book of nature; moreover, they standardized the language of classification by producing textual restraints on description and naming which facilitated communication within the network.[102] This facilitation was a valuable reform for collectors both institutional and private. Seen from this standpoint, the claims of Linnaean supporters to have created order from a botanical science which was previously in chaos take on a new meaning. It is clear that the act of ordering was not a natural result of inevitable scientific progress, but a product of the social structures of European botany in the second half of the eighteenth century.[103] In these cultural circumstances, order was a moral imperative which stretched into the realms of finance and administration, into the manufacture of dictionaries, encyclopedias, and projects for the reform of language, and even into the naturalist's own work space, as in the numbered drawers with which Thouin created an interface between the classified seeds and the correspondents.[104]

100 BCMHN, MS 1384: "Catalogue des Graines qui ont été semé le 19 Avril 1770 au Jardin du Roy . . . ."

101 BCMHN, MS 1974: Courset to Thouin, January 19, 1793, Courset near Samer. Here the authority relationships expressed in the act of naming are clearly delineated. Courset avoided the widely acknowledged botanical faux pas of applying his own name to the nondescript, offering it, instead, as a patronage gift for the Parisian botanists to take priority in the appropriative act of naming.

102 Jean-Baptiste-Pierre-Antoine de Monet, chevalier de Lamarck, *Flore françoise ou Descriptions succinctes de toutes les plantes qui croissent naturellement en France, disposées selon une nouvelle méthode d'analyse, et précédées par un Exposé des Principes élémentaires de la Botanique,* 3d ed., 6 vols. (1779; Paris: Desaudray, 1815). This successful work, which had been supported by Buffon, contained a dichotomous key to enable identification of all known plant species.

103 The Adamic model of Linnaeus as drawing forth order from confusion, an important part of his quasi-religious self-construction as classifier, is discussed by Sten Lindroth, "The Two Faces of Linnaeus," in *Linnaeus: The Man and His Work,* ed. Tore Frängsmyr (Canton, MA: Science History Publications, 1994), 1–62; Stafleu, *Linnaeus and the Linnaeans.*

104 Foucault, *Order of Things.* On the interlocking relations of material, textual, and classificatory concerns in the eighteenth century, see Müller-Wille, *Botanik und weltweiter Handel,* chapter 6; Anke te Heesen, *Der Weltkasten: Die Geschichte einer Bildenzyklopädie aus dem 18. Jahrhundert* (Göttingen: Wallstein Verlag, 1997); on scientific projects for language reform, see Duris, *Linné et la France,* 126–132.

Collecting and classifying, accordingly, had profound implications for the self-perceptions of natural historical practitioners and for their situated status within a world of interconnecting networks. The physical demands of the transmission and ordering of plants and animals in the late eighteenth century shaped and constrained the conventions of classification. When in 1773 Antoine-Laurent de Jussieu published his first memoir on the classification of a group of plants, based upon the new natural method which he was later to offer as *Genera plantarum,* he used the term *rapports* to describe the natural, visible relationships between plants.[105] De Jussieu, at this time involved in the reorganization of the Jardin's botany school and in the formation of a single general herbarium, was also faced with the problem of physically bringing species together. This act of physical juxtaposition of species likewise warranted the term *rapporter.* But there was another sense in which naturalists in the late eighteenth century commonly used the term. They needed to be able to control how specimens and knowledge were *rapporté* from distant places—both in the sense of being "brought back" and in the sense of being "reported." One of the primary concerns of individuals at the center of networks was to load travelers to whom they had access with literature. When Lapérouse came to Buffon to ask how he should go about collecting on the voyage, Buffon's first response was to give Lapérouse a complete set of the *Histoire naturelle.* Thouin preferred to rely upon a traveler who was already trained. He prepared journals and lists for his traveling gardeners, and recommended that both they and untrained collectors read suitable books while on the voyage, usually works by Linnaeus and travel accounts.[106] Reading these texts was a way of inducting travelers into the language of precise description which botanists used to communicate with one another.

105 Antoine-Laurent de Jussieu, "Examen de la famille des Renoncules," *MARS* 1773/1777, 214.

106 In AN, F/10/201: "Liste de Plantes Arbres et Arbrisseaux les plus rares et les plus agréables qu'on peut demander dans l'Amérique Septentrionale," Thouin recommended Linnaeus's *Amoenitates academicae, seu Dissertationes variae physicae, medicae, botanicae, antehac seorsim editae nunc collectae et auctae cum tabulis aeneis,* 7 vols. (Leiden: C. Haak, 1749–1769); Pehr Kalm, *Travels into North America; containing its natural history, and a circumstantial account of its plantations and agriculture in general, with the civil, ecclesiastical and commercial state of the country, the manners of inhabitants, & several curious & important remarks on various subjects,* trans. J. R. Forster, 2 vols. (London: Forster, 1770–1771); Mark Catesby, *The Natural History of Carolina, Florida, and the Bahama Islands/Histoire Naturelle de la Caroline, la Floride, & les Isles Bahamas* (London: Catesby, 1731–1743); and William Stork, *An Account of East-Florida: With remarks on its future importance to trade and commerce* (London: Woodfall, 1766). For the simultaneous classification of books, species, and naturalists by Linnaeus, see John Lewis Heller, *Studies in Linnaean Method and Nomenclature* (Frankfurt am Main: Peter Lang, 1983), 115–126.

Naturalists also wrote memoirs with the specific aim of directing the efforts of travelers. Le Monnier wrote a "Recueil de ce qu'on doit ramasser dans un pays Nouvellement decouvert" ('Collection of what one should gather in a newly discovered country'), and Daubenton prepared a "Notice des Oiseaux qu'on desire recevoir de Cayenne vivans" ('Notice of the living birds which one wishes to receive from Cayenne').[107] As the efforts of the naturalists to exert more control over their "income" became more pronounced in the 1780s, even more of such guides appeared, mostly in manuscript form. By arming travelers with lists and catalogues of plant names and descriptions, as well as training them to read, describe, and distinguish between plants, the naturalists were able to extend their vision over the whole world, as it were. Many projects to direct the traveling naturalist were specifically written as guides of what (and how) the traveler should learn to see in order to be a good naturalist: "It is in the researches and the observations which can only be made on the spot, that it is particularly necessary to enlighten, or facilitate, the traveling mineralogist."[108] Besides making travelers familiar with the translations of plants into text, these documents were concerned with the question of disciplining the voyagers. In what manner was it possible to convince such a traveler, often one with no financial or legal ties to the naturalist in question, to act properly at a distance from the center?

The travelers' instructions written by the members of the Société d'Histoire Naturelle de Paris in 1791, for the voyage in search of the lost Lapérouse expedition, primarily used two different tactics. First, they portrayed the traveler in the familiar guise of romantic explorer, dedicated to the pursuit of truth at all costs. This heroic model was evident in the instructions for traveling mineralogists prepared by Alexandre-Charles Besson, a subinspector of mines: "It is those hard-working men, who, for love of science and knowledge, choose to brave the intemperacies, the exhaustion and the dangers of voyages, who find no mountain too high, when they wish to make a discovery there, in whom we must interest ourselves." Such men alone would yield truth—in other words, men who were prepared to forsake their own personal safety and comfort in order to carry out the naturalists' wishes. "If on

107 BCMHN, MS 357, piece VIII, undated; BCMHN, MS 352, undated. See also Lorelai Kury, "Les instructions de voyage dans les expéditions scientifiques françaises, 1750–1830," *Revue d'histoire des sciences* 51 (1998): 65–91; Pierre Huard and Ming Wong, "Les enquêtes scientifiques françaises et l'exploration du monde éxotique aux XVIIe et XVIIIe siècles," *Bulletin de l'école française d'Extrême-Orient* 52 (1964): 143–155.

108 AN, AJ/15/565: Louis-Claude Richard, "Instruction partielle pour les voyageurs naturalistes," February 18, 1791.

the contrary the spirit of system or of fabrication *[création]* should become involved . . . , we shall remain where we are and we shall never be instructed; for we shall only see by way of the systems adopted by the voyagers and their particular prejudices."[109] For the central naturalist, the traveler should be an instrument lacking individual opinions and concerns, a telescope through which the central naturalist could see things at a distance. It was the central naturalist's privileged role to interpret what was seen. The more successful the naturalist in controlling the traveler at a distance, the more transparent the traveler became. The good, "scientific" traveler was the individual who behaved properly "out there."[110]

The stress on "empirical observation" in the late-eighteenth-century natural sciences is often claimed to be evidence of the development of the modern scientific method. It can, however, be read as a concern with controlling events and people on the edge of a rapidly expanding world. The point of "empirical observation" was not seeing as an unbiased individual at all, but rather being seen *through*. Far from being free to pursue truth, the traveling naturalists were subject to continual attempts to ensure their adherence to a rigidly structured, unvarying, universally agreed method of describing or inscribing. Ultimately, this language served to facilitate the vision of the central naturalists, to render the reduction of specimens to text easier, and thereby to increase the efficiency of communication between powerful centers. The value of the text accompanying the moving elements in the correspondence network, whether consignments of specimens or travelers, was as a way of controlling and changing the direction of those movements so as to maximize the returns from each translation.

Other factors were also essential to the successful movement of specimens into and out of natural historical centers, however. The second major concern of the Société d'Histoire Naturelle in 1791, and of Thouin throughout the 1780s, was the training of travelers to preserve specimens properly, in other words to make them not just mobile but

109 AN, AJ/15/565: Besson, "Observations sur le choix des minéralogistes et leurs recherches pendant ce voyage projetté pour la recherche de Mr. de Lapérouse, lues à la Societé des Naturalistes de Paris le 18 fevrier 1791."

110 The distinctions between this model of the traveler and that of the autonomous romantic explorer are evident. See E. C. Spary, "L'invention de l' 'expédition scientifique': Histoire naturelle, empire, et Egypte," in *L'invention scientifique de la Méditerranée,* ed. Marie-Noëlle Bourguet, Bernard Lepetit, Daniel Nordmann, and Maroula Sinarellis (Paris: Éditions de l'École des Hautes Études en Sciences Sociales, 1998), 119–138; Outram, "New Spaces in Natural History"; Michael Dettelbach, "Humboldtian Science," in *Cultures of Natural History,* ed. Jardine, Secord, and Spary, 287–304; Outram, *Cuvier,* 60–68.

immutable.[111] Hence a large amount of space in the journals and memoirs was devoted to the practices involved in the collection, packaging, care, and preservation of specimens en route. It was essential to the whole enterprise that the specimens should arrive at the centers in a condition good enough to permit them to be "legible" to the central naturalist. The physician Philippe Pinel's memoir for the instruction of traveling zoologists suggested that "if the animals are of small size, he could easily cause the flesh to be consumed by having them boil in a kettle with a strong solution of soda or potash," and similar instructions for the gruesome and malodorous process of preserving animals. In Thouin's case, seeds, despite their useful properties of being small, not given to decay, and capable of reproducing new individuals, did not always suffice as immutable mobiles. "As many of our plants which may be useful to the inhabitants of the countries which will be traversed do not produce seeds there, or are varieties which do not have the power of propagating themselves by seed without deteriorating, it is useful to try and transport them in nature."[112] "In nature" meant as developed living plants. Sending these was a considerably more expensive and tricky matter for a correspondent in the botanical network. Plants needed particular care and packaging on voyages, such as could be given by the expert gardeners whom Thouin trained.

> In order that the living plants . . . should not suffer any accident from the jolting of the vehicles, it will be necessary to close the vents of the greenhouses after having previously put the wire grilles and the glazed frames in place. The two little windows on the extremities of each of these greenhouses should be left open, to establish a current of air which, in renewing itself, prevents fermentation and the decay of the objects which are enclosed in them . . . Organize it so that these chests are placed in a large vehicle at the two extremities farthest from the axle, so as to avoid the roughest jolts.[113]

With increasing government involvement in the plant network in the late 1780s, such practices increasingly fell within the bounds of possibility. More gardeners were sent on voyages, and ministers often allotted funds to the gardener in question for the purchase and care of living

111 The expression "immutable mobile" derives from Latour, "Visualization and Cognition."

112 AN, AJ/15/565: Thouin, "Mémoire historique Rélatif à la partie d'Agriculture de l'expédition de Mr. le chevalier de La Peyrouse," extract of letter to Lapérouse, May 29, 1785.

113 BCMHN, MS 47: Thouin, "Instructions pour le Sieur Joseph Martin rélativement à son voyage à l'Isle de France et à la conservation des plantes qu'il transporte d'Europe et à celles qu'il doit y rapporter," quoted in Letouzey, *Le Jardin des Plantes,* 184.

plants to be transmitted to other centers, in particular the colonies.[114] Thouin promoted such arrangements heavily. "On the representations that he [Thouin, writing in the third person] made to the minister of the marine that it would not suffice to send the plants to fulfill the beneficent aims of His Majesty, but that it would be necessary to cultivate them during the voyage . . . Mr. Thouin was authorized to make the choice of a subject who would be fit to carry out this object."[115] In the second half of the eighteenth century, decades prior to the invention of the Wardian case, trainee gardeners were thus voyaging around France and the oceans of the world alongside miniature greenhouses filled with plants.

For a naturalist, the next best thing to traveling oneself was the use of a trained protégé as emissary. The language of patronage could be used by the naturalist in Europe to control the traveler effectively at a distance, as was the case in the exchanges between European centers. Thus the possibility of future financial and intellectual support acted to control what Dombey did with his collections, and how he addressed his patron Buffon from abroad. Increasingly, naturalists like Thouin began to use their own protégés as travelers, and to exert all their power as patrons over these travelers in order to render them more transparent to the central naturalist. Thouin wrote a number of memoirs for the guidance of his traveling gardeners which served precisely these functions of control through patronage. At the end of the 1780s, one of Thouin's protégés, Joseph Martin, set out on a voyage to the Cape of Good Hope, traveling by order of the king on a ship captained by Fournier, and accompanying a cargo of living plants to the Isle de France as part of a Crown correspondence between colonial centers and the Jardin du Roi.[116] Thouin provided for his *Elève cultivateur* a journal which still survives. It is a deceptively small and shabby leather-bound volume, entirely handwritten, which yet might be described as the material presence of authority in Martin's life.[117] As Martin turned the pages to reach the instructions for plant care and lists of duties, the models for observations and records which were to be completed on shipboard, he would always have to traverse the text of his official

114 BCMHN, MS 56. Martin received 172 livres, 18 sols to purchase fruit trees for the island's Monplaisir garden.

115 BCMHN, MS 56: "Journal d'un voyage fait par ordre du Roi a l'isle de France en 1788 et 1789," f°. 1.

116 BCMHN, MS 56: "Pieces rélatives au Voyage de M. Joseph Martin à l'Isle de France en 1788 et 1789."

117 BCMHN, MS 308: "Pièces rélatives au projet d'un correspondence agriculto-botanique entre les differentes colonies françaises et le Jardin du Roi."

passeport at the beginning, denoting his duties toward the Crown. Interleaved among the pages of botanical advice was Thouin's account of the two alternatives awaiting Martin on his homecoming: either a warm welcome, honor and glory from Thouin, Martin's family, and the ministers, if he carried out his task successfully, or disgrace and obscurity if he did not. These repeated references to home and to Thouin's thoughts about Martin while on the voyage served to reassert and make more immediate the ties of obligation, and they were interspersed with guides to the appropriate behavior of the naturalist-gardener: collecting, recording, describing, and preserving specimens. This supports Latour's contention that travelers never leave home, but merely extend the limits of their world by taking their concerns and apparatus for interpreting the world along with them.[118] A traveler is a traveler only by virtue of coming back to the center; thus the strength of Thouin's control over his protégé depended upon the anticipation of Martin's return: "Mr. Thouin has been authorized to promise him that his place as *garçon jardinier* will be preserved, his income will be continued, and [he will receive] a gratification proportionate to his labors and to the zeal which he will bring to bear in fulfilling the aim of his journey, when he returns to France."[119] In addition, the gardener was expected to solicit new potential correspondents for Thouin and to deliver desiderata lists or catalogues to existing correspondents.

> As soon as he arrives at the Cape one of the first cares of the gardener must be to distribute the letters with which he is charged to the Persons for whom they are destined. These letters have as an object to facilitate the means of collecting the most interesting productions of the country which must be taken on the return voyage of the vessel from the Isle de France to be brought back to the Jardin du Roi in Paris . . . If there are to be found at the Cape any agricultural botanists and even any merchant who would be pleased to enter into a correspondence with the Jardin du Roi, be it to make exchanges of seeds, of dried plants or of plants in nature or be it for money, Mr. Joseph could propose to him a correspondence with me, take his name and his address and send them to me. He will leave mine for him with the little instruction relating to the manner in which to write to me and to get the consignments to me.[120]

118 Latour, "Visualization and Cognition," 6; Mary Louise Pratt, *Imperial Eyes: Travel Writing and Transculturation* (London: Routledge, 1992), chapter 2.

119 BCMHN, MS 56, f°. 2.

120 BCMHN, MS 56, ff. 23–25. Martin was in fact awarded a gold medal on December 28, 1789, by the Société Royale d'Agriculture for his efforts. See François Regourd, "La Société Royale d'Agriculture de Paris face à l'espace colonial, 1761–1793," *Bulletin du centre d'histoire des espaces*

Thouin was endeavoring to discipline his protégés to act in particular ways through the inscriptions which they received from him. The journal served both to coax and to dominate, and even the record forms inside served to standardize what Martin saw at a distance.

CONQUERING CENTERS

The sort of control that was being exerted by the use of patronage ties also served to some degree to subjugate other centers in France. Throughout the 1780s, Thouin had made several attempts to recruit the state in his effort to centralize power over the movement of specimens into France around the Jardin du Roi, as did other groups and individuals with natural historical and agricultural concerns. In 1784 the Société d'Agriculture de Paris solicited its members for names of potential correspondents. Under the guidance of the abbé Jean-Laurent Lefebvre, the Société established a correspondence network in the following year, aiming to distribute agricultural information and possibly also seeds to cultivators. To do this, it primarily used the local curés, whom Lefebvre could recruit and control, owing to his post as procurator general of the order of Sainte-Geneviève, to which a large number of these curés belonged.[121] Thouin exploited the agricultural interests of ministers. The *économiste* model of nature as the source of ever-renewing wealth for the state, developed in the 1760s, was central to the legislative concerns of a succession of ministers, including the *contrôleur-général des finances,* Anne-Robert-Jacques Turgot, the minister in charge of agriculture, mining, and manufactures from 1763 to 1780, Henri-Léonard-Jean-Baptiste Bertin; the minister of the navy, César-Guillaume de La Luzerne; d'Angiviller; and Malesherbes.[122] In 1785 Thouin engaged in correspondence with Lubert, the secretary of Charles Gravier de Vergennes, the head of the Conseil Royale des Finances, who had just established a Comité d'Administration d'Agriculture to advise on matters of agricultural reform. This body included

atlantiques, new series, 8 (1998): 165; *Mémoires d'agriculture, d'économie rurale et domestique, publiés par la Société Royale d'Agriculture de Paris,* 25 vols. (Paris: Buisson, 1786) fall 1789, xiij.

121 Gillispie, *Science and Polity,* 385.

122 Ibid., 378–383. On the physiocrats see, among others, Larrère, *L'invention de l'économie;* Georges Weulersse, *Le mouvement physiocratique en France de 1756 à 1770* (Paris: Félix Alcan, 1910); id., *La physiocratie sous les ministères de Turgot et de Necker, 1774–1781* (Paris: Presses Universitaires de France, 1950); id., *Les physiocrates à l'aube de la révolution, 1781–1792,* revised by Corinne Beutler (Paris: EHESS, 1985); Ronald L. Meek, *The Economics of Physiocracy: Essays and Translations* (London: Allen and Unwin, 1962); Elizabeth Fox-Genovese, *The Origins of Physiocracy: Economic Revolution and Social Order in Eighteenth-Century France* (Ithaca, NY: Cornell University Press, 1976).

numerous members of the core group of the Académie Royale des Sciences, such as Lavoisier, Tillet, Darcet, and the *économiste* Pierre-Samuel Dupont de Nemours, although Thouin himself initally refused membership. Its creation heralded a period of renewed ministerial interest in agricultural improvement, culminating in the reform and improvement in status of the Société d'Agriculture de Paris, which received royal letters patent in 1788.[123]

The initiating factor in Thouin's correspondence with Lubert was apparently a forthcoming visit to China by an individual on government business, for Lubert had requested Thouin to inform the administration which Chinese species would be of use to the Jardin du Roi. Thouin sent copies of an instructive memoir which he had prepared for distribution to voyagers, one for each part of the world.[124] In succeeding letters he maximized the degree of control that he was able to obtain over these government-connected travelers: "If you approve, Monsieur, I shall take the liberty of sending you several lists of the most interesting plants growing in the European establishments of the two Indies and which are unknown in our gardens. It will hardly cost you any more to join them to the copies of the Memoir which you are sending; that will direct the goodwill of the persons to whom you are transmitting them and will procure for us more interesting consignments."[125] These lists were laced with references to textual sources for the names and descriptions of the species desired. Asking Lubert to have the offices make several copies of the lists, Thouin added : "One must sow plenty of things of this nature, and yet one still harvests very little."[126] Lubert's diplomats were the social superiors of men such as Thouin, and thus it was impossible to discipline them. Even accredited traveling naturalists such as Dombey were active participants in the interpretative enterprise that fashioned the meaning of the specimen en route from its source to the botanical center; this could be advantageous for central naturalists, but could also yield knowledge which conflicted with the concerns of the recipients. Such categories of travelers could not be made to behave.

123 On the Comité d'Administration d'Agriculture see Jean Boulaine, "Les avatars de l'Académie d'agriculture sous la Révolution," in *Scientifiques et sociétés pendant la Révolution et l'Empire: Actes du 114e congrès national des sociétés savantes (Paris 3–9 avril 1989), Section histoire des sciences et des techniques* (Paris: Éditions du CTHS, 1990), 211–227. On Lavoisier's role see W. A. Smeaton, "Lavoisier's Membership of the Société Royale d'Agriculture and the Comité d'Agriculture," *Annals of Science* 12 (1956): 267–277.

124 AN, F/10/201: Thouin to Lubert, December 13, 1785.

125 AN, F/10/201: Thouin to Lubert, December 15, 1785.

126 AN, F/10/201: Thouin to Lubert, December 12, 1785.

Ultimately, Thouin's favorite solution, and one for which he pressed continually through the 1780s and 1790s, was to increase the number of locally trained, dependent travelers. However, his attempts to interest the Crown in such projects were relatively unsuccessful. In 1786 he proposed the setting up of an agricultural training college with "a chief site at Paris with a cabinet of rural natural history" and "a course of pupil-travelers in different parts of Europe and even in other parts of the world with a temperature analogous to our climate" (to enable the plants to be grown throughout France). Other peripheral sites for colleges were to be set up in the extreme corners of France. A closer analogy to the ideal situation for perfecting Thouin's enterprises at the Jardin du Roi can hardly be imagined. As Thouin acknowledged, however, the accomplishment of plans on this scale was dependent upon political support and institutional reform: "these great changes cannot take place until after the government has constantly exercised its will and establishments proportionate to the task [have been formed]."[127]

Clearly, the correspondence network could benefit enormously from enlisting the state. In late 1787 and early 1788 the new minister of the navy, the comte de La Luzerne, sent Thouin two letters containing an outline of a plan for starting a regular correspondence between the Jardin and the colonies.[128] During early 1788 Thouin elaborated upon the sketch of this plan. Initially, he wrote to de Vaivre, the intendant general of the colonies, outlining La Luzerne's suggestions. De Vaivre replied, asking Thouin to visit him, and wrote again soon after requesting a nomenclature for the plants and instructions on transportation. Thouin wrote, and had printed, a "Mémoire instructif pour le Transport des Vegetaux," as well as preparing a "Note relative a la Correspondance d'Economie rurale, [politique] et de Botanique qui doit etre établie entre les Colonies françaises et le Jardin du Roi." The word *politique* is crossed out in the draft. The most striking feature of the last-mentioned document is the way in which Thouin rendered both himself and the Jardin central and indispensable to the project. The first part of it was mainly concerned with the reorganization and restocking of colonial gardens whose resources were poor, or which had

127 BCMHN, MS 308: Thouin, "Projet de Memoire pour un College d'Agriculture," January 1786. This same phrase was used by Thouin in a memoir read at the first public meeting of the Société d'Agriculture de Paris on March 30, 1786, a month after the Société had been formally presented to the king and queen by the *contrôleur-général des finances.* See Thouin, "Mémoire sur les avantages de la Culture des Arbres étrangers pour l'emploi de plusieurs terrains de différente nature abandonnés comme stériles," *Mémoires d'agriculture, d'économie rurale et domestique, publiés par la Société Royale d'Agriculture de Paris,* spring 1786, 46–47.

128 BCMHN, MS 308: La Luzerne to Thouin, January 5 and 7, 1788.

gone to rack and ruin. Once they had been restored, however, their culture was to be "confided to the hands of instructed gardeners who will correspond assiduously with the Jardin du Roi," in other words, individuals who could be controlled at a distance. Moreover,

> It is necessary that [the gardens] should be directed and inspected by an instructed man who joins knowledge of botany and of the natural history of plants to knowledge of practical agriculture.
>
> It is necessary, if possible, that this direction should be effected at the Jardin du Roi, under the supervision of the Minister of the navy.[129]

The control over the flow of plants between the colonies was thus to be in Thouin's hands, under the auspices of his patron La Luzerne, whose "zeal will merit him the recognition of his contemporaries and that of future generations."[130]

These centralizing concerns reappear time and again in the manuscripts written by Thouin between 1785 and 1793. He sent lists of questions to the administrators of each colony concerning the nature of the plants grown at each place, as well as the Jardin du Roi's desiderata lists, and organized the port gardens as "entrepôts" for specimens on their way to, or from, the colonies. The outgoing specimens were paid for by the Crown; the incoming ones enriched the Jardin's collections and could usually be cultivated in the greenhouses there. Here Thouin was uniting state interests with the flow of plants into and out of the Jardin. Controlling the plant economy in the late eighteenth century conferred considerable political, as well as botanical, power. The close link between the plant economy and the interests of political economists in the French government ensured that managers of plant economies became state experts. As an expert in the germination and cultivation of exotic seeds, Thouin was invaluable in introducing new plants to France. By the mid-1780s he had already become a relatively well-known name in the savant world and at court—after all, several of his correspondents possessed court connections, besides those ministers and administrators who corresponded with Thouin for governmental purposes. But his expertise itself was political, in the sense that his success in cultivating and accounting depended upon success in managing other people, both above and below him in social rank. Without that success Thouin could not have reached a position where he was

129 BCMHN, MS 308: Thouin to de Vaivre, January 7, 1788; de Vaivre to Thouin, January 7 and 10, 1788.

130 BCMHN, MS 308: "Projet d'Etablissement de Correspondance entre les Colonies françaises et le Jardin du Roi."

able to manage so many specimens; on the other hand, he was able to exploit his wealth in terms of specimens to advance his own social position. There was a considerable increase in the total number of Thouin's correspondents overall up to 1786, followed by a slight decline when the Jardin enlargement plan reached its conclusion. All kinds of correspondents—botanical, artisanal, the clergy, the nobility, and so on—increased in number between 1775 and 1785; but the *proportion* of correspondents who were nobles increased at the expense of the botanists. This suggests that more nobles were becoming interested in botany, but it also reveals that Thouin's own social status was rising, as he associated more and more with the nobility. At the same time, the solution to the problem of maximizing the natural historical income of the Jardin turned out not to involve the enlistment of polite correspondents, but rather the attachment of the Jardin's interests firmly to those of successive governments. It is thus not surprising that Thouin's activities slipped easily into an institutional politics of botanical resources after 1789.

During the Revolution the Jardin du Roi became the privileged center for French natural history. In 1792 and 1793 the minister of the interior, Roland, organized the stripping of plant assets from other important Parisian gardens—the Trianon, the Bellevue, the Bagatelle, and gardens of nobles, religious orders, and public institutions closed by order of the Assemblée Nationale.[131] Under the Convention this practice continued: Thouin and his colleagues at the newly founded Muséum d'Histoire Naturelle were ordered to travel not only to French gardens but also to the European cities conquered by the Revolutionary army, to remove natural historical specimens from collections and have them shipped back to the Jardin. The best-known case is the confiscation of the cabinet of the stathouder of Holland, carried out by Thouin and Faujas de Saint-Fond in 1794.[132] In 1795 and 1796 Thouin traveled to Italy, visiting botanical gardens at Mantua, Padua, Parma, and elsewhere.[133] Among these gardens were centers which had corresponded with the Jardin for many years. A succession of governments seized and then sold off the properties of émigrés and *condamnés,* most of whom were former nobles and several of whom had previously corre-

131 André Guillaumin, "André Thouin et l'enrichissement des collections de plantes vivantes du Muséum aux dépens des jardins de la liste civile, des émigrés et condamnés: D'après ses notes manuscrits," *Bulletin du Muséum National d'Histoire Naturelle,* series 2, 16 (1944): 483–490.

132 Ferdinand Boyer, "Le Muséum d'Histoire Naturelle et l'Europe des sciences sous la Convention," *Revue de l'histoire des sciences et de leurs applications* 20 (1973): 251–257.

133 AN, AJ/15/836.

sponded with Thouin. Thouin was, therefore, well placed to know exactly what plants of value existed in the gardens of national properties. In an attempt to raise cash, the Convention began to sell off the *biens nationaux* in the late summer of 1793. One of the properties threatened was the Roule nursery, run by the abbé Nolin, one of Thouin's most assiduous correspondents since 1765. Thouin appealed to agronomically minded deputies in the Convention, such as the deputy to the Drôme, Joseph-Antoine Boisset, and the former abbé Henri Grégoire, a member of the Société Royale d'Agriculture and of the Convention's Comité d'Instruction Publique. On September 9, 1793, Thouin sent the latter a copy of his "Notes Relatives à l'Etablissement des Jardins de Botanique dans les departemens."[134] This document formed the basis for Grégoire's report on the establishment of rural economy institutions, given to the Convention on October 4, 1793. A second report followed on 16 brumaire year II (November 6, 1793).[135] Both of these were concerned with the setting aside of important sites from the *domaines nationaux* for the maintenance of natural history and rural economy establishments in every department.

Thouin continued to send information to the deputies. A copy of the "Notes" on botanical gardens which he had sent to Grégoire in the preceding September was sent to Joseph-Antoine Boisset in early 1794. On 16 germinal year II (April 5, 1794), the question of the Roule arose again. On the intervention of Charles Lacroix, the Convention voted to place Thouin in command of the valuable plants remaining in national properties, and annexed the Roule to the new Muséum d'Histoire Naturelle, as the Jardin was now known. It further ordered a review of the departmental botanical gardens, to be organized by the Comité d'Agriculture. On 6 floréal (April 25, 1794), the Convention voted to improve the departmental botanical gardens for the benefit of the arts and humanity. On 15 floréal (May 4), Boisset read a discourse, based on Thouin's notes, on "the advantages which the establishment of

134 BCMHN, MS 308.

135 Henri Grégoire, "Rapport et projet de décret, sur les moyens d'améliorer l'agriculture en France, par l'établissement d'une maison d'économie rurale dans chaque département. Présentés à la séance du 13 du premier mois de l'an deuxième de la République française, au nom des comités d'aliénation et d'instruction publique par le Citoyen Grégoire," and "Nouveaux développemens sur l'amélioration de l'Agriculture, par l'Etablissement de maisons d'Economie Rurale; Présentés par le citoyen Grégoire à la séance du 16 brumaire, l'an deuxième de la République une et indivisible," in *Oeuvres de l'abbé Grégoire,* 14 vols. (Paris: Éditions d'Histoire Sociale; Nendeln: Kraus-Thomson Organisation, 1977), 2:71–102, 119–140. See Harten and Harten, "Mustergüter und botanische Gärten: Ansätze einer sozialen Landwirtschafts- und Gartenbaupolitik," in *Die Versöhnung mit der Natur,* 33–41.

botanical gardens in the departments would procure," which was referred to the Comité d'Agriculture. Seven days later the Convention passed a decree establishing botanical gardens in all the departments. Possibly at Boisset's suggestion, Thouin wrote to his correspondents at public provincial gardens, and to the administrators of towns where he had no correspondent, asking for information about the size and condition of existing public gardens.[136] Then, on 11 prairial (May 30, 1794), Grégoire read a report to the Convention containing passages extracted verbatim from Thouin's notes, as well as information about the departmental gardens sent by Thouin and departmental administrators to the Comité d'Agriculture.[137] Grégoire's *projet de décret* was adopted; the Convention set aside 150,000 livres to pay for the maintenance and conservation of departmental gardens. Thouin's notes to the two deputies had included a cost estimate for the running and layout of an ideal establishment.

Over the summer of year II, the departmental administrations across France continued to receive requests for information on all botanical gardens within their districts from the Comité d'Instruction Publique and the Comité d'Agriculture. The more enthusiastic and detailed responses that flooded into the Commission d'Agriculture—some including plans and catalogues of the existing gardens, others proposing utopian schemes for the annexation of national properties to increase the size of the departmental gardens—were often from Thouin's own correspondents in the network of the 1780s.[138] Some even explicitly sought to model their garden upon the newly formed Muséum d'Histoire Naturelle. Moreover, a significant number of former Jardin protégés, whom Thouin had sent on voyages or found posts, were now in

136 BCMHN, MS 308: A. Thouin, "Notes pour le Répresentant du Peuple. Le Cen. Boisset," undated. This draft *projet de décret* formed the basis for Boisset's *Notes sur la necessité d'établir un jardin des plantes dans chaque département, faisant suite d'un rapport sur le même sujet* ([Paris]: Imprimerie Nationale, [1794]); and id., *Rapport et projet de décret relatifs à l'établissement des jardins des plantes dans les departemens, par Boisset, membre du Comité de commerce* (Paris: Imprimerie Nationale, [1794]).

137 BCMHN, MS 315: André Thouin, "Etat des Jardins Botaniques de la Republique d'après les renseignements adressés à la Commission d'Agriculture," undated; list of departmental gardens, undated; "Rapport fait au nom des Comités des Finances, des domaines, et d'Instruction publique, par Grégoire. Séance du 11 prairial, l'an 2e de la République une et indivisible." See *CIPCN,* 1:54, 509–511.

138 AN, F/12/1223–1224; "Etât." Harten and Harten, *Die Versöhnung mit der Natur,* 42, note that the formal decision by the Comité d'Instruction Publique to send a circular was taken on 3 messidor year II/June 21, 1794, but earlier responses from departmental administrations also exist; see AN, F/10/366. The exchanges continued for some five years.

charge of, or employed at, these departmental centers. They included Jean-Louis-Maurice Laurent, who owed his post as gardener and professor at the garden of the naval hospital at Brest to Thouin's intervention; Joseph Martin, who was trained by Thouin; Gaucherot, who was found a post by Thouin in the Orléans garden; and François-de-Paule Latapie, who had followed Desfontaines's botany courses. The gardener at Angers also owed his post to Thouin, as did the gardener Schweykert, whom Thouin had planted at Kew under William Aiton.[139]

Thouin's inventories of the plants in the provincial gardens were based upon the catalogues which his correspondents had provided to facilitate the movement of plants along the lines of correspondence. Many collections had suffered seriously during the revolutionary disturbances, and Thouin used the overflow in plants from the stripped "national" gardens to restock the gardens of his correspondents. The "Notes" on which the prairial project was based reveal the same concerns with increasing power over the network of correspondents that were to be seen in Thouin's reworking of the colonial correspondence project in 1788: "The garden of the Muséum d'Histoire Naturelle having become the central point for the reunion of the plants dispersed in the different parts of the world, its Administration will make a choice of those that could be useful to the different departments."[140] In other words, the Jardin would control which plants were available to which correspondents in France. By the mid-1790s the Jardin as a botanical center dominated the other French botanical centers, both in the eyes of the state and in terms of the control that Thouin had over what moved into and out of these peripheral centers, for such they had become. This situation was enabled by the increase in state intervention in the botanical correspondence network during the Revolution. But hardly any of the ambitious programs of the botanical reformers would be implemented.[141]

139 BCMHN, MS 1978, letter 1450: Laurent to Thouin, December 10, 1771, Brest; BCMHN, MS 56; BCMHN, MS 1975, letter 851: Gaucherot to Thouin, August 27, 1782, Orléans; BCMHN, MS 1977, letter 1427: Latapie to Thouin, February 10, 1789, Bordeaux; BCMHN, MS 1980, letter 1987: Merlet de la Boulaye to Thouin, 9 prairial year II/June 28, 1794, Angers; BCMHN, MS 1983: Schweykert to Thouin, September 16, 1782, Kew.

140 BCMHN, MS 308; Madeleine Ly-Tio-Fane, "A Reconnaissance of Tropical Resources during Revolutionary Years: The Role of the Paris Museum d'Histoire Naturelle," *Archives of Natural History* 18 (1991): 333–362.

141 Harten and Harten, *Die Versöhnung mit der Natur,* 50. Subsequent state support for organized botany and agriculture took the form of encouragement for private enterprise.

Conclusion

Thouin's success in constructing a network of correspondence which increased the inward flow of plants to the Jardin du Roi reflects his skill in capturing the "improving" interests of the state in the 1780s and 1790s. In effect, Thouin translated his interests and those of the state through the same channel, that of the increasing of the plant income of the Paris Jardin.[142] That obligatory passage point was the late-eighteenth-century concern with *économie.* Thouin was an expert economist in the eighteenth-century French sense: he was honest, laborious, and accountable; he was working for the public good by managing an economy of utility; and he was able to encourage a great number of people and seeds to behave in the ways that he wished. The inside of the seed was as inaccessible as the inside of the mind of a traveling naturalist, yet central naturalists promised that both, with correct knowledge of the social and natural laws that governed them, could be made to perform in specific ways. This was possible by recreating a specialized surrounding for each individual which mimicked the conditions under which they would normally perform. Whereas Thouin had to ensure that his protégés took France with them on shipboard, he also had to make certain that they would bring back the conditions under which exotic seeds could be made to feel "at home" in the Jardin du Roi.

Of course, many other naturalists, agronomists, and others were successful in capturing state interests for plant importation; the project of agricultural improvement in late-eighteenth-century France involved many individuals, both private and public. It is worth emphasizing that correspondence and seed exchange were directed, both explicitly and implicitly, toward the more efficient exploitation of colonial natural resources. The botanical networks were woven into the broader colonial projects of European nations in this period. Moreover, their study demonstrates with great clarity the degree to which such enterprises were a truly Europe-wide phenomenon, which, particularly in the eighteenth century, served the general hegemony of Western cultures over the rest of the world as much as the ends of individual nations. Likewise, it should be noted that a botanical network constructed by, or viewed

142 On obligatory passage points see Law, *Power, Action, and Belief,* esp. Michel Callon, "Some Elements of a Sociology of Translation: Domestication of the Scallops and the Fishermen of St. Brieuc's Bay," 196–233; Bruno Latour, *The Pasteurisation of France,* trans. Alan Sheridan and John Law (Cambridge: Harvard University Press, 1988).

from, the periphery would necessarily appear quite different. Although this study explored botanical networks, all "samples of the physical world" on which natural historical practice depended had to be moved from a distance to the centers of natural history, and preserved on the way. Natural history, in this sense, was a science of networks. The multiplicity of memoirs concerned with how to collect when at a distance from the center, and how to preserve—that is, how to create immutable mobiles—reveals that gardeners, naturalists, merchants, agriculturalists, and landowners faced similar problems.

In describing the structure of the correspondence network, I have elaborated upon the model of actor-network relationships developed by Latour and others, by discussing the ways in which the movement of inscriptions into and out of centers was initiated, controlled, and directed.[143] But Latour's model also has its shortcomings. The work put into rendering the objects of natural history both mobile and immutable must be investigated by historians, but so also must the work done in enforcing particular interpretations both on voyages and at home. Interpretations and inscriptions of objects are also acquired progressively. Latour's account supposes that there is one single interpretation which the center imposes upon objects when they are captured by travelers at a distance. Yet the objects themselves alter the interpretations that the center makes of the world "out there." They transform understandings of distant and invisible places—the past, the "other"—which are such important levers for activity at home. In other words, Latour's account does not allow for the results of networks—change. These vast circulating systems of specimens and people generated new natural historical knowledge, but also transformed the status of those who engaged in correspondence. Contemporary interest in Rousseau's "botaniste sans maître" suggests the fragility of botanists' efforts to control each other at a distance during the second half of the eighteenth century.[144] In particular, the French Revolution, by enabling attacks on the patronage system, would problematize that system for botanists who could not attach their interests to those of the new state.

Pierre-Paul Saunier, a young gardener who worked first at Montbard and then at the Jardin under Thouin, was the protégé selected to accompany André Michaux, a seasoned naturalist-voyager, to North

143 Esp. Law, *Power, Action, and Belief*; John Law, Michel Callon, and Arie Rip, eds., *Mapping the Dynamics of Science and Technology* (Basingstoke: Macmillan, 1986).

144 Jean-Jacques Rousseau, *Le Botaniste sans maître, ou manière d'apprendre seul la botanique: Fragments pour un dictionnaire des termes d'usage en botanique* (Paris: A. M. Métailié, 1983).

America to establish a botanical garden for the French crown.[145] While Michaux was originally to return after a few years, Saunier, who spoke no English, was out there for good. One of Michaux's letters complains of the periodic fits of depression experienced by the young man in the early years. In addition, Saunier barely received a subsistence wage, and repeated appeals from Saunier and Michaux to Thouin and d'Angiviller for an increase in the gardener's salary failed to have results. During the Revolution, the French government, concerned with more immediate issues, neglected the New York plot. In letters to his patrons Michaux noted that Saunier was using the New York garden to grow his own food, and complained that the gardener would no longer obey his orders. Effectively, the only solution for Saunier when existing patronage structures crumbled was to become independent. He had escaped from the controls of patronage which imposed laws of behavior upon young gardeners and naturalists; the mutual relationship, always hard to maintain at a distance, had broken down completely. When the French government at last turned its attention back to America, Saunier had earned enough from his "smallholding" to purchase a farm of his own and settle down with a wife. Thouin's exertions were directed toward limiting the traveler's power to change in this way. He sought to enforce particular kinds of behavior on protégés, and to prevent travelers from "escaping" and being transculturated, acquiring new values, and abandoning the social system which ensured right behavior at a distance. Yet the traveler was to be the agent of change in European society: the end result of traveling was to transculturate all of France. Thus there was a permanent tension between continuity and change, between preserving specimens and people and using them as levers for transforming society. This dialectic was, as we shall see, deeply implicated within contemporary accounts of the natural economy.

145 Letouzey, *Le Jardin des Plantes,* 197: Thouin to Poivre, August 20, 1785; *Correspondance,* 2:286, letter 586: Buffon to Thouin, June 10, 1785; William J. Robins and Mary Christine Howson, "André Michaux's New Jersey Garden and Pierre Paul Saunier, Journeyman Gardener," *Proceedings of the American Philosophical Society* 102 (1958): 351–370; Henry Savage Jr. and Elizabeth J. Savage, *André and François André Michaux* (Charlottesville: University Press of Virginia, 1986).

CHAPTER THREE

Naturalizing the Tree of Liberty: Generation, Degeneration, and Regeneration in the Jardin du Roi

Shall the tree of liberty be the only one which cannot be naturalized at the Jardin des Plantes?

"Seconde adresse des officiers du Jardin des Plantes à l'Assemblée nationale"

Natural history as defined by Buffon and Daubenton was broad in scope. But it was in the study of the processes of change within living beings that *histoire naturelle* related most closely to the concerns of improving landowners, legislators, physicians, and many others involved with the centralized management of French institutions and, ultimately, the French nation. I suggested in chapter 2 that much French natural historical inquiry was centered around the study of the natural economy—the cycle of processes which governed the perpetuation of nature, as the totality of organized bodies upon earth. Explanations of the operation of the natural economy were also driven by the needs of naturalists to account for change in the natural world to their patrons. Naturalists' ability to interpret the useful qualities of living beings allowed them to serve as consultants to improving legislators. However, contemporary distinctions between pure and applied science have ensured that such activities have appeared in histories of agriculture, rural economy, and animal husbandry rather than natural history.[1] The recent appearance of studies on natural economy and government for Sweden, Weimar, and Britain has further underlined the absence of a narrative connecting French naturalists' roles as managers of the natural

1 Octave Festy, *L'agriculture pendant la Révolution française,* 2 vols. (Paris: Gallimard, 1947–1950); Bourde, *Agronomie et agronomes.*

economy with the political function of their explanations of change in the natural world.[2]

Modern attempts to write a history of biology have thus constructed boundaries between parts of natural historical practice which reflect the development of nineteenth-century scientific specialization.[3] While one prominent study seeks the origins of "modern evolutionary biology, biogeography, ecology, physical anthropology, historical geology and cosmology" in eighteenth-century natural history, others have focused upon taxonomy.[4] But the attempt to relate past natural history to modern biology is problematic, for it has led historians to interpret natural historical writings selectively. Some aspects of the eighteenth-century discipline do not even appear to be "real science" to modern eyes, and so have been ignored; others are portrayed as examples of the feckless anthropocentricity of the Enlightenment. It has thus become very difficult to construct the links between the apparently diverse practices that composed eighteenth-century natural history and the social and political meanings that this discipline possessed for contemporaries. In this chapter, I shall suggest how a history of natural history might appear if such connections are taken as a central part of the discipline during the eighteenth century.

During the second half of the eighteenth century, Parisian naturalists explored the effects of a variety of external factors upon the characteristics of living beings. The two main determinants of change in the characteristics of a species were the climate—a general term drawn from the classics, which might embrace soil quality, air composition, temperature, topography, and humidity—and man, via the processes of domestication.[5] Discussions about climate, and interest in the relative im-

2 Koerner, "Purposes of Linnaean Travel"; Jackson, "Natural and Artificial Budgets"; James L. Larson, *Interpreting Nature: The Science of Living Form from Linnaeus to Kant* (Baltimore: Johns Hopkins University Press, 1994); Limoges, "Économie de la nature"; Carolus Linnaeus, *L'équilibre de la nature,* introduction by Camille Limoges (Paris: Vrin, 1987).

3 Paul L. Farber, "Research Traditions in Eighteenth-Century Natural History," in *Lazzaro Spallanzani e la biologia del settecento: Teorie, esperimenti, istituzioni scientifiche,* ed. Walter Bernardi and Antonello La Vergata (Florence: Leo Olschki, 1982), 397–403; Joseph Caron, "'Biology' in the Life Sciences: A Historiographical Contribution," *History of Science* 26 (1988): 223–268.

4 Lyon and Sloan, *From Natural History to the History of Nature,* 3. See also Jacques Roger, "Buffon et la théorie de l'anthropologie," in *Enlightenment Essays in Honour of Lester G. Crocker,* ed. Alfred J. Bingham and Virgil W. Topazio (Oxford: Voltaire Foundation, 1979), 253–261; Victor Hilts, "Enlightenment Views on the Genetic Perfectibility of Man," in *Transformation and Tradition in the Sciences: Essays in Honour of I. Bernard Cohen,* ed. Everett Mendelsohn (Cambridge: Cambridge University Press, 1984), 255–271.

5 For a good introduction, see Theodore S. Feldman, "The Ancient Climate in the Eighteenth and Early Nineteenth Century," in *Science and Nature: Essays in the History of the Environmental Sciences,* ed. Michael Shortland (Oxford: Alden Press, 1993), 23–40.

portance of culture and nature in determining the characteristics of living beings, rose to particular prominence in Revolutionary debates over the best way of managing society. While climatic theories of regeneration were fundamental to *histoire naturelle* after 1749, in the hands of revolutionaries they acquired a more utopian implication, as a means of implementing practical programs for the future perfection of the human species. As a royal establishment and a center for acclimatization, the Jardin des Plantes was closely involved with the study of generation, degeneration, and regeneration from the beginning of Buffon's intendancy. Although the role of degeneration in Buffon's work has recently been discussed, the connection between the natural historical and political meanings of the term has not been explored in the secondary literature, since it was principally and ephemerally elaborated after Buffon's death.[6] Degeneration and regeneration have found a place in the historiography of the French Revolution predominantly as political metaphors.[7] Concerns to explain the characteristic phenomena of living species prompted naturalists' growing interest in the potential of new natural philosophical entities, including electricity, magnetism, airs, and even animal magnetism. In the radical culture of the French Revolution, the revised model of climate was implemented in national programs for the reform of the physical and moral condition of humanity.

French natural historical improvers moved beyond utility and patriotism to implement utopian programs in which natural productions played a formative role in fashioning the future of republican France. As savant administrators of the sites of natural historical and agricultural practice, naturalists daily experimented with the relations between the natural and the social. Under the new regime, they claimed that their expert knowledge of the operations of the natural economy could contribute to an understanding of the proper government of society. At the Jardin des Plantes naturalists promised naturalized accounts of revolution and regeneration through their privileged understanding of the effects of cultivation upon living beings: in effect, they promised to

6 See Lyon and Sloan, *From Natural History to the History of Nature;* John H. Eddy Jr., "Buffon, Organic Alterations, and Man," *Studies in the History of Biology* 7 (1984): 1–45; Rheinberger, "Buffon."

7 Similar criticisms, which emphasize the medical significance of these terms, have been advanced by Nina Rattner Gelbart, "The French Revolution as Medical Event: The Journalistic Gaze," *History of European Ideas* 10 (1989): 417–427; Jeffrey Horn, "The Revolution as Discourse," a review of Baker, *Inventing the French Revolution, History of European Ideas* 13 (1991): 623–632. See also Peter Gay, "The Enlightenment as Medicine and as Cure," in *The Age of the Enlightenment: Studies Presented to Theodore Besterman,* ed. W. H. Barber, J. H. Brumfitt, R. A. Leigh, R. Shackleton, and S. S. B. Taylor (Edinburgh: Oliver and Boyd, 1967), 375–386.

convert the whole of France into a well-cultivated garden. This highly political agenda for natural history was to be progressively dismantled in Cuvier's program for a depersonalized, deductive, politically neutral natural history; but that is another story.[8]

Generation, Degeneration, Regeneration

The most famous writings of the Enlightenment made explicit reference to the social significance of climate. Jean Le Rond d'Alembert's *Encyclopédie* article "Climat" identified Charles-Louis de Secondat, baron de Montesquieu, magistrate, political theorist, and landowner, as the reviver of the view that climate was the principal determinant of the social institutions of different countries.[9] In *De l'Esprit des Lois* of 1748, Montesquieu had appealed to climate to justify his assertion that French society should be reformed in line with natural laws. According to d'Alembert, no one could doubt "that the climate influences the habitual disposition of the body, and thus the character; that is why laws must conform to the nature of the climate in unimportant things, and on the other hand, combat it in its negative effects."[10] But, notwithstanding the long Hippocratic tradition of climatic medicine, it was to the realm of natural history that, d'Alembert claimed, the study of the effects of climate on the constitution of living beings properly belonged. The results of just such a study appeared in successive volumes of Buffon's *Histoire naturelle.*

8 Admirably told in, among others, Outram, *Cuvier;* Corsi, *Age of Lamarck;* Toby A. Appel, *The Cuvier-Geoffroy Debate: French Biology in the Decades before Darwin* (New York: Oxford University Press, 1987); Goulven Laurent, *Paléontologie et évolution en France, 1800–1860: Cuvier-Lamarck à Darwin* (Paris: Éditions du CTHS, 1987); Adrian Desmond, *The Politics of Evolution: Morphology, Medicine, and Reform in Radical London* (Chicago: University of Chicago Press, 1989), 1–100.

9 Jean Le Rond d'Alembert, "Climat (Géog.)," in *Encyclopédie, ou Dictionnaire raisonné des Sciences, des arts et des métiers,* ed. Denis Diderot and Jean Le Rond d'Alembert (Paris: Briasson, 1753), 3:532–534. On climate theory see James C. Riley, *The Eighteenth-Century Campaign to Avoid Disease* (New York: St. Martin's Press, 1987); Ludmilla J. Jordanova, "Earth Science and Environmental Medicine: The Synthesis of the Late Enlightenment," in *Images of the Earth: Essays in the History of the Environmental Sciences,* ed. Ludmilla J. Jordanova and Roy S. Porter (Chalfont St. Giles: British Society for the History of Science, 1979), 119–146; Glacken, *Traces on the Rhodian Shore,* esp. 551–596; Theodore S. Feldman, "Late Enlightenment Meteorology," in *The Quantifying Spirit in the Eighteenth Century,* ed. Tore Frängsmyr, John L. Heilbron, and Robin E. Rider (Berkeley: University of California Press, 1990), 143–177. Parts of an earlier version of my argument have been reproduced in Richard H. Grove, *Green Imperialism: Colonial Expansion, Tropical Island Edens, and the Origins of Environmentalism, 1600–1860* (Cambridge: Cambridge University Press, 1995), chapter 4, esp. 162–164.

10 Jean Le Rond d'Alembert, "Éloge de M. le baron de Montesquieu," in *Encyclopédie, ou Dictionnaire raisonné des sciences, des arts et des métiers,* ed. Denis Diderot and Jean Le Rond d'Alembert (Paris: Briasson, 1755), 5:iii–xviii.

The social and moral relativism of writers like Montesquieu and Buffon mirrored accounts of the effects of climate upon society contained in other literary genres, notably the widely read travel accounts, which detailed the political, religious, and social institutions of exotic human societies in tandem with descriptions of climate and natural productions, dwelling especially upon the habits and appearance of humans and animals. Such accounts compared European society with other cultures and countries, and often ascribed differences in custom and conformation to differences in climate.[11] Claims to observe moral and physical difference in other locations were thus normative moves which could support arguments for either the decline or the superiority of Europeans.[12] Different civilizations were mapped onto a gradient of progress according to an implicit classification in which certain physical and moral features demarcated perfection.[13] Cultural contrasts were thus not static and inalterable, but the outcome of processes which were the combined work of nature and humanity, known in France as degeneration and regeneration. In Buffon's hands the study of these processes was formalized into an experimental program which spanned nearly half a century. Like Montesquieu, Buffon linked the degree of physical degeneration within a given society to the state of enlightenment or civilization of that society. The two aspects were integrated: a nation might learn to counteract the degenerative effects of the climate by becoming more enlightened, while an improvement in the climate might counteract the degeneration of society.[14] Thus, while contempo-

11 Ira O. Wade, *The Intellectual Origins of the French Enlightenment* (Princeton: Princeton University Press, 1971); Percy G. Adams, *Travelers and Travel Liars, 1600–1800* (Berkeley: University of California Press, 1962); Numa Broc, "Voyages et géographie au XVIIIe siècle," *Revue d'histoire des sciences et de leurs applications* 22 (1969): 137–154. An example, written by one of the Jardin's voyager-naturalists, was Pierre Sonnerat's *Voyage à la Nouvelle Guinée, dans lequel on trouve la description des lieux, des observations physiques et morales, et des détails rélatifs à l'histoire naturelle dans le règne animal et le règne végétal* (Paris: Ruault, 1776).

12 Henry Vyverberg, *Human Nature, Cultural Diversity, and the French Enlightenment* (Oxford: Oxford University Press, 1989). Daniel Pick, *Faces of Degeneration: A European Disorder, c. 1848–c. 1918* (Cambridge: Cambridge University Press, 1989), traces the later history of degeneration.

13 Elizabeth A. Williams, *The Physical and the Moral: Anthropology, Physiology, and Philosophical Medicine in France, 1750–1850* (Cambridge: Cambridge University Press, 1994); Glacken, *Traces on the Rhodian Shore,* chapters 12–14; David Spadafora, *The Idea of Progress in Eighteenth-Century Britain* (New Haven: Yale University Press, 1990); A. O. Lovejoy, *The Great Chain of Being* (Cambridge: Harvard University Press, 1970), 246 ff.

14 See, e.g., *Époques,* 212–213; Blanckaert, "Buffon and the Natural History of Man," 36; Phillip R. Sloan, "The Idea of Racial Degeneracy in Buffon's Histoire Naturelle," in *Racism in the Eighteenth Century,* ed. Harold E. Pagliaro (Cleveland: Press of Case Western Reserve University, 1973); Michèle Duchet, *Anthropologie et histoire au siècle des Lumières: Buffon, Voltaire, Rousseau, Helvétius, Diderot* (Paris: Maspero, 1971), 240; Amor Cherni, "Dégénération et dépravation: Rousseau chez Buffon," in *Buffon 88,* ed. Beaune et al., 143–154; Christian Marouby, "From

rary medical writers and even Buffon himself represented man as relatively immune to the degenerative effects of climate, it was still the most widely invoked factor in explanations of racial difference and the relations between nature and culture in humans.[15]

Buffon's concern to determine the causes of degeneration was reflected in his interest in studying living beings which changed climate—a topic which his contemporaries, both within and outside the savant world, confronted with considerable trepidation. The dangers of traveling to exotic climates were widely recognized in this colonial age. For many, the debate centered around whether displacement from the site at which nature had given rise to the originals of the species caused degeneration. This view of the specificity of locality for different natural kinds was still current in the 1780s.[16] The local failings of nature could cause degeneration in several ways: through variations in the temperature, altitude, wind direction, and constitution of the soil, but also through diet, disease, and breeding. As Buffon argued:

> It is principally via food that man receives the influence of the earth that he inhabits, that of the air and of the sky acts more superficially . . . In animals, these effects are faster and greater, because they are far more dependent upon the earth than man is, for as their nourishment is more uniform, more constantly the same, and is not prepared in any way, its qualities are more prominent and the influence is stronger; also, because animals cannot clothe themselves, nor protect themselves, nor make use of the element of fire to warm themselves, they remain nakedly exposed, and fully open to the action of the air and to all the intemperacies of the climate.[17]

Early Anthropology to the Literature of the Savage: The Naturalisation of the Primitive," *Studies in Eighteenth-Century Culture* 14 (1985): 289–298.

15 See articles such as "Le Lion," *Histoire naturelle* (1761), 9:1–2. The debate over the causes of skin color, however, was very lively in the second half of the eighteenth century. See Claude-Nicolas Le Cat, *Traité de la couleur de la peau humaine en général, de celle des negres en particulier, et de la métamorphose d'une de ces couleurs en l'autre, soit de naissance, soit accidentellement* (Amsterdam: n.p., 1765); Londa Schiebinger, *Nature's Body: Sexual Politics and the Making of Modern Science* (London: Pandora, 1993), chapter 4; Duchet, *Anthropologie et histoire,* 267–269.

16 E.g., see Pierre-Jean-Claude Mauduyt de La Varenne, "Conclusions," in "Ornithologie," in *Histoire naturelle des animaux* (part of the *Encyclopédie méthodique*), 423 (Paris: Panckoucke, 1782). Such discussions of the geography of Creation versus distribution were widespread across Europe: see Larson, *Interpreting Nature,* chapter 4; id., "Not without a Plan: Geography and Natural History in the Late Eighteenth Century," *Journal of the History of Biology* 19 (1986): 447–488; Jean-Marc Drouin, *Réinventer la nature: L'écologie et son histoire* (Paris: Flammarion, 1993), chapter 1.

17 *Histoire naturelle* (1766), 14:319–322. For Buffon's experiments on the effects of domestication upon the dietary habits of wild animals, see, e.g., Buffon, *Histoire naturelle* (1757), "Le Blaireau," 7:104–110; "La Fouine," 161–165.

Climate influenced animals in specific ways: in warm countries they were stronger and larger, in cool countries, weaker and smaller. These concerns depended upon Buffon's much-contested cosmogony, in which the earth was a rapidly cooling blob of matter formerly scattered from the sun by a passing comet. In *Les Époques de la nature*, published in 1778, Buffon would chart the different climatic epochs of the earth's history, claiming that living Nature could only have arisen on the globe once it had cooled to a certain point, and would come to an end once the earth became too cold. Three years before, in the "Hypothetical Part" of the second volume of the *Supplément* to the *Histoire naturelle*, he had carefully calculated the cooling rates of all the bodies in the solar system, suggesting in which order and for how many years "living Nature" was able to manifest itself on each, and drawing "the general conclusion of the real existence of organized and sensible beings in all the bodies of the solar system, and the more than probable existence of these same beings in all the other bodies which compose the systems of the other Suns, which increases and multiplies the extent of living Nature almost to infinity, and simultaneously erects the greatest of all monuments to the glory of the Creator."[18]

Buffon's cosmology was partly founded upon the view that two fundamental categories of matter, the organic and the brute, had existed *ab origine*. In much-publicized experiments in the 1740s with Daubenton and the English naturalist John Turberville Needham, Buffon demonstrated that generation was the result of the integration of organic molecules into an interior mold, or *moule intérieur*. At death, living bodies disintegrated once more into their component molecules. Diet was preeminent among the physical factors which Buffon advanced as causes of degeneration, for, in its food, the living being absorbed inorganic molecules from the local soil, which distorted its *moule intérieur*, irreversibly marking the individual with the characteristics of its locale.[19] The most serious form of degeneration, however, was experi-

18 *Histoire naturelle, Supplément* (1775), "Partie hypothétique," "Premier mémoire. Recherches sur le refroidissement de la Terre & des Planètes," 2:509, 515.

19 On the experiments with Needham, see Lyon and Sloan, *From Natural History to the History of Nature*, 165–209; Phillip R. Sloan, "Organic Molecules Revisited," in *Buffon 88*, ed. Beaune et al., 162–187. Buffon's distinction of all nature into organic and inorganic matter, while not universally rejected, was not widely accepted by other naturalists. See Eddy, "Buffon, Organic Alterations, and Man"; Sloan, "The Idea of Racial Degeneracy"; Jacques Roger, *Les Sciences de la vie dans la pensée française du XVIIIe siècle* (Paris: Armand Colin, 1963), 527–584; Vartanian, "Trembley's Polyp"; Elizabeth Gasking, *Investigations into Generation, 1651–1828* (London: Hutchinson, 1967). Less attention has been paid to writers who explicitly or implicitly accepted the existence of organic molecules. See [Ane (possibly Panckoucke)], *De l'homme, et de la reproduction des differens individus*. Félix Vicq d'Azyr, "Table pour servir à l'histoire anatomique et naturelle

enced by living beings lacking "liberty."[20] By this, Buffon meant the freedom to choose the most suitable conditions of existence—in short, freedom to act to prevent oneself from degenerating, either morally or physically. Later, the Jardin naturalists would explicitly interpret Buffonian liberty as a naturalized version of the political term.[21] But even in the 1760s, the political point was clear: a good society should allow man to act in all possible ways to counteract the degenerative effects of climate. "It is from [society] that man obtains his power, it is by means of it that he has perfected his reason, exercised his mind and gathered his forces; beforehand, man was perhaps the wildest and least redoubtable of all the animals."[22] Natural history became, in Buffon's hands, a means to investigate the past, and to predict the future, development of mankind.[23] Buffon was prominent among those arguing that humanity was a single species; indeed, only thus could accounts of the positive and negative effects of climate on exotic societies appear relevant to European society.[24] The original type of man had been the European; other varieties had emerged "after centuries had gone by, continents had been traversed, and generations had already been degenerated by the influence of different lands."[25] Degeneration, however, had both natural and social causes, and the powers of scientific inquiry could potentially be pitted against the forces of nature in order to combat it.

des corps vivans ou organiques," *Observations sur la physique, sur l'histoire naturelle et sur les arts,* series 2, 4 (1774): 479, left space for organic molecules in the order of nature.

20 Buffon, "Les animaux domestiques," in *Histoire naturelle* (1753), 4:169.

21 Hamy, "Les derniers jours du Jardin du Roi," 97–98.

22 Quoted in Roger, *Buffon,* 343–345.

23 In the *Discours sur l'origine et les fondements de l'inégalité parmi les hommes* (1755), Jean-Jacques Rousseau took Buffon's account of the natural history of animals and man to be fundamental for his own argument. See Rousseau, *Oeuvres complètes,* 3 vols. (Paris: Seuil, 1971), 2: 248–251. Among the literature on Rousseau's borrowings from Buffon, see Francis Moran III, "Between Primates and Primitive: Natural Man as the Missing Link in Rousseau's Second Discourse," *Journal of the History of Ideas* 54 (1993): 37–58; Asher Horowitz, "'Laws and Customs Thrust Us Back into Infancy': Rousseau's Historical Anthropology," *Review of Politics* 52 (1990): 215–241; id., *Rousseau, Nature, and History* (Buffalo: University of Toronto Press, 1987); Gilbert F. LaFreniere, "Rousseau and the European Roots of Environmentalism," *Environmental History Review* 14 (1990): 41–72; Étienne Géhin, "Rousseau et l'histoire naturelle de l'homme social," *Revue française de sociologie* 22 (1981): 15–31. The writings of eighteenth-century naturalists concerning a "science of man" were highly gendered, thus I deploy the gendered translation of "homme" purposely; see Sylvana Tomaselli, "The Enlightenment Debate on Women," *History Workshop* 20 (1985): 101–124.

24 *Histoire naturelle* (1749), 3:371–530. On Buffon's place in the debate over the unity of mankind, see Duchet, *Anthropologie et histoire,* 235–237.

25 Buffon, "De la dégénération des animaux," *Histoire naturelle* (1766), 14:311. For a more detailed discussion, see Eddy, "Buffon, Organic Alterations, and Man."

> To carry out an experiment on the change of color in the human species, it would be necessary to transport some individuals of that black race of Senegal to Denmark, where, man having commonly white skin, blond hair, blue eyes, the difference in the blood and the opposition of color is the greatest. It would be necessary to shut these Negroes in with their women, and scrupulously conserve their race without allowing them to interbreed; this is the only means one could use in order to know how much time would be necessary to reintegrate the nature of man in this aspect; and in the same way, how long was necessary to change [man's nature] from white to black.[26]

Under absolute rule, most middling educated individuals and many of the nobility were not able to partake in political, in the sense of "governmental," life in France, nor even to engage in explicit commentary on king and church. With climate theory Buffon and Montesquieu were able to convert the naturalization of social institutions into a political move, countering royal displeasure over adverse reflections upon French society with the implicit claim that the theory of climate was a natural law, thus beyond the reach of monarchs. The *Histoire naturelle* was singularly free of references to the social and moral status quo. Instead, it contained proposals for a morality based upon nature, couched in an evidential style resembling that of travel accounts. By using such styles, Buffon could simultaneously appeal to the existing audience for travel accounts, and embed his claims about the moral condition of French society in what purported to be an incontrovertible description, since it was an account of nature. Thus it was significant that Buffon's new classification of knowledge, unlike even the radical *Encyclopédie,* jettisoned sacred history altogether: "One could thus divide all the Sciences into two principal classes, which would contain all that it is suitable for man to know; the first is Civil History, and the second, Natural History."[27] Authors of eighteenth-century travel accounts largely restricted themselves to these two categories. Travel literature as a genre thus offered the possibility of generating a new secular morality, in which questions of individual and social behavior would no longer be discussed within a framework of Scriptural prescriptions for human conduct.[28] Eurocentric cultural perceptions could

26 *Histoire naturelle* (1766), 14:314. Compare Martin S. Staum, *Cabanis: Enlightenment and Medical Philosophy in the French Revolution* (Princeton: Princeton University Press, 1980), 26.

27 Buffon, *Oeuvres,* 15–16; compare Darnton, *Great Cat Massacre,* chapter 5.

28 On the rise of secular history, see H. W. Frei, *The Eclipse of Biblical Narrative* (New Haven: Yale University Press, 1974).

be applied to other societies within the "factual" style of travel literature in such a way as to disguise the work of interpretation that had gone into the writing of the account.[29]

But both Buffon and Montesquieu were judged by some to be treading close to the limits of acceptability. The *Histoire naturelle* was attacked by the Jansenist periodical *Nouvelles ecclésiastiques* for its claim that moral truths were "only matters of decorum and probability."[30] This was a dangerous claim in an era when the whole structure of society purportedly rested upon inalterable moral laws laid down by the state and church. Certain passages in the *Histoire naturelle* were also condemned by the Sorbonne. Montesquieu suffered even more: his contentious work, published in more liberal Geneva, was almost immediately placed on the *Index* of prohibited books. Nevertheless, it went through twenty-two editions in under two years.[31] Works like these helped to make the theory of climate a common discursive resource for all enlightened readers.

Culture and nature were thus closely interwoven in *histoire naturelle.* Contemporaries were intensely interested in phenomena which demonstrated the relations between these two agencies, such as domestication. In volume 6 of the *Histoire naturelle,* Buffon sought to explain the process of domestication of useful animals as the combined work of man and the climate.

> But when, with time, the human species extended itself, multiplied, spread, and by way of the arts and society man was able to march in force to conquer the Universe, little by little he made the wild beasts draw back . . . he pitted animals against animals, and by subjugating some by means of his skill, others through force, or by keeping them at a distance . . . and attacking them with rational means, he managed to place himself in safety . . . [Wild animals] thus degenerated, if their nature was ferocity joined to cruelty, or rather they merely experienced the influence of the climate: under a gentler sky, their nature softened, that which they had in excess was tempered, and by the changes which they underwent, they only became more suited to the country which they inhabited.[32]

29 Pratt, *Imperial Eyes,* chapters 2–4.

30 Lyon and Sloan, *From Natural History to the History of Nature,* 242.

31 Paul Edwards, ed., *The Encyclopedia of Philosophy* (New York: Macmillan, 1967), s.v. "Montesquieu."

32 *Histoire naturelle* (1756), 6:55–58. Compare Condorcet's attack on the slave trade in his "Éloge de Camper," *HARS* 1789/1793, 45–52: if Europeans civilized Africa and enlightened instead of depopulating, ferocious beasts would vanish.

There is clearly a tension present in Buffon's interpretation of the consequences of domestication. He refers to it in one sense as the worst possible form of degeneration, more devastating even than the effects of climate; this claim was reiterated by Jean-Jacques Rousseau and his followers.[33] Yet at much the same period in his work, Buffon also portrays man's role in artificially introducing, domesticating, and acclimatizing new species as a way of "embellishing," of perfecting Nature.[34] Hybrids, for example, were monstrous from a natural perspective, but also examples of man's ability to transform Nature from something that was "brute and dying" to an earthly paradise.[35]

For both Buffon and his coauthor Daubenton, man's rational faculty, the Creator's unique gift, justified his dominion over other created beings, even to the extent of deforming them from their original plan to suit his requirements for domesticated animals and plants. As Daubenton put it, "Man surpasses all material beings in dignity, by the ray of Divinity which animates him and enlightens him. His soul, immortal, gives him empire over the earth, and the use of all its productions." Unlike Buffon, however, Daubenton treated man as distinct from the animals in his introduction to the *Histoire naturelle des Animaux* in the *Encyclopédie méthodique* of 1782. "Man cannot be confounded with any

33 Jean-Jacques Rousseau, *Emile, ou de l'education* (1762; Paris: Flammarion, 1966), 34: "Tout est bien sortant des mains de l'Auteur des choses, tout dégenère entre les mains de l'homme. Il force une terre à nourrir les productions d'une autre, une arbre à porter les fruits d'un autre; il mêle et confond les climats, les éléments, les saisons; il mutile son chien, son cheval, son esclave" ('Everything is good as it comes from the hands of the Author of things, everything degenerates in the hands of man. He forces one soil to nourish the productions of another, one tree to bear the fruits of another; he mixes and confounds the climates, the elements, the seasons; he mutilates his dog, his horse, his slave'). I discuss these issues further in Spary, "The Nut and the Orange."

34 Contrasting views are in "Premiere Vue!" in *Histoire naturelle* (1764), vol. 12, and "De la degénération des animaux," in *Histoire naturelle* (1766), 14:311–374. See also Richard W. Burkhardt Jr., "Le comportement animal et l'idéologie de domestication chez Buffon et chez les éthologues modernes," in *Buffon 88,* ed. Beaune et al., 569–582. On the problems of domestication in the nineteenth century, see, among others, Harriet Ritvo, "At the Edge of the Garden: Nature and Domestication in 18th- and 19th-Century Britain," *Huntington Library Quarterly* 55 (1992): 363–378; James A. Secord, "Nature's Fancy: Charles Darwin and the Breeding of Pigeons," *Isis* 72 (1981): 162–186.

35 "Premiere Vue!" Buffon ended this account with a plea to the Creator to end wars upon earth and bring about an era of peace and plenty—a telling concern in the wake of the Seven Years' War, but also a passage with distinctly utopian echoes. John H. Eddy Jr., "Buffon's Histoire Naturelle: History? A Critique of Recent Interpretations," *Isis* 85 (1994): 644–661, however, offers a predominantly pessimistic Buffon. For eighteenth-century experimentation on and concerns about hybridization, see Bentley Glass, "Heredity and Variation in the Eighteenth Century Concept of Species," in *Forerunners of Darwin,* ed. Bentley Glass, Owsei Temkin, and William L. Straus Jr. (Baltimore: Johns Hopkins University Press, 1959), 144–172; Larson, *Interpreting Nature,* chapter 3; Gunnar Eriksson, "Linnaeus the Botanist," in *Linnaeus: The Man and His Work,* ed. Tore Frängsmyr (Canton, MA: Science History Publications, 1994), 95–98.

of the three Kingdoms of Nature, because he is the King. His power is founded, not only on the conformation of his body, whose organs produce more effects than those of the animals, but even more on his intelligence, his reason, and his industry."[36] The divine ray allowed man to improve upon nature and counteract the degenerative effects of adverse climate. Such a view supported both Daubenton's activities in the 1760s and 1770s in "improving" the race of French sheep, and the Jardin's function as a center for the introduction and acclimatization of new useful plants from all over the world. In 1766, when the last volume of the *Histoire naturelle des Quadrupèdes* was in press, Daubenton, who had prepared anatomical descriptions for the volume, was approached by the minister Daniel-Charles Trudaine and his son, Jean-Charles-Philibert Trudaine de Montigny, to serve as a consultant to the government by undertaking a state-funded program to improve the quality of French wool. Over the next fifteen years Daubenton studied not only breeding, but also the kind of food, the geographical location, and the temperature which best suited the animals imported from different climates (Morocco, Spain, Holland, and England, among others).[37] The results of his work were presented to the Société Royale de Médecine and to the Académie Royale des Sciences, and published in six memoirs between 1770 and 1782.[38] Similar programs were carried out by other French naturalists from the 1760s. In the 1780s, for example, abbé Henri-Alexandre Tessier, one of Antoine-Laurent de Jussieu's oldest friends and a central figure in the Société Royale d'Agriculture, was commissioned by the king to make a study of all the agricultural productions of Europe, and to set up an experimental

36 Louis-Jean-Marie Daubenton, "Histoire naturelle de l'Homme," in *Histoire naturelle des Animaux,* 1:xix. It should be noted that this was a reaction against Linnaeus, rather than Buffon. In manuscript fragments on monkeys, Daubenton criticized Linnaeus's choice of characters for including man among the quadrupeds, and went on to mount a full-scale attack on the "errors" and "faults" of the Linnaean system (BCMHN, MS 216).

37 Bourde, *Agronomie et agronomes,* 2:857–878; Gillispie, *Science and Polity,* 165–168.

38 Louis-Jean-Marie Daubenton, "Mémoire sur le mécanisme de la rumination, et sur le tempérament des bêtes à laine," *MARS* 1768/1770, 389–398, read April 13, 1768; id., "Observations sur des bêtes à laine parquées pendant toute l'année," *MARS* 1772/1775(1), 436–444, read November 15, 1769; id., "Mémoire sur l'amélioration des bêtes à laine," *MARS* 1777/1780, 79–87, read April 9, 1777; id., "Mémoire sur les remèdes les plus nécessaires aux troupeaux," *HSRM* 1776/1779, 312–320, read December 3, 1777; id., "Mémoire sur le régime le plus nécessaire aux troupeaux, dans lequel l'auteur détermine par des expériences ce qui est relatif à leurs alimens et à leur boisson," *HSRM* 1777–1778/1780 (2), 570–578, read December 11, 1778; id., "Mémoire sur les laines de France," read November 6, 1779. These memoirs, or extracts from them, also appeared in various editions of his *Instruction pour les Bergers et pour les Propriétaires de Troupeaux* (1st ed., Paris: Ph.-D. Pierres, 1782).

farm on the royal estate of Rambouillet for the purpose.[39] Thus there was a close relationship between the duties attached to naturalists' posts in the Crown administration in the second half of the century, and their attention to the interactions of nature and culture.

Whereas Buffon's Nature was a constantly changing, organic whole, whose plan unfolded in time rather than in space, Daubenton's investigations privileged the relationship between conformation and fitness to carry out one's role in the economy of nature. As a result, Daubenton's work devoted less space to determining the extent and causes of degeneration in living beings.[40] Both naturalists, however, shared a common concern to investigate the *moeurs* of animals. *Moeurs* is a term impossible to translate by a single modern English equivalent, but it encompassed behavior, custom, character, morality, habits, and instincts. The application of a single expression to cover the moral phenomena of animals, humans, and entire nations bespeaks the simultaneously political, social, and scientific nature of the discourse of natural history for Buffon and Daubenton. A detailed examination of one long entry, "Le Lion," in volume 9 of the *Histoire naturelle* (1761), will demonstrate the ways in which accounts of animal species could be converted into commentaries on human society, and the different strategies deployed by Buffon and Daubenton to describe a species as a natural kind.

Buffon's "Histoire du Lion" began with an extended comparison of the effects of climate on man with its effects upon the animals. In man, the air could cause changes in skin color and other characteristics which denoted the different races, but it could not produce distinct species. By contrast, in animals, the many varieties and closely similar species seemed to have been produced by exposure to the effects of climate. Some species could exist only in certain climates, some in others: each species "has its country, its native fatherland to which each is limited by physical necessity." Only man appeared able to inhabit all climates; the reason Buffon gave for this was that animals' "nature is infinitely less perfected, less extensive than that of man."[41] The histories of individual

39 Henri-Alexandre Tessier, "Mémoire sur la manière de parvenir à la connoissance exacte de tous les objets cultivés en grand dans l'Europe, et particulièrement dans la France," *MARS* 1786/1788, 574–589. He was also involved in the colonial correspondence set up by Thouin and La Luzerne: see id., "Mémoire sur l'importation et les progrès des arbres à épicerie dans les colonies françoises," *MARS* 1789/1793, 585–596.

40 Paul L. Farber, "Buffon and Daubenton: Divergent Traditions within the *Histoire naturelle*," *Isis* 66 (1975): 63–74.

41 Buffon, "Le Lion," in *Histoire naturelle*, 9:2.

species thus fitted into Buffon's extended commentaries upon the relations between man and the natural world, and allowed the nature, powers, and particular qualities of man to be highlighted by comparison with particular animal qualities.

By prefacing his account of the lion with an extended discussion of the nature of man, Buffon ensured that his discussion of the *moeurs* of the lion would implicitly possess significance for those same moral qualities in men. He compared the lion in captivity unfavorably with the wild individual, "the proudest [and] most terrible of all [animals]": "One has often seen him disdaining petty enemies, spurning their insults and pardoning them for offensive liberties; one has seen him in captivity, reduced to being bored without becoming angry, taking on gentle habits instead, obeying his master, flattering the hand which feeds him." But Buffon's description extended well beyond an Aesopian concern with the moral qualities of particular animals to more general relations of man and the animals.[42]

> To all these individual noble qualities, the lion also joins nobility of species. By species which are noble in Nature, I mean those which are constant, invariable, and which one cannot suspect of having degraded themselves; these species are usually isolated and alone in their genus; they are distinguished by characters which are so clear that one cannot fail to recognize them, nor [can one] confound them with any of the others. To begin with man, the most noble being of creation: his species is unique, because men of all races, climates, and colors can mix and produce together, and . . . no animal has a natural kinship with man, whether close or distant. In the horse, the species is not as noble as the individual, because it has the species of the ass as a neighbor . . . ; because these two animals produce individuals together which Nature in truth treats as bastards, unworthy to form a race, incapable even of perpetuating one or the other of the two species from which they originate; but which, coming from the mixture of the two, nonetheless demonstrate their great affinity.[43]

In his famous attack upon the philosophical basis of all classification, in the prefatory discourse of the first volume of the *Histoire naturelle*, Buffon had defined the species as a succession of generative acts.[44] As

42 For similar arguments concerning other periods or places, cf. Peter Burke, "Fables of the Bees: A Case-Study in Views of Nature and Society," in *Nature and Society in Historical Context*, ed. Mikuláš Teich, Roy S. Porter, and Bo Gustafsson (Cambridge: Cambridge University Press, 1997), 112–123; Harriet Ritvo, *The Animal Estate: The English and Other Animals in the Victorian Age* (Cambridge: Harvard University Press), 1987.

43 Buffon, "Le Lion," 9.

44 BCMHN, MS 218; Buffon, "Premier discours." See also Jean-Baptiste-Pierre-Antoine de Monet, chevalier de Lamarck, "Espèce," in *Dictionnaire de Botanique* (1786), 2:395; Mauduyt de

his experiments on crossing different species of animal—the horse and ass, the goat and sheep, the wolf and dog—progressed between 1749 and 1766, Buffon continually revised his views on the original forms of the Creation. By the end of his life he was writing articles like "Des Mulets," in which he gave what his peers considered to be highly speculative discussions of the development of present-day species from past crosses.[45] His arguments for the unity of the human species were bolstered by an appeal to the limits of interbreeding, but the effects of climate and diet on humans were nonetheless profound. Little is known about Buffon's cross-breeding experiments apart from the references he gave in the *Histoire naturelle,* but the *Dictionary of Scientific Biography* refers to his practice of "marriage broking" among Montbard peasants[46]—presumably to test his theories of the regenerative effects of cross-breeding, since "if, in the same canton, you examine men who inhabit high ground, . . . and compare them to those that occupy the middle of the neighboring valleys, you will find that the former are agile, well-disposed, well made, spiritual, and that the women there are commonly beautiful; while on flat ground, where the soil is heavy, the air thick, and the water less pure, the peasants are thick, weighty, badly formed, stupid, and the *paysannes* are almost all ugly."[47] Human intervention could thus reverse the impoverishment of Nature, as well as counteracting the degenerative effects of changing climate. As Buffon claimed at the very end of his discussion on the varieties of the human species, in volume 3 of the *Histoire naturelle:*

> It seems that the model of the beautiful and the good is dispersed over the whole earth, and that in each climate only a portion resides which is constantly degenerating, unless it is united with another portion taken from far away; so that in order to have good seed, beautiful flowers, etc., it is necessary to exchange seeds and never sow them in the same terrain which produced them; and, in the same way, in order to have beautiful horses, good dogs, etc., it is necessary to give foreign males to the females of the country, and reciprocally to the males of the country, foreign females; failing that, seeds, flowers, animals will degenerate, or rather take on such a strong hue of the climate, that matter will dominate over

la Varenne, "Discours deuxième," in "Ornithologie," 1:372. On the philosophical definition of species for Buffon, cf. Phillip R. Sloan, "Buffon, German Biology, and the Historical Interpretation of Biological Species," *British Journal for the History of Science* 12 (1979): 109–153; A. O. Lovejoy, "Buffon and the Problem of Species," in *Forerunners of Darwin,* ed. Glass et al., 84–113.

45 *Histoire naturelle, Supplément* (1789), 7:1–38; Larson, *Interpreting Nature,* 78–84.

46 *DSB,* s.v. "Buffon."

47 *Histoire naturelle* (1749), 3:528.

> form and will seem to bastardize it . . . In mixing . . . the races, and above all in renewing them constantly with foreign races, the form seems to perfect itself, and Nature [seems] to revive herself and produce all that she can of the best.[48]

By the time of the publication of volume 9 of the *Histoire naturelle,* Buffon had developed his view on the moral significance of interbreeding still further. In "Le Lion" he claimed that lower species of animals, from the dog—which had the wolf, fox, and jackal as its related species—to the rabbits, rats, and other small animals, with a much larger number of collateral branches, were increasingly promiscuous families of interbreeders, within which it was hard even to find the common root or *souche commune.* In other animal types such as insects and birds, it might even be necessary to classify species as a block of related forms. These sorts of claims legitimated Buffon's relative inattention to such "minute species of Nature," so loaded with hybrids as to disguise the true original species. They also underpinned his classificatory concerns: to denote the lion as a "*cat with a ruff and a long tail,* is to degrade and disfigure Nature rather than describing or denominating her." This dig at Linnaeus also indicates the profoundly different stakes involved in classifying for Buffon; clearly, because classificatory relationships should be based upon past generative acts, the classifier should choose his relatives carefully: "That is the true origin of methods." By this point in the *Histoire naturelle,* Buffon's breeding experiments and his order of describing species mapped onto a hierarchy of natural nobility, at a time when debates over the importance of ancestry, intermarriage, and adulterous liaisons for the noble condition were producing a considerable literature.[49]

The article "Le Lion" also served as a political commentary on different issues concerning ministers, in this case, population and industry. Buffon addressed the broader effects that man had, or could have, upon nature: "The industry of man increases with number, that of the ani-

48 Buffon, "Le cheval," *Histoire naturelle* (1753), 4:215–217. Probably the source for the Jacobin deputy Jacques-Michel Coupé's model of breeding in *Des animaux de travail et de leur tenue* ([Paris]: Imprimerie Nationale, year III/1794). That such issues continued to preoccupy Daubenton and his successors in breeding experiments at the Jardin and elsewhere is shown by Huzard's note, 422, to the posthumous edition of Daubenton's *Instruction pour les Bergers et pour les Propriétaires de Troupeaux; Avec d'autres Ouvrages sur les Moutons et sur les Laines,* 3d ed. (Paris: Imprimerie de la République, year X/1802).

49 Buffon, "Le Lion," 9–10; Chaussinand-Nogaret, *French Nobility;* Bien, "Manufacturing Nobles"; E. C. Spary, "Codes of Passion: Natural History Specimens as a Polite Language in Late Eighteenth-Century France," in *Wissenschaft als kulturelle Praxis, 1750–1900,* ed. Hans Erich Bödeker, Peter Hanns Reill, and Jürgen Schlumbohm (Göttingen: Vanderhoek and Ruprecht, 1999), 105–135; Spary, "'Nature' of Enlightenment."

Figure 6 "Male Dog-mule, second generation." A product of Buffon's experiments in crossing two species—the domestic dog and the wolf—stands calmly upon a trunk surrounded by the prosaic paraphernalia of everyday life. Images such as this served to import Buffon's claims about nature's pliability into the home itself. Engraving by C. Baron, after De Sève, 1789. From *Histoire naturelle, Supplément* (1789), vol. 7, plate 46, opp. 175. By permission of the Syndics of Cambridge University Library.

mals always remains the same."[50] It was this growing power of human societies, proportional to population, which had enabled man to overcome and alter the nature of wild animals. Buffon was cited by various authors of demographic studies for his interest in population, a widespread concern in eighteenth-century France.[51] Behind his comments

50 Buffon, "Le Lion," 3, 7, 5.

51 Buffon, "Des probabilités de la durée de la vie" and "État général des Naissances, des Mariages & des Morts," in *Histoire naturelle, Supplément* (1777), 4:149–264, 265–288; Morand, "Mémoire sur la population de Paris, et sur celle des Provinces de la France, avec des Recherches qui établissent l'accroissement de la Population de la Capitale & du reste du Royaume. Depuis le Commencement du siècle," *MARS* 1779/1782, 459–478; Pierre-Simon Laplace, "Sur les naissances, les mariages et les morts à Paris, depuis 1771 jusqu'en 1784; & dans toute l'étendue de la France, pendant les années 1781 & 1782," *MARS* 1783/1786, 693–702.

about the importance of industry for the natural history of the human species and the specific effects of captivity upon the nature of man and the animals, moreover, lay a political concern with the moral nature of labor and liberty which induced him to ally himself with Necker's government in the 1780s. It is well known that in 1749 he had written in opposition to the slave trade in the European colonies in "Variétés dans l'espèce humaine": "I cannot write their history without expressing my sorrow over their condition; are [slaves] not unlucky enough to have been reduced to servitude, to be obliged to work all the time without ever earning anything? Is it necessary to overwork them, to beat them, and treat them like animals? Humanity revolts against this odious treatment which greed has rendered customary." This was a humanitarian argument, in our sense, but it was also an economic one, taking as a fundamental condition of liberty the freedom to reap the fruits of one's own labor. Such concerns were perfectly mirrored in Buffon's philanthropic activities on his estate at Montbard, which consisted of supplying paid work for the local poor, reflecting a wider move against traditional forms of charity which was also supported by the Necker family.[52] Accounts of *moeurs* and descriptions of the effects of domestication thus mapped the topography of Buffon's political views. His history of the lion tied animals to man, and all living beings to a cosmology centered around the duration and powers of "living Nature" which linked all his diverse natural historical concerns, from experiments on wild and domestic animals to statements on human races, from the constitution of matter to the temperature of the air, from the meaning of industry and the sciences and arts to the end of life itself.

Daubenton's role, which was to write the description of the species, seems at first sight more neutral and less ambitious. For "Le Lion" his commentary was restricted to a topographic description of the lion's body surface, indicating the color and texture of its fur, the proportions of its body parts, their color and position; describing an actual example seen by Daubenton at the Paris bullring in 1757; and noting the ways in which the lioness's body differed from the description already given. His method consisted of measuring body proportions, and relied upon detailed anatomical studies. Tables supplied precise measurements of lengths, thicknesses, circumferences, separations, and widths of different parts, fashioning a virtual lion in figures. The succeeding part of the description covered the internal organs, with numerous cross-

52 Buffon, "Variétés dans l'espèce humaine," in *Histoire naturelle* (1749), 3:469. On Buffon's support for Necker, see chapter 1; on his Montbard philanthropic activities, see *Correspondance*, passim.

references to the relevant illustrations, and then, similarly, supplied tables of measurements of the organs, giving their lengths, circumference, thickness, separation, and width. Finally, the same procedure followed for the skeleton, bone by bone. Contemporaries such as Grimm and Bonnet praised this approach as more useful, in the long run, than Buffon's cosmological speculations and sweeping portraits of nature.[53] In Daubenton's later writings, however, the social and political meanings of his natural historical enterprise are evident. The study of conformation underpinned any account of the physical and moral faculties of man and the animals; in man, conformation, character, and degree of civilization were closely linked. "The Calmucks, although situated lower, near the Caspian Sea, seem to offer the traits most charged with this lugubrious and frightening profile under which the human species presents itself here. These are, by Tavernier's report, the most hideous of all Men . . . . [The traits of] the other Tartars . . . civilize and soften themselves as one advances toward China, where we will find a race of Men less disgraced by Nature."[54] Investigations into the natural world by a new generation of naturalists, brought up on the fifteen volumes of the Buffon-Daubenton *Histoire naturelle* that appeared between 1749 and 1767, drew upon both traditions, investigating conformation in its relationship to moral and physical faculties, and the effects and causes of changes in conformation and constitution.[55] Such concerns were widespread among European naturalists; but the emphasis among the Jardin's naturalists was definitely upon the exploitation of the transformative powers of living beings, rather than upon the exploration of the limits to transformation. Naturalists debated the speed, reversibility, and inheritability of the transformations operated by changes in physical surroundings; and the best place to experiment upon them was the privileged site of the botanical garden, where central naturalists were continually confronted with the transformative powers of nature and culture.

Acclimatization

"*Gardens,*" opined Lamarck in his article "Jardin de Botanique" in the *Encyclopédie méthodique,*

53 Daubenton, "Description du lion," in *Histoire naturelle* (1761), 9:26–48; Tourneux, *Correspondance littéraire,* 3:112–113, 301–306, 4:131, 170, 5:55.

54 Louis-Jean-Marie Daubenton, "Seconde Variété," in *Histoire naturelle des Animaux,* 1:xxxii.

55 This dual tradition also appeared in other European countries. See Wood, "Natural History of Man in the Scottish Enlightenment."

> may be reproached for having the disadvantage of changing the true habit of plants a little; for only giving an imperfect and sometimes misleading idea of that habit; for almost always augmenting the dimensions of the parts of the plants cultivated there; and often for altering their fructificatory parts, by monstrous multiplications which take place in the flowers, at the expense of their most essential organs. These monstrous multiplications, doubtless produced by an abundant sap which gives rise to a faulty development or growth of the fructificatory parts, constitute what are called *double flowers,* semi-double flowers, and are a highly sought after object of pleasure for the Florists. But the Botanist, who has the knowledge of nature uniquely in view, can no longer find in these plants the true traits which characterize them and which he is looking for. Thus the parts of plants which have undergone these changes which denature them, are no longer susceptible of being studied. . . . Thus, in *Gardens,* the abundance of sap can deform and extinguish the essential organs of flowers; but the characters which remain do not deceive; whenever they can be found, they are what it is appropriate for them to be.[56]

The eighteenth-century garden was thus a site at which the interaction of nature and culture could be closely observed—the place of investigation into the relative importance of natural powers and social intervention in remodeling the living being, and into the limits of such reshaping. Lamarck's discussion of the effects of culture upon flowering plants captures all the aspects of the garden which made it at once useful for man and less valuable for the savant interested in the study of "true nature." While the double flowers that could be generated in a garden under cultivation were of value to commercial gardeners and hobbyists, they were nevertheless the products of art, and in natural terms were only degenerated versions of the "real" thing.

Thus naturalists such as Lamarck were endeavoring to demonstrate their power to distinguish between the natural and artificial in the physical and moral properties of living bodies. In practice, however, such distinctions were highly problematic, and naturalizing judgments were contested not merely within France but within the wider European natural historical community.[57] The characteristic practices of eighteenth-century natural history underlined the difficulties of defining the natural, for naturalists continually encountered a series of new plants which changed size, shape, color, and habit with bewildering rapidity. In the same work, Lamarck explained how "[s]eeds of the

56 Lamarck, "Jardin de Botanique," in *Dictionnaire de Botanique* (1789), 3:211–212.

57 Larsen, *Interpreting Nature;* Wood, "Natural History of Man in the Scottish Enlightenment."

same plant carried to two different places, exposed and cultivated in completely opposed circumstances, will necessarily produce, after several years, two plants which differ greatly, principally in their appearance; so that one can be vigorous, succulent, of a darker green, more leafy in all its parts, etc., while the other will be thin, tough, shorter, and less straight."[58] Such chameleonic behavior created problems both for determining which features of a living being were properly to be called "natural," and, potentially, for delimiting "natural" boundaries between classificatory groups. The ascription of change in gardens to "social" or "cultural" rather than "natural" causes permitted naturalists to define their own realms of expertise at the same time as they demonstrated their control over the shapes and qualities of natural bodies. As Daubenton put it, "The Agriculturalist fortifies Nature in the production of plants, by means of labor and fertilizer."[59] Many French botanists regarded the deformations observed in gardens as the result of excesses of "nutritive juices" in the soil of gardens, from the fertilizer that gardeners applied to cultivated plants. The nature of the fertilizing principle was under much discussion during the second half of the century. For Lamarck and others, such as the Swiss botanist Louis Reynier, writing in the *Journal d'Histoire naturelle* which Lamarck edited, the variations in plant structure caused by cultivation were the direct result of an excess of nutritive molecules, which the plant built up into extra structures such as petals.[60] An anonymous correspondent of the Revolutionary journal *Décade philosophique* drew upon Wallerius's view that the active principle of fertilizers was animal "emanations" in the form of animal fat, although "the first, the best of these fertilizers is that nutritive humidity which is in the air, and that Nature combines herself."[61] For all these writers, fertilizers were manifestations of a nutritive principle circulating through the natural economy between the mineral,

58 Lamarck, "Espèce," 2:395.

59 Louis-Jean-Marie Daubenton, "Introduction à l'Histoire naturelle," in *Histoire naturelle des Animaux* (1782), 1:ix.

60 Lamarck, "Jardin de Botanique," 212; Louis Reynier, "De l'influence du climat sur la forme et la nature des végétaux," *Journal d'Histoire naturelle* 1 (1792): 101–148.

61 "Agriculture: engrais," *La Décade philosophique* 4 (1794–1795): 455. Interest in the fertilizing principle was shared by chemists. Van Helmont and others claimed that plants could grow in water alone, but during the 1770s Lavoisier argued that the complex atmosphere near ground level contained nutritive principles as subtle fluids, which fed plants ("Mémoires sur la Nature de l'Eau, et sur les Expériences par lesquelles on a prétendu prouver la possibilité de son changement en terre," *MARS* 1770/1773, 73–89, 90–107). Compare Simon Schaffer, "The Earth's Fertility as a Social Fact in Early Modern England," in *Nature and Society in Historical Context*, ed. Mikuláš Teich, Roy S. Porter, and Bo Gustafsson (Cambridge: Cambridge University Press, 1997), 124–147.

vegetable, and animal kingdoms. Applying a fertilizer was merely one way of supplementing the nutritive principle in the soil, since animal dung was a concentrated form of the products of fermentation and putrefaction which "vegetables continually borrow from the air."[62]

The manipulability of living nature allowed savants to experiment with the care of exotic animals and plants in order to stop the process of degeneration that faced living beings when changing climate. This gave rise to elaborate programs to investigate the relations between climate, cultivation, and characteristics, such as Thouin's proposed experiments on exotic cereal species, where soil type, time of sowing, and plant growth were to be recorded in registers and related to tables of climatic conditions.[63] In determining the utility of exotic plants, the job of the botanist was to see if these incoming plants could be "acclimatized" in France. But at the heart of such enterprises was a paradox: "acclimatization" or "naturalization" was the process by which a *social* operator successfully changed the *natural* properties of a living being. Because of the intimate relation between conformation and physical or moral qualities, "acclimatized" productions were in need of culture, at least in the short term, in the form of an artificial environment which would prevent them losing their useful qualities. For Thouin, botanical gardens could serve as acclimatization centers. Reporting to the Convention on the role of provincial gardens for the Republic, he portrayed Bordeaux as "the only maritime city of the globe in the 49th degree of latitude, and this equal distance from burning and frozen climates indicates of itself [that it has] the most suitable climate for the experiments necessary to naturalize the most useful plants of Sweden, Russia, Tartary, North and South America, Africa, and the Indies, in France."[64] Likewise, the colonial gardens were naturalizing intermediaries, serving to soften the blow of changing climate for exotic species; greenhouses served a similar function (figure 7). Naturalization was a process that had to be accomplished with care, often in several stages. In his ornithological dictionary of 1782, part of the *Encyclopédie méthodique*, Pierre-Jean-Claude Mauduyt de La Varenne devoted many pages to a discussion of how to acclimatize useful birds that were to be established in France: "First, it is necessary to transport them to our northern

62 "Prix proposé par l'académie des sciences, pour l'année 1794," in *Journal d'Histoire naturelle* 1 (1792): 467–469.

63 BCMHN, MS 318: Thouin, "Observations et experiences à faire sur la division des Plantes Céréales," prairial year II/May 1794.

64 BCMHN, MS 315: Thouin, "État des Jardins Botaniques de la Republique, d'après les renseignements adressés à la Commission d'Agriculture," circa 1793.

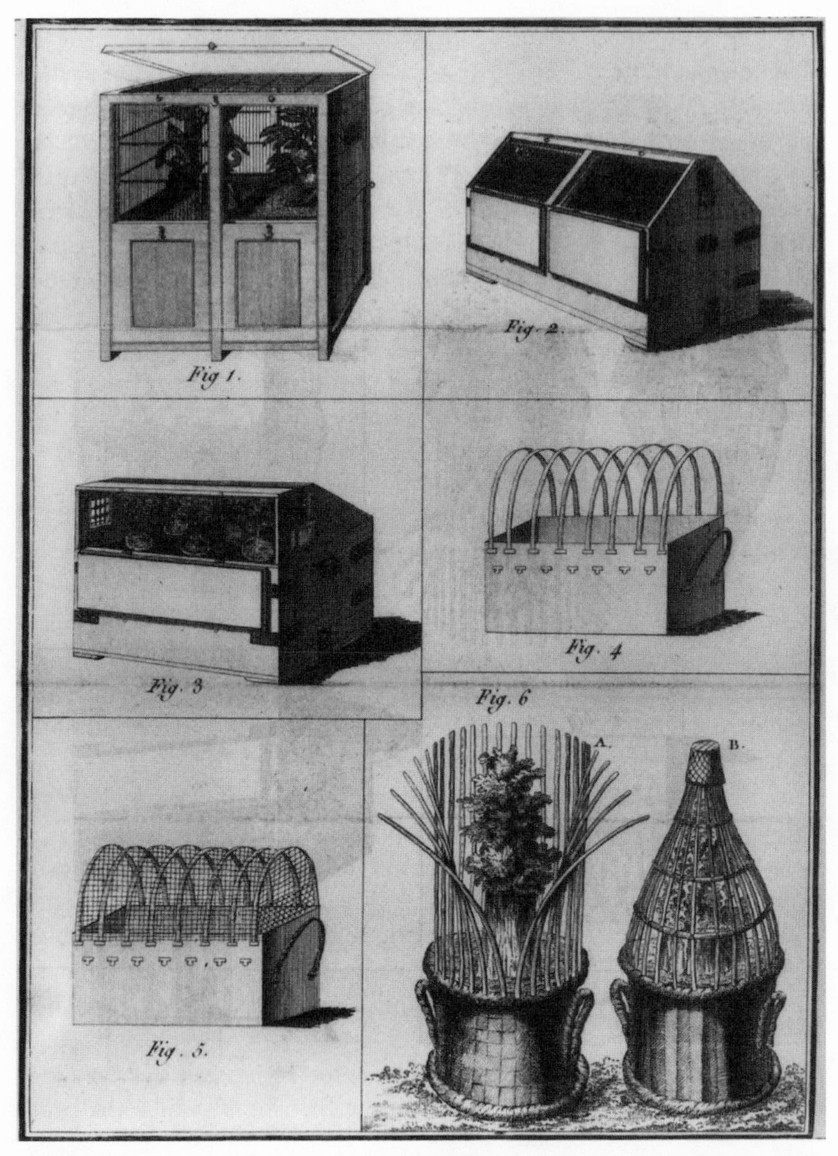

Figure 7 These containers and baskets, designed by François Lebreton, a correspondent of the Société Royale d'Agriculture de Paris, were specifically designed to assist in acclimatizing "the Mangostan, the Breadfruit, or any other precious tree from the great Indies or the Islands of the south Sea" (François Lebreton, *Manuel de botanique, à l'usage des amateurs et des voyageurs* [Paris: Prault, 1787], 385). From Lebreton, *Manuel de botanique,* plate 8. By permission of the British Library: shelfmark 452.b.15.

provinces or to those of the south, according to the temperature of the climate which they inhabit. France is happily situated for such a choice." The effects of an abrupt change in climate were very dangerous for the constitution of living beings; they could only be overcome by the greatest of care in transport, food, and extent of exposure to the climate of imported species. Once the species had bred in the new climate, however, "there will be less fear of losing it, it will be necessary to accustom the young ones by degree to the climate under which they are born; and by proceeding with prudence, they will be able to pass by degrees from the pen to the poultry-yard."[65]

The acclimatization of exotic living beings had great financial potential for the Crown. Thus it was important to assess whether imported beings retained their characteristics and could be naturalized, or whether they showed a tendency to degenerate, to the possible detriment of their valuable qualities. The Jardin du Roi, under the de Jussieu brothers and later under Le Monnier, had been among the first of the Parisian establishments to yield useful results for the state from the acclimatization of an exotic species, as Thouin and fellow naturalists of the following generation were to repeat over thirty years in their appeals to government on behalf of the institution. "For a century the greater part of the plants introduced into France have had as their nursery the Jardin du Roi and there are some which, at this moment, have become the objects of a considerable trade which has produced several millions for the State, such as that brought in by coffee cultivation."[66] France's many military enterprises in the first half of the century, coupled with the disastrous failure of French attempts to retain colonial possessions, created an unstable situation in which the naturalists could not form global correspondence networks. Different colonies changed hands several times between 1730 and 1760, during the period when Buffon and his colleagues might otherwise have been establishing networks. The nadir of French colonial ambition came with the surrender of the Indian colony of Pondichéry to the British, for which the com-

65 Mauduyt, "Quatrieme Discours," in "Ornithologie," 1:433.

66 AN, AJ/15/502: Thouin, draft project for the regulation of the Jardin, September 1788. Lavoisier estimated the quantity of goods consumed annually in Paris: spices to the value of 10,000,000 livres, cocoa worth 500,000, sugar worth 78,000,000, and coffee worth 3,125,000 (*AP*, vol. 12, March 15, 1791). Both coffee and the cedar of Lebanon, originally introduced to France and its colonies by the Jardin's botanists in the first half of the century, were well-known and material symbols of the promise of naturalization projects. See Marie-Noëlle Bourguet, "La collecte du monde: Voyage et histoire naturelle, fin XVIIe siècle–début XIXe siècle," in *Le Muséum au premier siècle de son histoire,* ed. Claude Blanckaert, Claudine Cohen, Pietro Corsi, and Jean-Louis Fischer (Paris: Muséum National d'Histoire Naturelle, 1997), 180–183.

manding officer, Lally-Tollendal, was executed. Only in 1763, after the Peace of Paris, did France regain many of her colonial possessions in India, North America, and the West Indies.[67] The Compagnie des Indes was forced to sell the Isle de Bourbon and the Isle de France to the Crown in 1764.[68] It was at about this time that concerted attempts began to be made by Parisian naturalists to use the Crown as a channel for the transmission of exotic plants to France and the Parisian gardens. Pierre Poivre, a traveling botanist and correspondent of Buffon—who referred to him several times in the *Histoire naturelle*—took up residence as head of the Pamplemousses botanical garden on the Isle de France in 1766. Louis-Guillaume Le Monnier, professor of botany at the Jardin between 1759 and 1786, formed a colonial correspondence network in the mid-1760s.[69] The botanist De Reine set up a colonial network in the late 1760s, centered around the garden at Petit Trianon.[70] It was not until the early 1770s, however, that Buffon and Thouin established state-recognized correspondence networks, justified by promises of national improvement and financial gain. Even outside physiocratic circles, the view that the wealth of a nation was founded upon its natural productions was widely supported by enlightened landowners, and was fueled by a corpus of improving agricultural texts of which the most famous were probably the works of Duhamel du Monceau. Major ministerial patrons of savants during the 1760s to 1780s, for example Turgot and Bertin, also interpreted natural history and agriculture in terms of the development of natural resources, and were highly important in supporting programs for the introduction of useful plants into acclimatization centers. A tree grown by Antoine-Laurent de Jussieu from a seed sent to the Jardin in 1778 by Turgot's cousin, who was a prominent member of the Société d'Agriculture de Paris, still survives.

Throughout the second half of the century, landowners and naturalists were engaged in testing the effects of climate upon a growing number of new exotic species arriving in France. Buffon's career in some

67 Paul Galliano, Robert Philippe, and Philippe Suissel, *La France des Lumières, 1715–1789* (Paris: Culture, Art, Loisirs, 1970), 171–177.

68 Grove, *Green Imperialism,* chapter 6. On the colonial situation in the last years of the old regime, see Paul Butel, "Revolution and the Urban Economy: Maritime Cities and Continental Cities," in *Reshaping France: Town, Country, and Region during the French Revolution,* ed. Alan Forrest and Peter Jones (Manchester: Manchester University Press, 1991), 37–51. Also *AP,* vol. 12, April 9, 1790; *AP,* vol. 24, March 15, 1791.

69 BCMHN, MS 357; BCMHN, MSS 1971–1985: Thouin's correspondence; Savage and Savage, *André Michaux,* 10–14.

70 BCMHN, MS 357.

ways typified the paternalist landowner's concerns for the advancement of agriculture and commerce in France. He worked throughout his life to make his estate at Montbard into a model example of the new agriculture, with plantations of trees, orchards, and vegetable gardens, as well as his famous ironworks.[71] Buffon's improving concerns are demonstrated in his efforts to enlarge the Jardin, but also in his choice of staff, many of whom slipped easily between natural history and its dependent arts in their roles as consultants to the Crown. Daubenton, Desfontaines, and Thouin were members of the Parisian Société d'Agriculture, chief among the network of agricultural societies initially established under the ministry of Bertin in the 1760s and 1770s.[72] The activity of these societies varied greatly, but there was a distinct decline in interest by members during the early 1770s, possibly as a consequence of Crown disapproval of the use of the societies as arenas for debates over agricultural reform. Local and central authorities prohibited the discussion of agricultural changes which would have involved legislative reform; such overtly political activity ran counter to the increasingly repressive moves of Louis XV's government in endeavoring to control the implications of the Damiens affair and *parlementaire* defiance.[73]

The accession of Louis XVI in May 1774 was widely perceived by the French educated elite to herald the dawn of an enlightened era for the French nation.[74] The concerns about climate that appeared in savant enterprises between the 1760s and 1780s were often expressed as part

71 Bourde, *Agronomie et agronomes,* 1:238–242. For another case study see Robin Middleton, "The Château and Gardens of Mauperthuis: The Formal and the Informal," in *Garden History: Issues, Approaches, Methods* (Washington, DC: Dumbarton Oaks Research Library and Collection, 1992), 219–242.

72 For the foundation of the agricultural societies, see Boulaine, "Les avatars de l'Académie d'agriculture"; id., *Histoire de l'agronomie en France* (Paris: Lavoisier Tec and Doc, 1992), chapter 6; Regourd, "La Société Royale d'Agriculture de Paris"; Bourde, *Agronomie et agronomes,* 2:1109–1116; Daniel Roche, *Le siècle des Lumières en province: Académies et académiciens provinciaux, 1680–1789,* 2 vols. (Paris: Mouton, 1978), 1:61–63. The periodical *Feuille du cultivateur,* 7 vols. (Paris: Bureau de la Feuille du Cultivateur, 1790–1798), edited by Dubois, the Société Royale d'Agriculture's perpetual secretary, Broussonet, and the abbé Lefebvre, and the *Mémoires d'agriculture, d'économie rurale et domestique, publiés par la Société Royale d'Agriculture de Paris,* 25 vols. (Paris: Buisson/Cuchet/Bureau de la Feuille du Cultivateur, 1785–1791), reveal the degree of overlap between contributors and Thouin's colonial or domestic correspondents. See esp. *Mémoires,* 1: 22–23.

73 Bourde, *Agronomie et agronomes,* 3:1195–1198; Dale Van Kley, *The Damiens Affair and the Unravelling of the Ancien Regime, 1750–1770* (Princeton: Princeton University Press, 1984); Keith Michael Baker, "Politics and Social Science in Eighteenth-Century France: The Société de 1789," in *French Government and Society, 1500–1850: Essays in Memory of Alfred Cobban,* ed. J. F. Bosher (London: Athlone, 1973), 208–230.

74 Simon Schama, *Citizens: A Chronicle of the French Revolution* (Harmondsworth, Middlesex: Penguin, 1988).

of a larger program for French agricultural reform. This was the case in both the Paris Société d'Agriculture and the Jardin du Roi.[75] At the same time, the French nation began to be portrayed as being in need of massive moral reform. Rousseauist writers held up the image of French degeneracy before the eyes of the urban rich. Natural history and agriculture appeared to many to offer solutions to the problems of moral and physical degeneration of the nation, and these sciences came also to embody the concerns for social reform of many individuals who were later to be involved in the French Revolution.[76] André Bourde's lengthy study of French agronomy identifies a program of "new agriculture" involving government officials, private landowners, gardeners, naturalists, and agriculturalists.[77] In their appeals to ministers between 1774 and 1789, French savants ascribed greater English prosperity to English agricultural supremacy—a powerful argument in the context of prolonged commercial and military competition with England.[78] These reformers based their arguments on the writings of English agricultural improvers such as Jethro Tull, which reached a French audience via Duhamel du Monceau's free translations after 1750.[79] In the hands of the French in the late eighteenth century, Tull's reforms diversified into

75 Falls, "Buffon et l'agrandissement du Jardin du Roi"; Yves François, "Buffon au Jardin du Roi, 1739–1788," in *Buffon*, ed. Heim, 105–124.

76 See Charlton, *New Images of the Natural*, esp. chapter 9; Schama, *Citizens*, chapter 1. Darnton, *Mesmerism*, 119–120, discusses Mesmerism as an alternative route to French regeneration, and see Mona Ozouf, "La Révolution française et l'idée de l'homme nouveau," in *The French Revolution and the Creation of Modern Political Culture*, vol. 2, *The Political Culture of the French Revolution*, ed. Colin Lucas (Oxford: Pergamon, 1988), 213–232. The religious origins of the term *regeneration* are explored in Michel Peronnet, "L'invention de l'*ancien régime* en France," *History of European Ideas* 14 (1992): 49–58.

77 Bourde, *Agronomie et agronomes;* Jean-Laurent Rosenthal, *The Fruits of Revolution: Property Rights, Litigation, and French Agriculture, 1700–1860* (Cambridge: Cambridge University Press, 1992).

78 Jeremy Black, *Natural and Necessary Enemies: Anglo-French Relations in the Eighteenth Century* (London: Duckworth, 1986), chapters 2 and 3; Doru Todericiu, "L'Académie Royale des Sciences de Paris et la mise en valeur des richesses minérales du Roussillon au XVIIIe siècle," *Comptes rendus du 106e congrès national des sociétés savantes, Perpignan, 1981, Section des sciences* 4 (1982): 153–159. Crown funding of the exploitation of natural resources was more common in France than in England. See D. G. C. Allan, "The Society of Arts and Government, 1754–1800: Public Encouragement of Arts, Manufactures, and Commerce in Eighteenth-Century England," *Eighteenth-Century Studies* 7 (1973–1974): 434–452; Hazel Le Rougetel, "Encouragement Given by the Society of Arts to Tree Planters: John Buxton's Work at Shadwell, Norfolk," *Journal of the Royal Society of Arts* 129 (1981): 678–681; Dulcie Powell, "The Voyage of the Plant Nursery, H.M.S. Providence, 1791–1793," *Economical Botany* 31 (1977): 387–431; Carter, *Sir Joseph Banks.*

79 Festy, *L'agriculture pendant la Révolution française*, vol. 1, chapter 2; André Bourde, *The Influence of the English on the French Agronomes, 1750–1789* (Cambridge: Cambridge University Press, 1953). See also Arthur Young, *Travels in France, during the Years 1787, 1788, and 1789*, ed. J. Kaplow (Garden City, NY: Anchor Books, 1969). Young himself was a correspondent of the Société Royale d'Agriculture.

a vast program of improving agriculture, including soil improvement, land clearance and drainage, and the introduction of new crops onto land previously considered to be unusable, such as fallow or wild land. Many improvers in Europe and America were able to experiment on various aspects of crop growing or animal husbandry on their own estates, often employing savant gardeners—sometimes Thouin's protégés—to introduce the practices and crops of new agriculture into their farms and plantations. By the 1770s the science of agriculture was highly fashionable, primarily among the wealthy and literate.[80] But there was a darker side to the claim that agricultural progress was to be achieved through increasing enlightenment, and naturalists' claims to be improving agriculture in order to bring happiness to the people should not be accepted at face value.[81] Members of agricultural societies sought to fashion a new peasantry which unquestioningly accepted the authority of savants over matters natural, and obeyed their dictates. Savant concerns with agricultural improvements were thus not politically neutral. Improvement, correspondence, and authority were closely linked within a particular politics of natural knowledge: increasing expertise in knowing the natures of things could be used to order and improve nature, people, and the arts. Thus, for example, Daubenton's sheep memoirs were presented as advice to shepherds, rather than as specialist texts for a scientifically literate readership. Simple shepherds were to achieve good husbandry, and make their herds profitable, through correct government of their beasts: in participating in the understanding of natural productions being developed by enlightened savants, shepherds could improve that nature and thus improve their own circumstances and those of the nation.

French writers on agriculture and natural history in the latter half of the eighteenth century exhorted enlightened readers to take into their own hands the task of improvement; agricultural reform would accomplish the economic and climatic amelioration of France. This discourse of national and personal improvement was used to elicit fund-

80 William Howard Adams, *The French Garden, 1500–1800* (London: Scolar Press, 1979), chapter 5, discusses the fashion for agricultural follies in the gardens of the wealthy during the late eighteenth century. See also Charlton, *New Images of the Natural,* chapter 2; Chrétien-Guillaume Lamoignon de Malesherbes, *Mémoire sur les moyens d'accélérer les progrès de l'économie rurale en France* (Paris: Ph.-D. Pierres, 1790); Laurent-Benoît Desplaces de Montbron, *Préservatif contre l'Agromanie, ou l'Agriculture réduite à ses vrais principes* (Paris: Hérissant, 1762).

81 Gillispie, *Science and Polity,* 244–256, and Robert Forster, "Obstacles to Agricultural Growth in Eighteenth-Century France," *American Historical Review* 75 (1970): 1600–1615, side with the reformers. A more critical view of the agronomists' activities is offered by Philip T. Hoffman, "Institutions and Agriculture in Old Regime France," *Politics and Society* 16 (1988): 241–264, and Rosenthal, *Fruits of Revolution,* esp. part 1.

ing and support from landowners, ministers, and the Crown. Savants, increasingly employed as consultants to an improving monarchy, linked claims about the effects of climate with political claims about the duties of government toward the citizen. Writing to the prince de Soubise to obtain royal support for a plant network centering on the Petit Trianon garden, the botanist De Reine promised the king "the advantage, without cost, of the sight of all the productions of the torrid regions, and this amusement is the more worthy of a great king because it contributes to the happiness of his people."[82] The regeneration of agriculture would lead to the happiness of the citizen—the true duty of government, according to many enlightened writers. During the 1780s new organizations mediated between savant supporters of the new agriculture and government officials. A subsection of the Société Royale d'Agriculture, known as the Comité d'Agriculture, was established in 1785 to liaise with the government's special committee of the same name.[83] One of its principal functions was the introduction of useful plants into France from abroad. In effect, the state was approving and supporting a practice which already had a long history in botanical centers. But these changes marked an increasing official interest in botanical networks which would confer greater power upon central naturalists.

But naturalization would acquire a more prominent role during the French Revolution. Since the mid-eighteenth century there had been an exchange between natural historical and political economical writers over the regeneration and perfectibility of living species, including man. The minister Turgot and his protégé Condorcet both argued that man's rationality, the Creator's special gift, allowed him to counteract the effects of natural degeneration and opened the possibility of infinite perfectibility. Writing on public instruction in 1791, Condorcet discussed how civilization was achieved through the pursuit of truth, and, since each truth led to other truths, "it is impossible to assign an end to this [process of] perfecting."[84] For Condorcet, like many of his con-

82 BCMHN, MS 357.

83 Bourde, *Agronomie et agronomes,* 3:1310; AN, F/10/201: letters, Comité d'Agriculture to Thouin, 1785.

84 Marie-Jean-Antoine-Nicolas Caritat, marquis de Condorcet, "Première mémoire: Nature et objet de l'instruction publique," in *Condorcet: Écrits sur l'instruction publique,* annotated and introduced by Charles Coutel and Catherine Kintzler (Paris: Edilig, 1991), 1:42; Condorcet, *Esquisse d'un tableau historique des progrès de l'esprit humain* (Paris: Agasse, year III/1794). Condorcet's patron, the minister Turgot, was the author of a Latin manuscript written in 1750 and afterward translated into French as "Tableau philosophique des progrès successifs de l'esprit humain" for circulation among his friends; it was published in English as *On the Progress of the Human Mind,* trans. McQuillin De Grange (Hanover, NH: Sociological Press, 1929). See also Michel Morange, "Condorcet et les naturalistes de son temps," in *Sciences à l'époque de la Révolution française,* ed. Roshdi Rashed (Paris: Albert Blanchard, 1988), 445–464.

temporaries, there was a close relation between cultivation and the utopian society; "it is not so chimerical as it might seem at first sight, to believe that culture can improve generations themselves, and that the perfection of the faculties of individuals is transmissible to their descendants."[85] The "culture" in question was as much natural historical as educational. Condorcet had been particularly excited by Daubenton's experiments in sheep breeding; they seemed to suggest that improvement in quality could be obtained after a very few generations. The slide to discussions about human perfectibility through breeding was easy, although Condorcet warned that other factors, such as type of government, might impinge upon the effects of breeding in humans.[86] It was precisely in these debates over the powers that breeding and naturalization conferred upon humanity that naturalists and agriculturalists possessed expertise, upon which legislators would repeatedly draw in preparing their reform proposals between 1789 and 1795.

Accounts of the operations of climate and cultivation blurred the distinction between the government of society and the government of nature, for naturalists claimed the ability to use social operations—introduction and acclimatization of exotica—to produce natural transformations. Revolutionary legislators would approach them as experts who could demonstrate the natural facts concerning these elisions between society and nature, in their attempts to ensure the success of further social operations. At the hands of botanical networkers, climatic theories were thus made central to social problems, both before and during the Revolution. After 1770 Thouin, at the nexus between the agricultural and natural historical worlds, was consistently occupied with plans for the introduction of food and other useful plants to France and to her colonies.[87] He portrayed this role as a "fortification" or improvement of Nature: "Nature has only given to the climate of France, tasteless vegetables, insipid fruits, and flowers of little beauty . . . all that we possess [which is] good and agreeable has been sent to us from foreign Countries."[88] One of the more important goals of acclimatization, for Thouin himself, was the introduction of the breadfruit as a crop plant into the West Indies. Thouin's view derived from his reading

85 Condorcet, "Premier mémoire," 1:44: "il n'est pas aussi chimérique qu'il le parait au premier coup d'oeil, de croire que la culture peut améliorer les générations elles-mêmes, et que le perfectionnement dans les facultés des individus est transmissible à leurs descendants." The term "culture" was often used for plants, but also described the development of children.

86 *HARS* 1777/1780, 16–17. Compare Staum, *Cabanis,* chapter 8.

87 Letouzey, *Le Jardin des Plantes,* 175.

88 BCMHN, MS 308: Thouin, "Nottes pour le Répresentant du Peuple. Le Cen. Boisset," second draft, autumn 1793.

Figure 8 From John Ellis, *A Description of the Mangostan and the Bread Fruit* (London: n.p., 1775), opp. 11. By permission of the Syndics of Cambridge University Library.

of John Ellis's account of the breadfruit of 1776, which had recently been translated into French (figure 8).[89] Since the Bougainville voyage to Tahiti, where breadfruit was a dietary staple, in the mid-1760s, Tahiti and all things Tahitian had become a potent symbol for the French elite.[90] Thouin's concern to naturalize the breadfruit in the mid-1780s, soon after the failed British *Bounty* voyage, reflected far more than mere nationalistic rivalry.[91] The breadfruit was the ultimate desideratum of

89 John Ellis, *Description du mangostan et du fruit à pain* (Rouen: P. Machuel, 1779).

90 Louis-Antoine de Bougainville, *Voyage autour du monde par la frégate du roi La Boudeuse et la flûte l'Etoile, en 1766, 1767, 1768, & 1769,* 2d ed., 2 vols. (Paris: Saillant et Nyon, 1772). See also Charlton, *New Images of the Natural;* Mornet, *Le sentiment de la nature en France;* Barbara Maria Stafford, *Voyage into Substance: Art, Science, Nature, and the Illustrated Travel Account, 1760–1840* (Cambridge: MIT Press, 1984); Bernard Smith, *European Vision in the South Pacific,* 2d ed. (New Haven: Yale University Press, 1985).

91 Mackay, *In the Wake of Cook,* 130–131.

improving acclimatizers—it was described as the perfect food for mankind, doing away with the need for other vegetable foodstuffs and balancing the constitution. If Tahitian society relied only on breadfruit for its physical (and, some said, moral) perfection, then the fruit itself must contribute to that utopian social state. Thouin's protégé Lahaye was the gardener sent on the d'Entrecasteaux expedition of 1791 to search for the missing Lapérouse expedition. In his instructions to his protégé, Thouin particularly stressed the value of the breadfruit tree.

> But one thing to which he must pay particular attention is to employ all the means that are in his power to procure himself the greatest number that he can take away of saplings of the best variety of the Breadfruit Tree and to do the impossible by cultivating them carefully to transport them alive to the Isle de France or other French or European colonies which he meets en route, and finally to bring them back to Europe . . . If he succeeds in enriching us with this tree, he will make the most useful of all presents to his fatherland, and by himself he will have done more for the happiness of mankind than all the savants of the world.[92]

The utopian status of the breadfruit as food for Thouin was shared by his colleagues in the Société d'Histoire Naturelle, whose report to the Assemblée Nationale on the d'Entrecasteaux expedition stressed the value of naturalizing the breadfruit "which by itself can replace all the vegetables necessary to life."[93] Breadfruit experiments continued throughout the Revolution; Guillaume-Antoine Olivier published a history of the discovery, varieties, and cultivation of the breadfruit tree in the *Journal d'Histoire naturelle,* which he and Lamarck edited.[94] A few years later, one of Thouin's and Olivier's recommendations was being carried out: the *Décade philosophique,* in fructidor year III (August–September 1795), reported the attempts of Bermond, "enlightened cultivator," to naturalize the plant along with other exotic species in the département des Alpes-Maritimes.[95] Naturalizers stressed the need for labor and the processes of art to be involved in the production of the ideal food. In 1778 Buffon had argued in his *Époques de la nature* that "the grain of which man makes his bread is not a gift from

92 Letouzey, *Le Jardin des Plantes,* 234: Thouin, "Instruction pour le Jardinier de l'Expédition autour du monde de M. D'Entrecasteaux," 1791.

93 AN, AJ/15/565: "Observations générales de la Société d'Histoire naturelle sur le voyage à entreprendre pour aller à La Recherche de Mr. de La Peyrouse," [1790–1791].

94 Guillaume-Antoine Olivier, "Observations sur la culture de l'arbre-à-Pain et des épiceries," *Journal d'Histoire naturelle* 2 (1794): 72–80.

95 "Essai tenté pour la naturalisation de plusieurs végétaux," *La Décade philosophique* 2 (1795): 340–344; 3 (1795): 199–202.

Nature, but the great, the useful fruit of his researches and of his intelligence in the foremost of arts; nowhere on earth has wild wheat been found, and it is evidently a plant perfected by his care," and went on to outline the precepts of the new agriculture.[96] Likewise, Olivier's memoir noted that "[i]t is highly probable that culture has rendered the fructificatory parts of [the breadfruit] tree gradually imperfect"—since cultivated breadfruit were all seedless and were propagated by cuttings.[97] The utopia of the new agriculturalists was not to be accomplished, or maintained, without labor.

In the context of a prolonged subsistence crisis during the Revolution, the efforts of Thouin and his fellows to introduce new food plants attracted attention from a succession of legislative bodies and Republican cultivators.[98] In the spring of year II Thouin was commissioned by the minister of the interior, Jules Paré, to organize the digging up of the luxury gardens of Paris—the Tuileries and the Luxembourg—and their replanting with potatoes.[99] However, central control over which reforms actually took place was far from perfect, since municipal governments sometimes arrested landowners who tried to grow fodder crops on their lands. For example, in pluviôse year II (February 1794) the marquise de Marbeuf, who had corresponded with Thouin in 1781, was executed for implementing a program of the new agricultural practices on her lands, rather than cultivating wheat on every available surface area.[100] Despite their promises of regeneration and plenty, the centralizing practices of the new agriculture were not viewed as transparently beneficial by all. Nonetheless, the breadth of improvers' claims and the wide appeal they possessed for many demonstrates the intimate associations that were being constructed between political, rural, and natural economy. During the 1770s and 1780s, however, a new impetus would be given to studies of the economy of Nature—primar-

96 *Époques,* 217–218.

97 Olivier, "Observations sur la culture de l'arbre-à-Pain et des épiceries," 79.

98 Susanne Petersen, *Lebensmittelfrage und revolutionäre Politik in Paris, 1792–1793: Studien zum Verhältnis von revolutionärer Bourgeoisie und Volksbewegung bei Herausbildung der Jakobinerdiktatur* (Munich: Oldenbourg, 1979). Letters from Thouin sending crop plants to cultivators still survive occasionally in provincial archives; e.g., Bibliothèque du Port de Brest, MS 1507, 4: f°. 102, 16 germinal year II/April 5, 1794.

99 *Actes CSP,* 12, 13 germinal year II/April 2, 1794; BCMHN, MS 308: Paré to Thouin, 18 ventôse year II/March 8, 1794; Harten and Harten, *Die Versöhnung mit der Natur,* 28–32.

100 Festy, *L'agriculture pendant la Révolution française,* 1:137. The Convention apparently supported the efforts of the new agriculturalists throughout its existence, in spite of *sans-culotte* and municipal opposition. The classic study of the *sans-culottes* remains Albert Soboul, *Les sans-culottes parisiens en l'an II: Histoire politique et sociale des sections de Paris, 2 juin 1793–9 thermidor an II,* 2d ed. (Paris: Clavreuil, 1960).

ily fueled by recent work on airs and atmospheres carried out by a number of European natural philosophers.

Revolution in the Air

Buffon's first entry onto the savant scene was as the translator of the English natural philosopher Stephen Hales's *Vegetable Staticks.*[101] This in itself was indicative of the primacy of the natural economy in the future intendant's agenda for the investigation of the natural world. Similar concerns were shared by Parisian naturalists, gardeners, and agronomists of the next generation, who assimilated into the program of agricultural reforms a series of recent chemical studies on the effects of plants on the air, sometimes portrayed as a tradition of research derived from Hales's well-known work.

The changing land use of the Jardin during Buffon's intendancy reveals the particular importance of trees for both Thouin and Buffon. After the expansion, the largest area was occupied by plantations of trees, including a large nursery for North American trees suitable for naturalizing in France. In contrast, the largest area of the Jardin before 1739 was taken up with the botany school (figures 9 and 10). These transformations reflect a widespread interest in promoting tree planting; wood supplied the principal material for building, especially shipbuilding, and fuel.[102] Even before becoming intendant at the Jardin in 1739, Buffon, a friend of Duhamel du Monceau, had had considerable interest in silviculture, which featured heavily in his earliest papers to the Académie Royale des Sciences.[103] In France, advocates of the new agriculture supported the planting of trees along the verges of the new road system. However, the magistrate Joly de Fleury noted that, although trees were supplied free by a system of royal nurseries estab-

101 Stephen Hales, *La Statique des Végétaux et l'Analyse de l'Air, Expériences nouvelles lues à la Société Royale de Londres* (Paris: Jacques Vincent, 1735).

102 E.g., Henri-Louis Duhamel du Monceau, *Du Transport, de la conservation et de la force des bois* (Paris: Delatour, 1767); Bourde, *Agronomie et agronomes,* passim. See also Andrée Corvol, "L'arbre et la nature, XVIIe–XXe siècle," *Histoire économie et société* 6 (1987): 67–82; id., *L'homme et l'arbre sous l'ancien régime,* preface by Pierre Chaunu (Paris: Economica, 1984); G. Buttoud and Y. Letouzey, "Les projets forestiers de la Révolution," *Revue forestière française* (1983): 9–20.

103 Georges-Louis Leclerc de Buffon, "Sur la conservation et le rétablissement des forêts," *MARS* 1739/1741, reprinted in *Histoire naturelle, Supplément* (1775), 2:249–270; id., "Moyen facile d'augmenter la solidité, la force et la durée du Bois," *MARS* 1738/1740, reprinted in *Histoire naturelle, Supplément* (1775), 2:185–203. See Lesley Hanks, *Buffon avant l'*Histoire naturelle (Paris: Presses Universitaires de France, 1966). For Duhamel du Monceau's interest in climate, see Guy Pueyo, "Duhamel du Monceau, précurseur des études climatiques et microclimatiques," *Actes du XIIe congrès international d'histoire des sciences, 1968* 12 (1971): 63–68.

lished across France in 1721, after sixty years their success was negligible.[104] Crown sanction for the establishment of tree nurseries across France and the colonies led to several projects for large-scale tree planting. In Burgundy, Marc-Antoine-Louis Claret de Fleurieu de La Tourette, Thouin's correspondent, was made director of a royal school in which one hundred orphan boys of good family were trained as apprentice silviculturalists. Over the period of the school's existence, approximately one million trees were grown to the sapling stage and distributed among local landowners. André Michaux, a correspondent both of Thouin and of Le Monnier, was sent out by the Crown on his successful return from a voyage to Persia, to found and direct a nursery for North American trees in Charlestown. During the 1780s he or his gardener Saunier sent back shipments of trees to be planted in the royal domain of Rambouillet, especially purchased for the purpose.[105]

From the 1760s onward experiments carried out throughout Europe and in the colonies indicated that trees had an important effect on the climate. Pierre Poivre's efforts to prohibit deforestation on the Isle de France, now Mauritius, are apparently the earliest example of such savant observations underpinning local administrative policy.[106] Poivre defended his policy with the claim that the removal of trees led to drought and soil erosion. In his *Époques de la nature* of 1778, Buffon pointed to the loss of trees in western Europe as the cause of the milder temperatures experienced there, as compared with Hungary, Poland, and other, less civilized places.[107] A milder climate would, of course, contribute directly to the gentler morals and greater enlightenment of the West. Claims like this were directly adaptable to revolutionary use. Fourteen years later, Louis Reynier carried out a careful analysis of references to the climate of ancient republican Rome in the *Journal d'Histoire naturelle,* and concluded "that the climate of Rome was approximately the same as the present climate of Paris," thus giving the climatic seal of approval to Paris as a seat of republican government.[108] The members of the Société Royale d'Agriculture were

104 Bourde, *Agronomie et agronomes,* 3:1548.

105 *NBU,* s.vv. "Michaux," "La Tourrette"; Savage and Savage, *André Michaux,* 39–44.

106 Grove, *Green Imperialism,* chapter 6.

107 *Époques,* 213–214.

108 Reynier, "De l'Influence du climat." See Feldman, "Ancient Climate." On Rome as a model city-state for the revolutionaries, see Outram, *The Body and the French Revolution.*

Figure 9 "Plan of the Jardin du Roi in 1640," by K. Collin. Numbers *2–4* denote the Cabinet, *7* denotes the formal medicinal garden, *12* the botany school, *13* the orchards, *14* the nursery beds, *10* and *16* tree plantations. *19* and *20* are the large and small mounds, or *buttes.* © Bibliothèque Centrale M.N.H.N. Paris 2000.

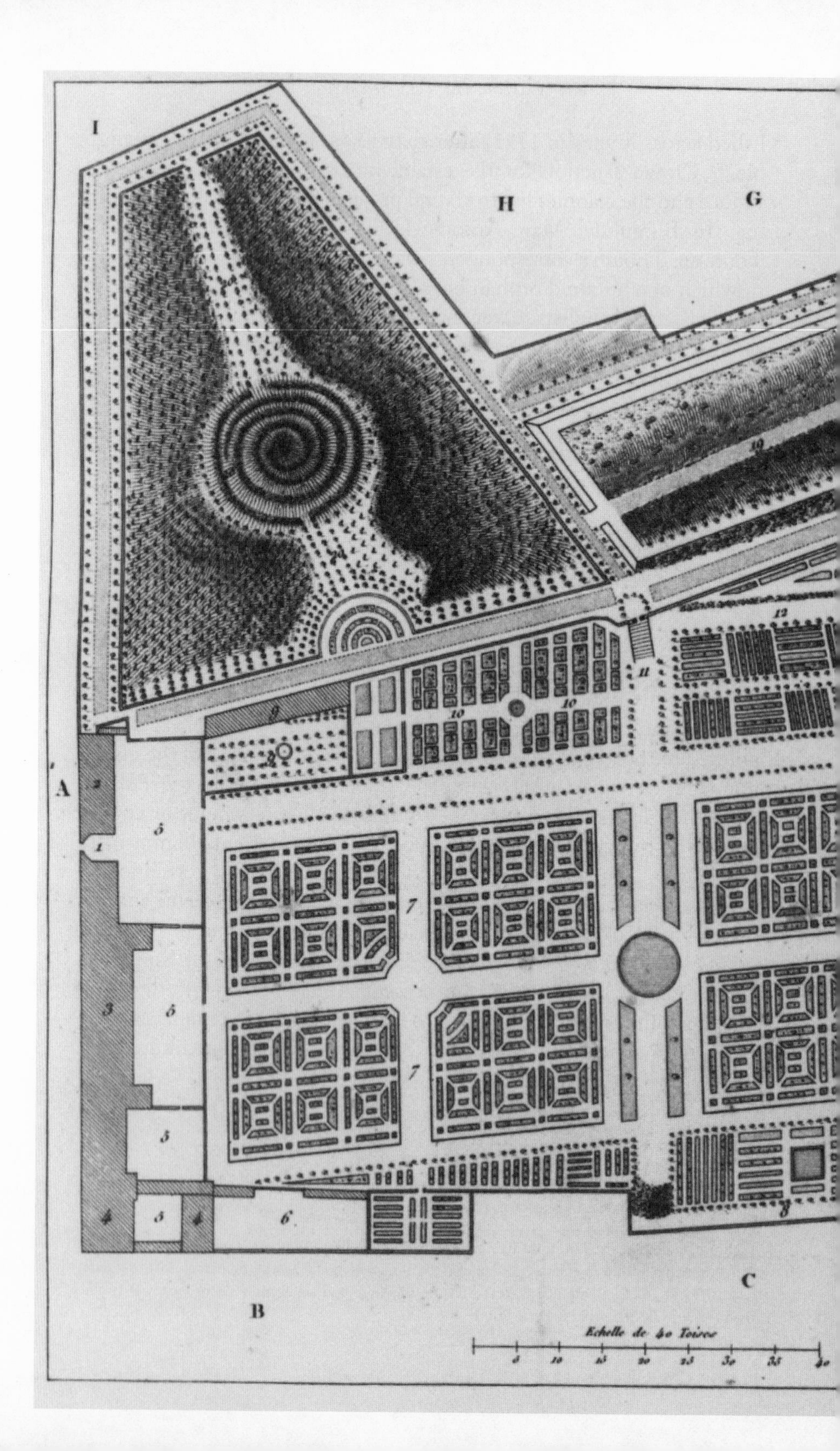

I
H
G
A
B
C
Echelle de 40 Toises

D
F
E
D
Gravé par E. Collin.

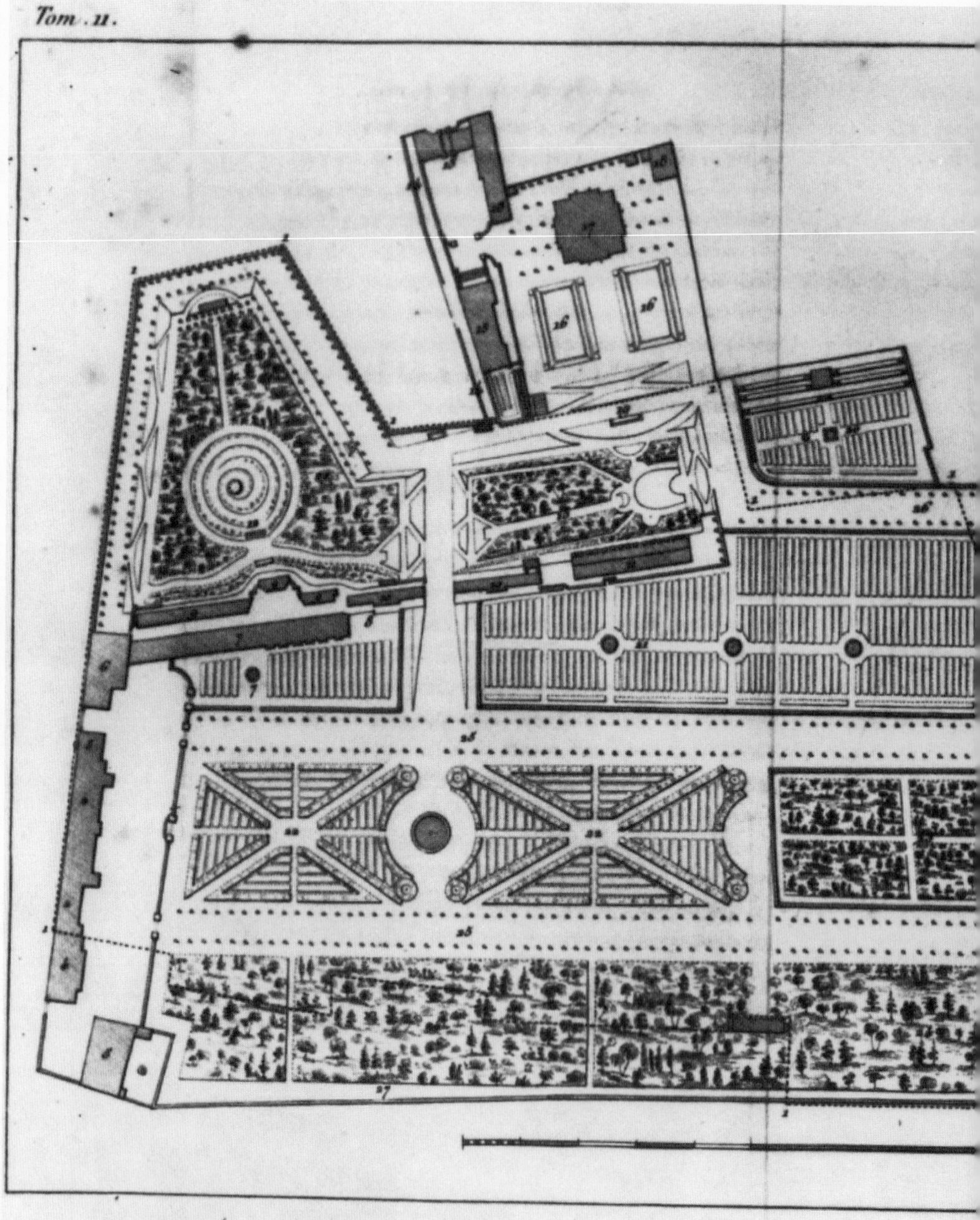

Figure 10 "Jardin des Plantes with its Additions in 1788," by T. Drouet, 1808. The line at *1* indicates the original boundaries of the Jardin. Numbers *2–6* are the buildings of the Cabinet; *12* and *13* are the two *buttes; 21* is the new, enlarged botany school; *22* denotes the formal beds, now given over to exotic plants; *23* and *27* are plantations of exotic trees; *29–33* contain trees for different purposes. The vast majority of the surface area was thus given over to tree plantations. © Bibliothèque Centrale M.N.H.N. Paris 2000.

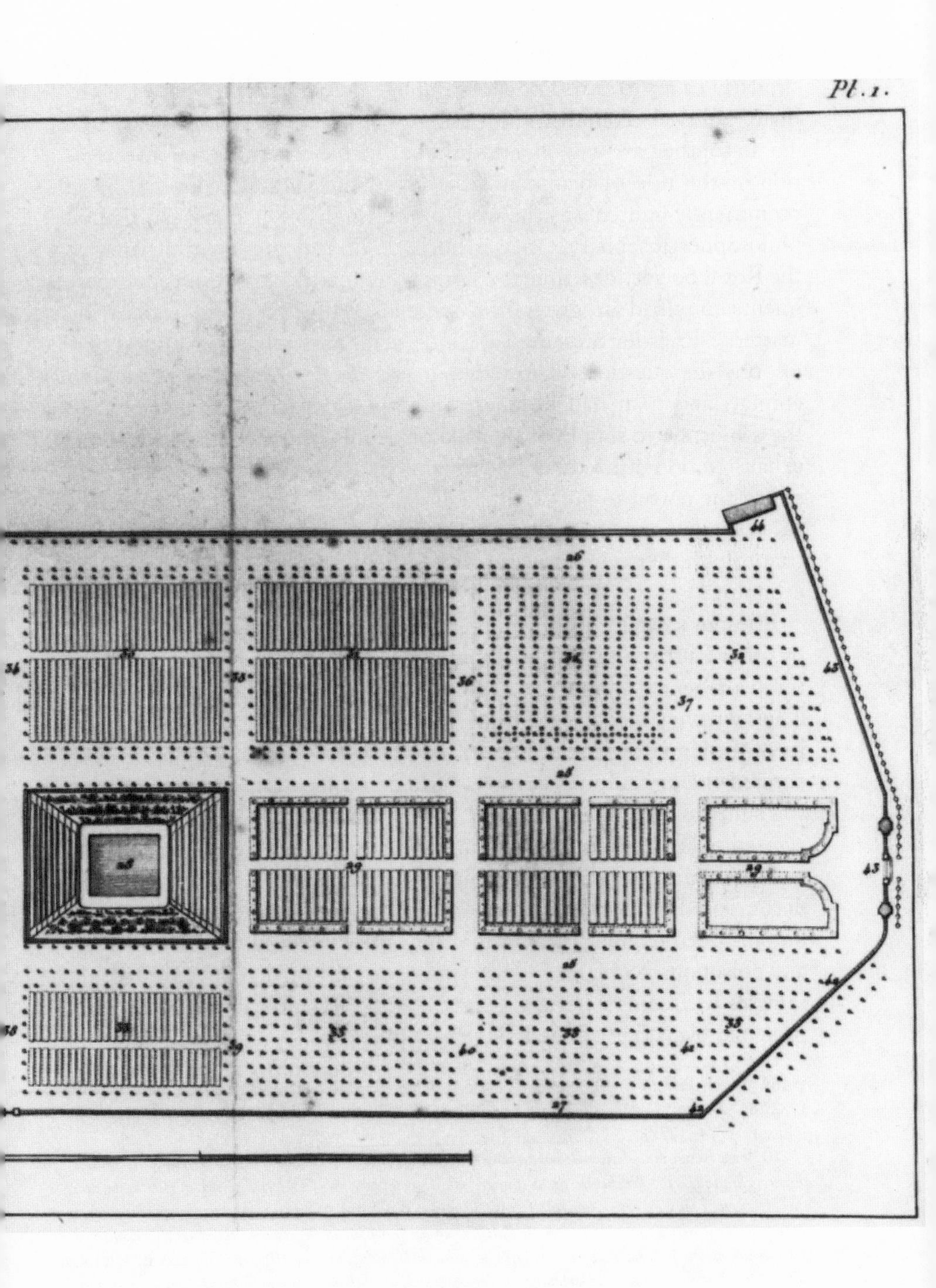

similarly eager to turn knowledge about the relationship between climate, political economy, and economy of nature to revolutionary use. To do so, they drew upon a tradition of British experimental investigation of the role of plants and the air in the natural economy. Most prominently figured was the work of the English Dissenter and natural philosopher Joseph Priestley, who in 1772 had presented a paper to the Royal Society describing, among other matters, the results of experiments on putrid air, given off by animals in breathing and by decaying matter.[109] Priestley presented as a long-standing problem of natural philosophy the question of how the air was renewed, since, even though animals and man had been breathing and decaying for many years, the atmospheric supply of purified or dephlogisticated air was not yet exhausted. He utilized his experiments to demonstrate that plants possessed the power to purify putrid air and make it breathable again. For Priestley the most important aspect of this work was that it revealed the true role, hitherto unknown, of the vegetable kingdom in the natural economy.

Priestley's experiments were replicated and given added significance by other European natural philosophers, particularly the Genevan pastor Jean Senebier and the Dutch experimenter Jan Ingen-Housz.[110] Their writings ascribed a central role to plants in determining the quality of the climate. This Protestant model of the vegetable economy also entered the agricultural writings of late-eighteenth-century France; it was summarized in the 1785 article "Air" of the *Cours complet d'Agriculture,* edited by a leading agriculturalist and member of the Société Royale d'Agriculture, François Rozier. In vegetating during the day, plants decomposed atmospheric air into two parts: fixed air, lethal to animals, but absorbed as nourishment by the plant itself; and dephlogisticated air, poisonous to plants. "This abundant rain of dephlogisticated air mingles with atmospheric air, and . . . augments the proportion of that principle over that of fixed air. From that [stems] the purity of air in

109 Joseph Priestley, "Observations upon different Kinds of Air," *Philosophical Transactions* 1772:156–247; id., *Experiments and Observations on different kinds of Air,* 3 vols. (London: J. Johnson, 1774–1777).

110 Jean Senebier, *Action de la lumière sur la végétation* (Paris: Didot, 1780); id., *Mémoires physico-chimiques, sur l'influence de la lumière solaire pour modifier les êtres des trois règnes de la nature, & surtout ceux du règne végétal,* 3 vols. (Geneva: B. Chirol, 1782); id., *Recherches sur l'influence de la lumière solaire pour métamorphoser l'air fixe en air pur pour la végétation. Avec des expériences & des considérations propres à faire connoître la nature des substances aëriformes* (Geneva: B. Chirol, 1783); Jan Ingen-Housz, *Expériences sur les végétaux, spécialement sur la Propriété qu'ils possèdent à un haut degré, soit d'améliorer l'Air quand ils sont au soleil, soit de la corrompre la nuit, ou lorsqu'ils sont à l'ombre; auxquelles on a joint une méthode nouvelle de juger du degré de salubrité de l'Atmosphère,* trans. Jan Ingen-Housz (Paris: Didot, 1780).

the countryside: an abundance of plants and trees, ceaselessly absorbing and consuming a quantity of fixed air, and disseminating streams of pure air on all sides, ceaselessly makes [the atmosphere] more suitable for breathing. Admirable compensation of nature! Masterpiece of wisdom by her author!"[111] Papers submitted to the Société Royale d'Agriculture in the 1780s employed the same set of chemical studies to support proposals for nationwide tree-planting schemes. What was more, these ambitious projects were expressed in an explicitly political way, in the context of calls for specific agricultural reforms which would reappear soon afterward in the legislation plans of successive revolutionary governments.[112] In 1789 the young Jean-Augustin-Victor Yvart, *fermier* to the archbishop of Paris and a correspondent of the Société Royale d'Agriculture, sent a paper to the latter which was read before the assembled company by Thouin and Mathieu Tillet, both at this time members of the Société's elite Comité d'Agriculture. Yvart's paper was on a subject that will seem familiar to any student of agriculture under the French Revolution, namely the necessity for land enclosure. However, he here presented trees as the ideal instruments of enclosure, making explicit reference to the work of Priestley and Ingen-Housz, whose experiments had shown "that [plants] possess the astonishing faculty of purifying the air contained in their substance, which they have no doubt absorbed from the atmosphere . . . that they sprinkle a kind of abundant rain, so to speak, of this vital and purified air, which, in expanding itself through the mass of the atmosphere, truly contributes to maintaining the salubrity of the air, and to rendering it more suitable to maintain the life of animals."[113] Yvart related this work to more recent studies by the Société Royale d'Agriculture's own correspondents, such as the baron de Tschudi, who "assures us that the fell-

111 "Air," in *Cours complet d'Agriculture Théorique, Pratique, Économique, et de Médécine Rurale et Véterinaire, suivi d'une Méthode pour étudier l'Agriculture par Principes; ou Dictionnaire universel d'Agriculture,* 10 vols., ed. François Rozier (Paris: rue et hôtel Serpente [Cuchet], 1785), 1:342–343; cf. also "Arbre," 1:627.

112 Priestley himself invoked his aerial discoveries in support of the Revolution, which he regarded as the prelude to an era of perfection in European society. See Jack Fruchtman Jr., *The Apocalyptic Politics of Richard Price and Joseph Priestley: A Study in Late Eighteenth-Century English Republican Millennialism* (Philadelphia: American Philosophical Society, 1983).

113 BCMHN, MS 318: Jean-Augustin-Victor Yvart, "Mémoire sur les bornes ou limites des Champs." The agricultural discussions of airs dating from the 1780s explicitly appealed to a Priestleyan, rather than a Lavoisierian, chemical model and experimental program. On the fate of Priestleyan chemistry in France, see Jan V. Golinski, *Science as Public Culture: Chemistry and Enlightenment in Britain, 1760–1820* (Cambridge: Cambridge University Press, 1992); Jonathan Simon, "The Alchemy of Identity: Pharmacy and the Chemical Revolution, 1777–1809" (Ph.D. dissertation, University of Pittsburgh, 1997).

ing of a forest of cedars on an island in the Pacific Ocean rendered the air so unhealthy, that it was necessary to replant them." Tschudi was one of Thouin's long-term correspondents, known to him since the late 1770s, and an expert on naturalization.[114] In addition, Yvart argued that the planting of trees, contrary to the old-fashioned view, actually improved the quality of the soil. "No one is ignorant of the fact that an infinity of insalubrious places, a very large number of marshes, formerly unhealthy, uncultivated and uninhabited, owe the salubrity of the air that can be breathed now to plantations, as well as the fertility that is manifested in all areas." He thus drew support for his claims about the effects of trees on climate both from the previous decade of studies on plants and the air, and from the tradition of agricultural reform within which he and his audience were situated.

Similarly, a memoir by the Société Royale d'Agriculture's correspondent Brunet, presented to the Société by Thouin and the abbé d'Harpicourt just twelve days before the fall of the Bastille, explicitly associated the deficits in the existing state of agriculture with a Rousseauist distinction between city and country, bewailing the fact that the *seigneurs* were often resident in Paris and not on their lands.

> It is a great shame that the landed rich abandon their inheritance and their lands. It is in the middle of these possessions that they would have found strength of temperament, a robust health, gentle morals, little disease, and a tranquility of soul far superior to the artificial pleasures that the blasé citizens of the towns and the great cities experience daily. These weak men have almost all distanced themselves from nature, in distancing themselves from their homeland. They were made for [that homeland], their organs acclimatized, so to speak, to the country which saw them born . . . The inconstancy of the air influences our age and our health. We correct it by our prudence. Why should we not act in the same way with regard to the influence which the air gives to our properties?[115]

Immediately prior to the outbreak of revolution, the members of the Société Royale d'Agriculture in Paris were attaching their models of climate and acclimatization to new natural philosophical entities in such a way as to support a particular physiology and politics of self-conduct for the French landowner. Fundamental to their claims was the forging of a connection between climate, national prosperity, and the condition of society; accordingly, many features of the "new agricul-

114 See Spary, "The Nut and the Orange."
115 BCMHN, MS 318.

ture" became converted into efforts to counteract local adversities of climate. The soil was to be fertilized and cultivated in areas which had hitherto lain infertile and deserted. Projects to drain marshes were aimed at improving air quality in regions where there was a high risk of epidemic diseases.[116]

Revolutionary legislators drew heavily upon the expertise of savants such as Thouin and his Société d'Agriculture colleagues in agricultural matters. Climatic measurement became a central part of the assessment of a nation's wealth, as is indicated by Constantin-François Chassebeuf de Volney's "Questions d'Economie Politique" of 1794, which included an entire section on the climate or "State of the Sky."[117] Supporters of the new agriculture argued that enlightenment and education could counteract the degenerative effects of climate; indeed, by inducting cultivators into the principles of the new agriculture, the climate of France could be improved, as the report on the plan of work for the committee of agriculture and commerce suggested in May 1790: "Spread enlightenment, and you will fertilize the soil. The agricultural societies will produce this happy effect; and a better moral and physical education, given to the children of colonists, would be a second advantage which would accelerate this much-desired change in our morals."[118] For these Revolutionary reformers, agriculture and natural history were thus to be an important route to the improvement of the French nation. During the early 1790s naturalists in Paris described their activities in an increasingly utopian tone, claiming to be able to regenerate French nature through enlightened teaching, and thus to transform French society. The expertise which they claimed to possess inextricably linked knowledge about cultivation and the natural economy with the good of the nation. A combination of climatic and legislative reform would accomplish the project of Revolutionary regeneration. Hence the dual significance of the appeal made by the Jardin naturalists before the assembled deputies in 1790 and again in 1793, soliciting the foundation of a national Muséum d'Histoire Naturelle. Their introductory address

116 Jean-Paul Desaive, Jacques-Philippe Goubert, and Emmanuel Le Roy Ladurie, eds., *Médecins, climat, et epidémies à la fin du XVIIIe siècle* (Paris: Mouton, 1972); Jean-Baptiste-Louis-Théodore, baron de Tschudi, *De la transplantation, de la naturalisation et du perfectionnement des végétaux* (London/Paris: Lambert and P. F. Didot le jeune, 1778).

117 *Magazin Encyclopédique, ou Journal des Sciences, des Lettres et des Arts, rédigé par Millin, Noel et Warens* 1 (1795): 352–362.

118 *AP,* 15:437; Grégoire, "Rapport . . . sur les moyens d'améliorer l'agriculture." Andrée Corvol, in her summary and preface to *La nature en Révolution,* notes that, although discussions of regeneration continued after the Revolution, the more ambitious claims common before and during the revolutionary period were no longer widely supported.

concluded by asking, "Shall the tree of liberty be the only one which cannot be naturalized at the Jardin des Plantes?"[119]

French aerial reform projects dating from the latter half of the eighteenth century have often been examined principally as the source for nineteenth-century hygiene policies. However, as Corbin's work on the history of odor demonstrates, public hygiene reforms were underpinned by medical arguments about the effects of atmospheres on the human constitution. Schaffer has shown how efforts to measure and reform air quality in the European state were driven by concerns for the moral management of society.[120] Atmospheres and airs were prominent not just in medical or chemical discourses but also, of course, in natural history. In 1784 Antoine-Laurent de Jussieu produced a report on his experiences as a member of the Société Royale de Médecine's commission, one of two charged with the investigation of Mesmerism.[121] De Jussieu was the only member of his commission to disagree with the conclusions of the main report. Certain observations seemed to the physician botanist to contradict the report of his fellow Société Royale de Médecine commissioners, Mauduyt, Andry, and Caille, which followed Lavoisier and the sister commission from the Académie Royale des Sciences in its denunciation of animal magnetism as the product of the imagination, and so entirely worthless.[122]

119 "Première adresse des officiers du Jardin des Plantes et du Cabinet d'Histoire naturelle, lue à l'Assemblée Nationale le 20 août 1790," in Hamy, "Les derniers jours du Jardin du Roi," 97–100. Leading members of the Jacobin agricultural committees subscribed to the same climatic models as the naturalists. In Grégoire's address "Des arbres de la liberté" of year II/1794, the deputy had explicitly portrayed trees of liberty as potential purifiers of corrupted air (Harten and Harten, *Die Versöhnung mit der Natur,* 23; Grégoire, "Essai historique et patriotique sur les arbres de la liberté. Par Grégoire, membre de la Convention nationale," in *Oeuvres,* 12:5–72, esp. 55). Agricultural *projets de décret* written by the Montagnard deputy for the department of the Oise, Jacques-Michel Coupé, and published by order of the Convention Nationale, reveal that he was familiar with the various atmospheric and climatic effects ascribed to trees by agriculturalists. See Coupé, *De l'amélioration générale du sol français dans ses parties négligées ou dégradées* (Paris: Imprimerie Nationale, year III/1794), 8; id., *De la tenue des bois* (Paris: Imprimerie Nationale, year III/1794), 26–27. Coupé was a member of the Comité d'Instruction Publique during the summer of 1793 and of the Comité d'Agriculture in year III/1794–1795.

120 Schaffer, "Measuring Virtue"; Alain Corbin, *The Foul and the Fragrant: Odor and the French Social Imagination,* trans. Jonathan Mandelbaum (Leamington Spa: Berg, 1986). See Owen Hannaway and Caroline C. Hannaway, "La fermeture du cimetière des Innocents," *Dix-huitième siècle* 2 (1977): 181–191; and Ludmilla J. Jordanova, "Urban Health in the French Enlightenment," *Society for the Social History of Medicine Bulletin* 32 (1983): 31–33, for the Académie des Sciences' hygienic projects of the 1770s and 1780s, in which Daubenton and Fourcroy were both involved.

121 Antoine-Laurent de Jussieu, *Rapport de l'un des Commissaires chargés par le Roi de l'examen du Magnétisme animal* (Paris: Herissant et Barrois, 1784).

122 Pierre-Isaac Poissonnier, Pierre-Jean-Claude Mauduyt de La Varenne, Caille, and Charles-Louis-François Andry, *Rapport des Commissaires de la Société royale de Médecine, nommés*

Although de Jussieu was acting in a medical role, his report is directly relevant for understanding his natural historical concerns, for it reveals the nature of his views on the relationship between the atmosphere, the animal economy, and the economy of Nature in a way that the only other work he published during this period, the *Genera plantarum,* does not. De Jussieu listed the experiments he had personally carried out; most of his patients had experienced a feeling of warmth, some had had more violent reactions such as fits, helpless emotional or physical spasms, and occasionally cures. In the case of the more spectacular effects, the "positive facts" appeared, for the most part, to support the claim that the effects of magnetism were ascribable to the imagination. In a few cases, however, de Jussieu himself was able to produce some physical results even when the patient was unaware of his activities. "[These results] suffice to admit the possibility of the existence of a fluid or agent, which carries itself from man to man, and sometimes exercises a sensible action on the latter . . . From this union of facts and consequences, it results that the human body is subject to the influence of different causes, some internal and moral, such as the imagination; others external and physical." The three processes of magnetization—rubbing, contact, and the action of fluid emanating from a similar body—all produced the same effects, and so could all be ascribed to one cause. "[But] what is the principle which insinuates itself thus into bodies?" In animate bodies, de Jussieu explained, there were two primary principles: matter and movement, the latter being the cause of all animal functions. "Principle of movement in the entirety of Nature, it becomes that of animal heat in living bodies; hence that marked correspondence between the variations of the atmosphere and the state of our organs." Under another name, he added, this principle of animal magnetism might be identified with electricity, "known by its effects, present in bodies, and exerting a sensible action."[123]

Reflecting the commonality of language and natural economical concerns between medicine and natural history, de Jussieu's report equated

---

par le roi, pour faire l'examen du magnétisme animal (Paris: Imprimerie Royale, 1784). The existence of the animal magnetic fluid had previously been supported by Charles-Louis-François Andry, who, with a medical colleague, Michel-Augustin Thouret, published "Observations et Recherches sur l'usage de l'aimant en médecine, ou Mémoire sur le Magnetisme animal," *HSRM* 1779/1782, 531–688.

123 Jussieu, *Rapport,* 25–26. See also de Thourry, "Mémoire qui a remporté le Prix proposé par l'Académie des Sciences, Belles-Lettres & Arts de Lyon, sur cette Question: L'Electricité de l'Athmosphère a-t-elle quelque influence sur le corps humain? quels sont les effets de cette influence?" *Observations sur la physique, sur l'histoire naturelle et sur les arts,* series 2, 9 (January–June 1777): 401–437.

Mesmeric atmospheres with chemical ones of the sort that natural philosophers across Europe were investigating. Toward the end of the century, French naturalists increasingly described matter and activity in the world as a finite quantity, continually cycling through the living and inanimate world, so that the natural economy was balanced like the financial resources in an administrator's accounting book.[124] De Jussieu's description of the operation of the "electric principle" in the economy of nature was probably derived from a number of different contemporary accounts of the role played by electricity in the natural economy, in particular the writings of Pierre-Joseph Bertholon and Jean-Antoine Nollet, and perhaps those of Mauduyt and Bernard-Germain-Étienne de Lacepède.[125] Indeed, de Jussieu regarded this account of the natural economy as so unproblematic that he gave no sources for his claims; they were, he claimed, "known to everyone." Since the electric fluid tended toward equal distribution throughout the world, it would pass

124 Wise, "Work and Waste"; Mikuláš Teich, "Circulation, Transformation, Conservation of Matter, and the Balancing of the Biological World in the Eighteenth Century," in *Lazzaro Spallanzani,* ed. Bernardi and La Vergata, 363–380.

125 Pierre-Joseph Bertholon, *De l'electricité du corps humain dans l'état de santé et de maladie; ouvrage couronné par l'Académie de Lyon, dans lequel on traite de l'électricité de l'Atmosphere, de son influence & de ses effets sur l'économie animale, des vertus médicales de l'électricité, des découvertes modernes & des différentes méthodes d'électrisation* (Paris: P. F. Didot, 1780); id., *De l'électricité des végétaux. Ouvrage dans lequel on traite des effets de l'électricité de l'athmosphère sur les plantes, de ses effets sur l'économie des végétaux, de leurs vertus médico- & nutritivo-électriques, & principalement des moyens de pratique de l'appliquer utilement à l'agriculture, avec l'invention d'un électro-végétometre* (Lyon: Bernuset, 1788); id., *De la salubrité de l'air des villes, et en particulier des moyens de la procurer* (Montpellier: J. Marcel aîné, 1786); Jean-Antoine Nollet, *Essai sur l'electricité des corps* (Paris: Frères Guerin, 1746); id., *Leçons de physique expérimentale,* 6 vols. (Paris: Frères Guerin, 1745–1768); Bernard-Germain-Étienne de La Ville sur Illon, comte de Lacepède, *Essai sur l'électricité naturelle et artificielle,* 2 vols. (Paris: Imprimerie de Monsieur, 1781); Pierre-Jean-Claude Mauduyt de La Varenne, "Mémoires sur l'électricité, considérée relativement à l'économie animale et à l'utilité dont elle peut être en médecine," *HSRM* 1776/1779, 461–513, 514–528; id., "Mémoire sur le traitement électrique administré à quatre-vingt-deux malades," *HSRM* 1777–1778/1780 (2), 199–431; id., "Mémoire sur les effets généraux, la nature et l'usage du fluide électrique, considérée comme médicament," *HSRM* 1777–1778/1780 (2), 432–455. The Jardin chemistry professor Pierre-Joseph Macquer, reviewing Lacepède's *Essai sur l'électricité* in the *Journal des Sçavans* 52 (13) (November 1781): 115, claimed that "aucun Physicien n'a jeté jusqu'à présent, sur l'électricité, un coup-d'oeil aussi étendu, aussi général & aussi hardi que M. le Comte de la Cepède. Il considere cette matiere d'une subtilité & d'une mobilité extrême, comme un des plus grands ressorts de la nature" ('until now, no physicist has presented such an extensive, general and daring account of electricity as the comte de la Cepède. He considers this substance to be one of the great springs of nature, possessing extreme subtlety and mobility'). See also W. Cameron Walker, "Animal Electricity before Galvani," *Annals of Science* 2 (1937): 84–113; Geoffrey V. Sutton, "Electric Medicine and Mesmerism," *Isis* 72 (1981): 375–392; Marcello Pera, *The Ambiguous Frog: The Galvani-Volta Controversy on Animal Electricity* (Princeton: Princeton University Press, 1992); John L. Heilbron, *Electricity in the Seventeenth and Eighteenth Centuries: A Study of Early Modern Physics* (Berkeley: University of California Press, 1979), 353–354.

"with impetuosity" into bodies deprived of it. "Distributed through the air without uniting with it, having the greatest affinity with water, this fluid is seized by vapors above the ground; condensed in clouds, it forms great meteors; brought back to the earth with rainwater, it penetrates [the earth] and brings life and fecundity." Similarly, in animals the principle formed an equilibrated atmosphere around each individual "sometimes relatively easy to recognize with the sense of smell, when it is charged with odoriferous particles." Personal atmospheres, de Jussieu suggested, might be measured by the odoriferous particles they contained.[126] Plants had living atmospheres as well, so that a blind man could detect the presence of a tree.[127] Every living being was a "true electric body," and the quantity of the active principle in each organ determined its activity. Malignant changes could be detected by a very delicate touch; and two bodies in imbalance would equilibrate via a conducting finger—this, de Jussieu argued, explained the phenomena of animal magnetism.

Plants and animals, de Jussieu reminded his readers, were influenced by physical and external causes such as the nature of the soil, the local exposition, and variations in the atmosphere. In animals the effects of climate were confounded by the more complex organization. Like Daubenton, he ascribed the characters of animals to their conformation: nervous systems and musculature made the possession of voluntary movement and sensibility or imagination possible. Under the influence of the imagination, the "active principle" could react variably "and so determine salutary or destructive effects." In man, as well, the imagination had variable effects upon the constitution, and different groups of individuals, depending upon the amount of imagination they possessed, contained a proportionate quantity of active principle. Ultimately, de Jussieu concluded that heat was the true principle "which establishes the physical influence of man over man," and the effects of animal magnetism could be ascribed to the effects of transfer of heat.[128] "One cannot in truth deny the existence of a principle identified with fire, with the electric fluid, penetrating the human body, and bringing heat to

126 This was a view shared by Daubenton; see BCMHN, MS 219.

127 Probably a reference to the English mathematician Nicholas Saunderson (1682–1739), also mentioned in Daubenton's "Histoire naturelle de l'homme," in *Histoire naturelle des Animaux* (1782), 1:xix–lxxxij, liij.

128 Jussieu, *Rapport,* 35. Lamarck also investigated the subtle fluids present within the earth. See Richard W. Burkhardt Jr., *The Spirit of System: Lamarck and Evolutionary Biology* (Cambridge: Harvard University Press, 1977), 65–69. The importance of atmosphere and climate in these debates of the 1770s to 1790s about change in the animal body suggests reasons for Lamarck's turn toward meteorology in his publications of the 1790s and beyond.

it." The therapeutic value of Mesmerism, like electricity, therefore, lay in its ability to stimulate animal heat in people. This was the assimilation of the eighteenth-century electrical body to a long tradition of climatic medicine derived from the ancients, and underpinning the past half-century of natural historical inquiry. But above all, this report related physical and moral causes and their effects very directly to the world's active principle—be it manifested as electricity or animal heat—which circulated throughout the economy of nature, from sky to plant to animal to atmosphere, and between individual humans. In prerevolutionary France atmospheres were not merely hazardous to health, they were what bound members together in a society, determined their physical and moral qualities, and tied them to the economy of nature. The cohesion of society itself was at stake in the improvement of the air.

De Jussieu's use of electricity as the mediator for moral/physical interactions was thus not unusual in the 1780s. But his world fluid resembled the universal Mesmeric fluid in many ways. Moreover, his account was taken up by the radical Jean-Louis Carra as the basis for his revolutionary political theory, as described by Darnton: "Moral causes, like unjust legislation, disrupted one's atmosphere and hence one's health, just as physical causes produced sickness; and conversely, physical causes could produce moral effects, even on a broad scale. 'The same effects take place, every moment, in society; and one has not yet ventured to acknowledge their importance, I believe, because one has not yet sufficiently connected the moral to the physical.' "[129] The investigation of that connection between the moral and the physical within the economy of nature characterized much of the activity of savants during the Revolution. In addition, the economy of nature became increasingly important as a site of scientific investigation during the Revolution. In 1792 the Académie des Sciences proposed as its prize essay competition the topic of the economy of nature, since "it was time that the attention of savants was fixed on the solution of this great problem." "By what processes," it asked, "does nature operate this circulation between the three kingdoms? . . . The cause and mode of the phenomena have, up to now, been enveloped in an almost impenetrable veil."[130]

At the new Muséum d'Histoire Naturelle, the concern to investigate the moral-physical relationship manifested itself in public dissections, such as Mertrud's dissection of the Versailles rhinoceros on September 26, 1793, before the members of the Comité de Salut Public. A detailed

129 Darnton, *Mesmerism,* 108, quoting from Carra.

130 "Prix proposé par l'académie des sciences, pour l'année 1794," *Journal d'Histoire naturelle* 1 (1792): 467–469.

account of this, including a brief description of its *moeurs* by Daubenton, survives in manuscript.[131] The study of animal *moeurs* was also more actively advocated by naturalists during the Revolution, and received considerable state support. In 1791 Pinel's "Recherches à faire par les voyageurs pour concourir efficacement aux progrès de la zoologie," a memoir written for the Société d'Histoire Naturelle when its members were engaged in deciding how to instruct the voyagers traveling in search of Lapérouse, lamented the inadequacy of naturalists' knowledge of varieties, "an object of natural history which can as yet only be sketched." Pinel's zoology was to be "anthropocentric" in the same way that Buffon's had been twenty or thirty years before: namely, the investigations of zoologists were to contribute to understanding the effects of climate on man. Above all, he appealed for travelers to "note with care all that can exert a more or less powerful influence" upon species, such as "the position of places, their temperature, the variations of the seasons, the qualities of the soil or the inequalities which can make them Low or Mountainous, the principal vegetable productions which grow there naturally, the Nature of the foodstuffs that result from them for animals and man, the more or less savage and rustic state of the latter and the more or less advanced state of civilization of the individuals of the human species."[132] Such an inquiry would transform the practice of natural history, in Pinel's view. The Société d'Histoire Naturelle's report to the Assemblée Nationale on the Lapérouse expedition similarly stressed the importance of examining the effects of climate.[133]

Prior to becoming intendant of the Jardin for six months in 1793, the writer Jacques-Henri Bernardin de Saint-Pierre had already attracted attention with a proposal for a menagerie to be established at the Jardin.[134] Aubin-Louis Millin, Pinel, and Alexandre Brongniart wrote a favorable report on his memoir of 1792 to the Société d'Histoire Naturelle.[135] However, the menagerie came into being only in 1794. The

131 BCMHN, MS 219. The animal had died three days earlier and smelt appalling, but "anatomists are used to it" (L. C. Rookmaaker, "Histoire du rhinocéros de Versailles, 1770–1793," *Revue d'histoire des sciences* 36 [1983]: 307–318).

132 AN, AJ/15/565: Pinel, "Recherches," [1790–1791].

133 AN, AJ/15/565: "Observations générales de la Société d'Histoire naturelle sur le voyage à entreprendre pour aller à La Recherche de Mr. de La Peyrouse," [1790–1791].

134 AN, AJ/15/512, piece 573: "Mémoire sur la nécessité de joindre une ménagerie au Jardin National des Plantes de Paris par Jacques Bernardin Henri de Saint-Pierre [*sic*]," [1792]. See the discussion in Masumi Iriye, "Le Vau's Menagerie and the Rise of the Animalier: Enclosing, Dissecting, and Representing the Animal in Early Modern France" (Ph.D. dissertation, University of Michigan, 1994), 193–202.

135 AN, AJ/15/512, piece 572: "Rapport fait à la Société d'Histoire Naturelle de Paris, sur la nécessité d'établir une ménagérie; par A. L. Millin, Pinel et Alex. Brongniart," Paris, Decem-

Commission des Travaux Publics, reporting on the professors' request for a menagerie in floréal year II (May 1794), produced a thrilling account of *moeurs* to persuade the Comité de Salut Public of the value of a menagerie.

> This garden can become the most beautiful in the universe; one will see here all the earth offers that is most curious, the marvels of the three Kingdoms, the rarest animals, plants from all climates; each animal would be placed here in the situation which most suited it; close by would be the trees that it loves, the plants that are appropriate to it. The Lion of Africa would have his den here in a Rock near the shade of Palms and Coconut trees together; the ferocious Panther, the bloodthirsty Tiger would inhabit the same regions; but the white bear of Siberia sheltered by some high rock from the Winds and from the heat of the south, would be placed in the coldest spot.[136]

The commissioners took for granted not only the necessity to use the Jardin's surface area as a series of sites of culture, but also the valuable effects upon French citizens that the sight of the climate-*moeurs* relationship might have. The *Décade philosophique*'s description of the menagerie in vendémiaire year III (October 1794) included an extract from the Muséum librarian's history of the menagerie's lion, which had as its closest companion a dog (figure 11). The lion had been raised from a cub in Senegal and given to the director of the Compagnie des Indes. In the hands of the Muséum's staff, he became the symbol of the ability of republican society to reconcile the natural and the artificial, and to improve upon both. "The lion of the menagerie has conserved all the primitive traits of his species. Returned to the plains of Africa, he would still reign there by the sentiment of his force which he owes to Nature. Society has not destroyed his instinct, but rather perfected it."[137] Animals revealed the ways in which life under a republican government resulted in the perfecting of society—assuming that proper measures were taken against the deleterious effects of climate.

---

ber 14, 1792. For a detailed discussion of the foundation of the menagerie, see Richard W. Burkhardt Jr., "La ménagerie et la vie du Muséum," in *Le Muséum au premier siècle de son histoire,* ed. Blanckaert et al., 481–508.

136 BCMHN, MS 457: "Second Raport fait par la Commission des Travaux publics au Comité de Salut public relativement au Jardin des Plantes."

137 "Description de la ménagerie du Muséum d'Histoire naturelle. Suite de l'Histoire du Lion et du Chien," *La Décade philosophique* 3 (1794): 129–138, 193–199. Based on Georges-Louis Toscan, *Histoire du Lion de la Ménagerie et de son chien, par le citoyen Toscan, bibliothécaire du Muséum* (Paris: n.p., year III/1794).

Figure 11 "The Lion of the Muséum d'Histoire Naturelle with His Dog. Drawn from nature." The "nature" being invoked here was that of Republican society; the iron bars which always penned in the Muséum's lion were entirely lacking in this sensible scene. From *La Décade philosophique* 3 (1794): opp. 129. By permission of the Syndics of Cambridge University Library.

Conclusion

There was a continuous tradition in Buffonian natural history which sought links between the physical and moral natures of man and the animals, and associated both closely with the effects of nature and of culture on living beings. From this connection naturalists and agriculturalists derived radical arguments about the need to reform French society, which also drew ammunition from the writings of individuals such as Bernardin de Saint-Pierre and Jean-Jacques Rousseau, even if there were evidently different models of Nature at work in the latter. The fascination with Buffon's writings as the germ of modern biological disciplines, however, has tended to obscure some of the ways in which contemporaries interpreted the natural historical enterprise upon which he and his contemporaries were engaged. Projects such as Antoine-Nicolas Duchesne's investigation of the Versailles strawberry and Michel Adanson's experiments in cross-breeding wheat have been seen, instead, as precursors to Mendelian theories of heredity. In other cases, studies of climate and degeneration have been interpreted as early strugglings toward theories of evolution or genetics.[138] The search for the origins of modern science has effectively masked the centrality of acclimatization and cultivation programs to the concerns of naturalists after the 1770s. In fact, the situation of many eighteenth-century naturalists, including those at the Jardin du Roi, within national programs of improvement, directed their attention toward the material and epistemological problems associated with the introduction and improvement of useful plants and animals. Their accounts of acclimatization may superficially resemble later models of evolution and heredity, to which they have been compared. But, aside from the presentism embodied in identifying these discourses as "the same as" nineteenth- and twentieth-century accounts of the natural world, there are more serious differences which must alter our view of the meaning of eighteenth-century climatic theories. Most prominent among these is the timescale employed by contemporaries to describe the changes operating: from their vantage points in the garden, naturalists and agriculturalists could observe transformations within the lifetime of an individual, or at the very least within a generation or two. Even Buffon concluded his *Époques de la nature* of 1778, in which he called for an expansion in the age of the earth to a number of years very much greater than that agreed upon by scriptural experts, with a dramatic call to his readers to recognize the enormous transformative power that man possessed through the sciences and arts: "And what is he not able to do to himself, I mean to his own species, if will is always directed by intelligence? Who knows to what point man can perfect his nature, whether morally or physically?"[139] Here was a powerful weapon for Revolutionary projects of a total transformation of society. In chapter 5 I shall pursue in detail the question of the relations between the teaching of natural history and the transformation of the citizen.

Eighteenth-century climatic concerns led naturalists, physicians, and agriculturalists, as well as others, to study the effects of external physical

138 Antoine-Nicolas Duchesne, *Histoire naturelle des fraisiers, contenant les vues d'économie réunies à la botanique; & suivie de remarques particulières sur plusieurs points qui ont rapport à l'histoire naturelle générale* (Paris: P.-F. Didot le jeune, 1766); Michel Adanson, "Examen de la question: si les espèces changent parmi les plantes; nouvelles expériences tentées à ce sujet," *MARS* 1769/1772, 71–78. For "early genetics" interpretations of such work, see, e.g., Conway Zirkle, "An Overlooked Eighteenth-Century Contribution to Plant Breeding and Plant Selection," *Journal of Heredity* 59 (1968): 195–198; J.-P. Nicolas, "Adanson: Ses travaux sur les blés, ses observations sur l'orge miracle," *Journal d'agriculture tropicale et de botanique appliquée* 11 (1964): 231–249.

139 *Époques*, 253; see also Ehrard, *L'idée de nature.* Secondary sources suggest that *Époques* was crucial in introducing a new timescale for natural events: see Wolf Lepenies, "De l'histoire naturelle à l'histoire de la nature," *Dix-huitième siècle* 11 (1979): 175–182; Rheinberger, "Buffon"; Lyon and Sloan, introduction to *From Natural History to the History of Nature.*

conditions upon living bodies. Their inquiry has implicitly been related to the environmental theories of the nineteenth and twentieth centuries. However, the term "environment" has become so loaded with modern biological and ecological implications that its use risks creating a false understanding of the particular purposes that theories of climate served for eighteenth-century writers. Thus I have avoided its use here, despite the several excellent studies which have applied the term to the eighteenth century.[140] In particular, and as recent usage of the expression "medical topography" demonstrates, climate theory was characterized by its medical origins and implications. Disease was the most evident of a whole range of potential disruptions of the animal economy arising from climate, diet, and social institutions. I have endeavored to construct naturalists' concerns with climate as part of a broadly conceived Hippocratic project, which scarcely resembles modern environmentalism either in its content or in its social uses. The rapidity with which transformations were taken to occur within bodies, and the widely held view that changes could become irreversible, even inheritable, accounts both for contemporaries' proportionately greater fear of climate, and for their optimistic engagement in projects which now seem unimaginably distant, such as the naturalization of the breadfruit in temperate countries. Naturalists offered their contemporaries the power to transform both nature and society.

Such arguments, as I have shown, possessed considerable appeal during the Revolutionary period. But the study of natural history also demands a new historiographical approach to the understanding of the French Revolution. Neither Marxist views of the Revolution as an event driven by socioeconomic factors, nor more recent intellectual histories of the Revolution as expressed in a series of symbolic and linguistic struggles for power, adequately contain the meanings of Revolution as employed by naturalists, nor, presumably, those utilized by the members of successive Revolutionary governments who actively promoted programs of acclimatization and climatic reform in France. Regeneration, liberty, and improvement were among the *practical* consequences anticipated as following from proper Revolutionary government, and

140 E.g., Jordanova, "Earth Science"; Riley, *Eighteenth-Century Campaign to Avoid Disease.* Most recently, Ludmilla J. Jordanova, "Environmentalism in the Eighteenth Century," in *Nature and Science: Essays in the History of Geographical Knowledge,* ed. Felix Driver and Gillian Rose, special issue of *Historical Geography Research Series* 28 (1992); Grove, *Green Imperialism;* Drouin, *Réinventer la nature;* Corvol and Richefort, *Nature, environnement, et paysage.* My reflections are provoked by the caution of Peter J. Bowler, "Science and the Environment: New Agendas for the History of Science?" in *Science and Nature,* ed. Shortland, 4.

they were terms implying a natural and physical process of transformation in living bodies. Such transformations could be observed every day in natural historical establishments, in exotic imports, on voyages, and even in one's own garden. They were recognized by most educated people, and discussions of them circulated widely in print. Naturalists' claims to expertise in the control of acclimatization appeared to offer the possibility of limiting or reversing such changes, just as physicians gave copious advice on how to protect the individual body against the intemperacies of climate. Both agricultural reform and botanical networking reveal powerful Crown support of antidegenerative measures even before the Revolution. For the naturalists the Revolution conferred a new legitimacy upon naturalization projects. In year II (1793–1794) the *Décade philosophique* summed up the role that naturalizing activities had in Revolutionary terms: "The culture of plants, envisaged in terms of its relationship with national prosperity, becomes, for the citizen, the practice of a moral *vertu,* founded upon love of the fatherland."[141]

The secondary literature has addressed acclimatization from the standpoint of its perceived utility in supplying new natural productions for food and for the arts.[142] This chapter supports such a reading, but nonetheless indicates a need to go beyond it. For utility is, as others have noted, a relative term, and in this period was a politically loaded expression denoting a particular self-presentation, one of numerous ways in which naturalists recruited the support of the regimes which succeeded one another between 1774 and 1799. The oft-retold story that Daubenton declared his occupation to be that of *berger* (shepherd) before the *section des sans-culottes* in order to obtain a certificate of civism, even if apocryphal, illustrates the extent to which Jacobinism and coarsely utilitarian models of the sciences were to become synonymous.[143] Utility alone, however, is not enough to explain the foundation

141 *La Décade Philosophique* 4 (1794): 130 (nivôse year III/December 1794).

142 Michael A. Osborne, *Nature, the Exotic, and the Science of French Colonialism* (Bloomington: Indiana University Press, 1994), which principally concerns the nineteenth century; id., "Applied Natural History and Utilitarian Ideals: 'Jacobin Science' at the Muséum d'Histoire Naturelle, 1789–1870," in *Recreating Authority in Revolutionary France,* ed. Bryant T. Raglan Jr. and Elizabeth A. Williams (New Brunswick: Rutgers University Press, 1992), 125–143, esp. 129–132; Gillispie, "The Encyclopédie and the Jacobin Philosophy of Science"; L. Pearce Williams, "Science, Education, and the French Revolution," *Isis* 44 (1953): 311–330, esp. 321–322.

143 This account is rarely documented when retold, but apparently originates in the memoirs of Henri Grégoire. See *Mémoires de Grégoire, ancien évêque de Blois, député à l'Assemblée Constituante et à la Convention Nationale, Sénateur, membre de l'Institut; suivies de la Notice historique*

of the Muséum at a time when other scientific establishments were being suppressed. Nor can it suffice to explain the success of the Jardin's naturalists in promoting their projects for autonomy when the "useful" activities of other naturalists outside the establishment did not save them from being impoverished or persecuted. A case in point is Tessier's acclimatization garden at Rambouillet, which was closed by the Convention Nationale. In the next chapter, therefore, I will consider the strategies of political agency adopted by the Jardin's naturalists in their response to the problems posed by Revolutionary life. Discourses about the natural economy, nature, and culture within the Parisian naturalist community were inescapably political, both in their use by naturalists addressing improving state officials, and in their use during the Revolution by those same naturalists as a way of talking about the regeneration of society. As the role of the cultivator acquired increasingly utopian connotations during the Republic, the importance of the Jardin naturalists for the Revolutionary state grew. In contrast with the many utopias, literary, religious, or political, which functioned principally in the realms of the imagination, revolutionaries often portrayed the ideal social and natural state as almost within reach, even perhaps embodied within certain privileged sites such as gardens and farms.[144] The transformative agendas of naturalists supported such reifications of gardens as ideal spaces. Revolution and regeneration were to encompass not merely the reform of political, moral, and social life, but the reshaping of physical nature as well. Addressing the deputies of the Assemblée Nationale in 1790, the naturalists even declared *themselves* regenerated: they were "new men."[145] Only in the wake of the Revolution would the

sur Grégoire d'Hippolyte Carnot, preface by Jean-Noël Jeanneney, introduction by Jean-Michel Leniaud (Paris: Editions de Santé, 1989), 61.

144 Among the substantial literature on millenarianism and utopias, see, in particular, Bronislaw Baczko, "Lumières et utopie: Problèmes et recherches," *Annales: Économies, sociétés, civilisations* 26 (1971): 355–386; Ernest L. Tuveson, *Millennium and Utopia: A Study in the Development of the Idea of Progress* (Berkeley: University of California Press, 1949); Clarke Garrett, *Respectable Folly: Millenarians and the French Revolution in France and England* (Baltimore: Johns Hopkins University Press, 1975); Karl Mannheim, *Ideology and Utopia: An Introduction to the Sociology of Knowledge,* preface by Louis Wirth and Bryan Turner (London: Routledge, 1991), esp. 173–236; Franco Venturi, *Utopia and Reform in the Enlightenment* (Cambridge: Cambridge University Press, 1971); Timothy Kenyon, "Utopia in Reality: 'Ideal' Societies in Social and Political Theory," *History of Political Thought* 3 (1982): 123–155; Judith N. Shklar, "The Political Theory of Utopia: From Melancholy to Nostalgia," *Daedalus* 94 (1965): 367–381; Mona Ozouf, "La Révolution française au tribunal de l'utopie," in *The French Revolution and the Creation of Modern Political Culture,* vol. 3, *The Transformation of Political Culture, 1789–1848,* ed. François Furet and Mona Ozouf (Oxford: Pergamon, 1989), 561–574.

145 Hamy, "Les derniers jours du Jardin du Roi," 64–68.

scope of such transformations, as envisaged in the 1780s, be seriously questioned.[146]

146 See, for example, Marie-Noëlle Bourguet, "L'image des terres incultes: La lande, la friche, le marais," in *La nature en Révolution,* ed. Corvol, 15–29. Whereas Williams (*The Physical and the Moral,* 104) argues that pre- and postrevolutionary medical doctrines of perfectibility both emphasized the limited nature of progress, I would suggest that the events and experiences of the Revolution were themselves central in helping to problematize the notion of progress.

CHAPTER FOUR

Patronage, Community, and Power: Strategies of Self-Presentation in New Regimes

"Greatness," said Pangloss, "is dangerous, according to all the philosophes. After all, Eglon, king of the Moabites, was assassinated by Aod; Absalom was hung by the hair and pierced with three darts; King Nadab, son of Jeroboam, was killed by Baasa; King Ela, by Zambri; Ochosias, by Jehu; Athalia, by Joiada; King Joachim, King Jechonias, and King Sedecias were enslaved. Do you know how Croesus, Astyages, Darius, Dionysius of Syracuse, Pyrrhus, Perseus, Hannibal, Jugurtha, Ariovistus, Caesar, Pompey, Nero, Otho, Vitellius, Domitian, Richard II of England, Edward II, Henry VI, Richard III, Mary Queen of Scots, Charles I, the three Henris of France, and the emperor Henri IV perished? Do you know . . . ?"
"I also know," said Candide, "that we have to cultivate our garden."

François-Marie Arouet de Voltaire, *Candide, ou l'optimisme*

Ever since the publication of François Furet's *Penser la Révolution française* in 1978, the French Revolution has been contested territory.[1] The success of Furet's attacks on Marxist histories of the Revolution caused historians on both sides of the debate to revise the old model of 1789 as the moment of transformation of a socially divided, industrially backward kingdom into an egalitarian, capitalist nation.[2] However, Furet

1 François Furet, *Penser la Révolution française,* translated by Elborg Forster under the title *Interpreting the French Revolution* (Paris: Éditions de la Maison des Sciences de l'Homme; Cambridge: Cambridge University Press, 1988).

2 François Furet and Denis Richet, *French Revolution* (London: Weidenfeld and Nicolson, 1970). On the reception of Furet's work, see Baker, introduction to *Inventing the French Revolution;* Joseph I. Shulim, *Liberty, Equality, and Fraternity: Studies in the Era of the French Revolution and Napoleon* (New York: Peter Lang, 1989), chapter 9; R. Emmet Kennedy Jr., "François Furet: Post-Patriot Historian of the French Revolution," *Proceedings of the 11th Annual Meeting of the Western Society for French History* 11 (1984): 194–200; Norman Ravitch, "On François Furet and the French Revolution," *Proceedings of the 11th Annual Meeting of the Western Society for French History* 11 (1984): 201–206; James Friguglietti, "Interpreting vs. Understanding the Revolution: François Furet and Albert Soboul," *Consortium on Revolutionary Europe, 1750–1850: Proceedings*

and many historians who cast themselves in the Furetian mold have continued to regard the Revolution as the point of origin of "the modern State."[3] While Marxists struggled to contain the causes of the event within a framework of socioeconomic explanation, Furet and his followers have spoken of the need to "rediscover . . . the political in the historiography of the French Revolution."[4] The meaning of the term "political" is, however, very different for these competing historiographies. For followers of Furet, such as Keith Michael Baker or Mona Ozouf, political agency is principally played out at the level of symbolism. The capture of the linguistic domain, in particular, could ensure the possession of legislative power.[5] This insistence on treating the language of the French Revolution as the central object of historical interpretation, rather than as a means to recover the "real" historical events of the period, has encouraged attention to the development of a political language and a political space for the people of France.[6]

Those emerging from the patronage system of the Old Regime were now faced with a society with a rapidly changing locus of sovereignty. The problems experienced by the Jardin naturalists during the Revolution were similar to those faced by many other groups which struggled for control of their own sphere or even of the political realm as a whole after 1789. The successful translation of the interests of an individual or group during the Revolution depended heavily upon the ability of that individual or group to monopolize the language of legitimacy. In this chapter I suggest that the naturalists of the Jardin du Roi were highly skilled at the adoption and reworking of Revolutionary language

17 (1987): 23–36; Jack R. Censer, "The Coming of a New Interpretation of the French Revolution?" *Journal for Social History* 21 (1987), 295–309.

3 See esp. *The French Revolution and the Creation of Modern Political Culture,* vol. 1, *The Political Culture of the Old Regime,* ed. Keith Michael Baker; vol. 2, *The Political Culture of the French Revolution,* ed. Colin Lucas (Oxford: Pergamon, 1987–1988).

4 Baker, *Inventing the French Revolution,* 3. Many historians of the Revolution are indebted to Furet's approach, including Mona Ozouf, Keith Michael Baker, Robert Darnton, Roger Chartier, and Dorinda Outram, although not all subscribe to his model in its entirety. Classic Marxist revolutions are in Albert Soboul, *La Révolution française, 1789–1799* (Paris: Éditions Sociales, 1948); Georges Lefebvre, *La Révolution française,* 3d ed. (Paris: Presses Universitaires de France, 1963); Rudé, *French Revolution.*

5 Baker, introduction to *Interpreting the French Revolution;* Lynn Hunt, *Politics, Culture, and Class in the French Revolution* (Berkeley: University of California Press, 1984).

6 Mona Ozouf, *Festivals and the French Revolution,* trans. Alan Sheridan (Cambridge: Harvard University Press, 1988); *French Caricature and the French Revolution, 1789–1799* (Chicago: University of Chicago Press, 1988); Noel Parker, *Portrayals of Revolution: Images, Debates, and Patterns of Thought on the French Revolution* (New York: Harvester Wheatsheaf, 1990); Baker, *Inventing the French Revolution;* Outram, *The Body and the French Revolution;* Robert Darnton and Daniel Roche, eds., *Revolution in Print: The Press in France, 1775–1800* (Berkeley: University of California Press, 1989); Hunt, *Politics, Culture, and Class;* id., *Family Romance.*

and behavior to justify their claims to reform and control the establishment, both before and after its transformation into the Muséum d'Histoire Naturelle on June 10, 1793. In so doing, however, I argue for the value of historical agency, against both Marxist historians who have pursued a French Revolution driven by social and economic forces, and certain readings of Furet which treat discourse, symbols, and ideology as the containers of power without exploring the strategies of construction and utilization by which their meaning and power is made and sustained.

Language, after all, is functional, not merely symbolic. Those who participated in the making of Revolutionary discourse did so because it offered them the possibility of achieving diverse goals. It is the very fact that the lives and intellectual activities of all savants were subject to considerable disruption and danger during this period that makes the reform of the Jardin at the height of the Terror seem such a striking event. Far from being the passive victims of Revolutionary political circumstance, the staff of the Jardin des Plantes preserved not only their lives, but also their institution and their posts throughout the nadir of French scientific institutionalization.[7] The Jardin was unique among the major scientific institutions of France in that its reform was based entirely upon the naturalists' own joint proposals. This was, in part, a result of the naturalists' ability to define the meaning of the Jardin for the future of France by alternating between different styles of addressing potential sources of sovereignty. The history of the Jardin's reform thus refutes Furet's claim that "no coalition nor any institution could last under the onslaught" of revolutionary ideology.[8] Naturalists incorporated concerns from the Old Regime within their self-fashioning in the Revolution; moreover, they actively appropriated and manipulated languages of the Revolution to suit their own ends.

Particularly important for the naturalists as they attempted to gain autonomy over their institution was the development of a rhetoric of community and the use of communal approaches to the negotiation of autonomy which conflicted with patronage-style behavior. The transformation of the Jardin into the Muséum marked the resolution of several possible strategies for preserving the institution. Instead of the power distribution of the Old Regime's patronage system, in which personal power and status were effected by personal vouchsafing and

7 See, e.g., Georges Kersaint, "Antoine-François de Fourcroy, 1755–1809: Sa vie et son oeuvre," *Mémoires du Muséum d'Histoire Naturelle,* series D, 2 (1966): 32; Hamy, "Les derniers jours du Jardin du Roi"; Letouzey, *Le Jardin des Plantes,* 323. Compare Outram, "Ordeal of Vocation."

8 Furet, *Interpreting the French Revolution,* 25.

guarantee, in the new Republic the naturalists of the Jardin could recruit far more power by representing themselves as members of a community, parts of a whole. The Jardin/Muséum was portrayed as an establishment of sensibility, a location within which human existence approached the ideal state of agricultural utility and Rousseauist social harmony. That vision contributed to naturalists' understanding of the proper function of the Jardin in a republic, and of how to regulate themselves and each other.

The Breakdown of Patronage

At the Jardin, patronage relationships were controlled by numerous conventions of communication, manifest in written and verbal exchanges. These tacit rules defined who could solicit whom for financial, intellectual, political, or other advantages. Buffon's patrons numbered several ministers, including the comte d'Angiviller, the *contrôleur-général des bâtiments du roi;* Amelot de Chaillou, the *ministre de la maison du roi;* and the baron de Breteuil, *ministre des finances.* These men, directly or indirectly connected with financing the Jardin, or otherwise in a position to affect events there, interceded for Buffon when he requested money, land, or *pensions* from the king for the Jardin. Likewise, Buffon himself stood as patron to his naturalists, and exerted his considerable power in the savant world to obtain their appointment to the Jardin, salary increases, sponsorship to other institutions such as the Académie Royale des Sciences, and diverse smaller privileges. The Jardin's naturalists were dependent upon their patrons, whose voices were the only ones they could use to address those with supreme power over their fate. In their turn, however, the naturalists themselves patronized correspondents, relations, and friends.

The French Revolution created a widespread dislocation of the patronage system upon which the naturalists' own posts in the scientific world of eighteenth-century France depended.[9] This period of disruption, associated with the loss of power of the pinnacle of all patronage relationships, the king, was characterized by a plethora of attempts by savant organizations to obtain autonomy over their internal affairs in their relationship with the state. The early years of the Revolution provided a political theater in which a considerable number of scientific practitioners participated in the ongoing debate over the redistribution of power, not least because the Assemblées Nationales, Constitutive

9 See Outram, *Cuvier,* 47.

and Legislative, contained numerous savant deputies.[10] Seeking to demonstrate that scientific projects were central to proper government, the Académie des Sciences' members involved themselves with such enterprises as the Weights and Measures Commission, setting new standards for Revolutionary society.[11] The Paris Société d'Agriculture attempted to demonstrate its usefulness to deputies by disseminating agricultural information to peasants and provincial landowners.[12] There were large rewards awaiting those who successfully captured the interests of the new state. Thus the Société d'Histoire Naturelle de Paris, entirely lacking state funding, was instrumental in the organization of a voyage in search of the missing expedition of Lapérouse, whose principal beneficiaries were to be naturalists.[13] No less than any other section of the educated administrative elite of France, savants were intensely politically active during this period, concerned with social organization, self-presentation, and the legitimacy of authority in the new regime.

Such concerns were not new at the Jardin in 1789. Already, at Buffon's death, the head gardener André Thouin had experienced the darker side of patronage when the new intendant, the marquis de La Billarderie, had taken up his post in April 1788. Serious difficulties could arise in the patronage system when there was an abrupt change of actor at a high level. Buffon had tried to avoid such a situation in 1771, by giving the survival of his own post to the comte d'Angiviller. D'Angiviller had proved a reliable patron of the Jardin enterprise, successfully arranging the transfer of specimens from many countries to the Jardin's collections. A few days before Buffon's death, however, pleading that his post of *contrôleur-général des bâtiments du roi* left him no time for other occupations, d'Angiviller had passed the prospective intendancy to his eldest brother, the marquis de La Billarderie.[14] The marquis was a man unfamiliar to the majority of the Jardin's employees; thus he lacked the bonds of social loyalty that had joined Buffon and his protégés.

In the Jardin under Buffon, Thouin had occasionally made suggestions for the improvement of certain aspects of the establishment.[15] These were processed through the proper channels—from Buffon to

10 Rudé, *French Revolution,* 36–41.

11 Hahn, *Anatomy of a Scientific Institution,* 162.

12 Gillispie, *Science and Polity;* Bourde, *Agronomie et agronomes.*

13 AN, AJ/15/565; *AP,* vol. 17, séance of August 5, 1790.

14 AN, AJ/15/502.

15 *Correspondance,* vol. 2, passim.

the ministers, from the ministers to the king, followed by a letter from the ministers to Buffon giving the king's final decision. As Buffon's ambassador to the ministers, Thouin associated with many of the ministers and *commis* who were involved with the Jardin. On the basis of this familiarity, Thouin attempted to obtain the approval of the new intendant for a project to reform the Jardin's regulations. He sent a memoir to La Billarderie which proposed alterations in the running of the establishment as a whole and the garden, his own sphere of power, in particular.[16] To justify his intervention, Thouin appealed to Buffon, "that man of genius," whose long tenure, he argued, had "accustomed all the members of the establishment to know only the law of his wishes, particularly since these always concerned the honor of the establishment and above all its utility." The recent enlargement of the Jardin's facilities had made the old legislation defunct, since it did not encompass the Jardin's new functions. The proposals that followed defined Thouin's own role and his importance in the running of the garden.

Unbeknownst to Thouin, La Billarderie promptly sent this document to the baron de Breteuil, and shortly afterward the gardener received a message acknowledging receipt from one of the Jardin's regular visitors and a fellow naturalist, Malesherbes, sometime minister to the king and later to be his ill-fated defending lawyer in the trial which preceded his execution on January 21, 1793. Thouin then made use of a typical patronage device, sending Malesherbes a copy of the memoir with a request to examine it and, if he approved, to submit it to the baron "without compromising me with M. de la Billarderie or in his offices" (presumably de Breteuil's).[17] Thouin was requesting Malesherbes to become a spokesman for his views before authority, but he was not doing it in the approved manner. Instead of waiting for the intendant to be the Jardin's spokesman in matters concerning the reform and the laws of the establishment, Thouin had attempted to bypass the legal authority and negotiate directly with the next step up the patronage ladder. Malesherbes returned the project, warning Thouin that "I am too much your friend, my dear Confrère, to give your memoir to the baron de Breteuil . . . You have given me some very good reasoning in natural history. I shall give you some that is very important where ministerial politics is concerned, which is that one should never send a memoir to a busy minister if one is afraid of its falling into other

16 Letouzey, *Le Jardin des Plantes,* 250–253. This was the reform project, discussed in chapter 2, in which Thouin attempted to lay down moral regulations for his subordinates.

17 Ibid., 257: Thouin to Malesherbes, July 22, 1788.

hands. I have seen dangerous results. Besides when you send a memoir know to whom you are sending it. Can one count on the security of ministerial life."[18] Malesherbes's letter reveals that Thouin's actions could be interpreted as subversive. In 1788, the patronage system as social structure was still intact. This episode has been portrayed by Thouin's biographer as one example of attempts by Thouin's superiors to suppress his talents for administration. What Thouin was in fact defeated by were the rules for advancement in old-regime society. His proposals and the way in which they were addressed to those in authority conflicted, after all, with the same patronage system that he was trying to utilize to get his views heard. This is one example of why the issue of spokesmanship—of who had the right to speak on another's behalf, and how—became so important for the middling sort in the French Revolution. Thouin would never again speak on his own behalf in putting forward politically sensitive proposals; as we saw in chapter 2, his later suggestions were channeled to successive governments via favorably inclined deputies such as Boisset or Grégoire.

Another of Buffon's protégés, the *géologue* Barthélémi Faujas de Saint-Fond, was evidently standing by with a copy of Thouin's project to be sent to d'Angiviller in the provinces. His letter, dated July 15, 1788, throws light upon the timing of their solicitations.

> Monseigneur
>
> you have loved M. de Buffon at all times, you cherish his memory, and you have Created Jointly with him a garden which fixes the attention of Europe, and attracts a Great number of foreigners to Paris.
>
> I recall having had the honor of hearing you say to M. de Buffon, that one would always respect His intentions and what he had done. I have learned that someone is already concerning themselves with asking you for a new regulation for the garden and the Cabinet of natural History. Mr. Thoin *[sic]* and I are absolutely ignorant of its tenor; I therefore take the liberty of asking you for permission, to have the honor of presenting to you the plan that [the] forty Five years of experience Of M. de Buffon had made him consider the most Suitable to maintain This useful establishment in all its splendor.[19]

The cold winds of change were blowing around the Jardin du Roi in the months after Buffon's death. The difficulty of addressing patrons in this way is underlined both by Faujas's linguistic contortions as he

18 Ibid., 257: Malesherbes to Thouin, July 23, 1788, headed "6 h du matin."
19 AN, AJ/15/510, piece 322.

attempted to make his act more acceptable to d'Angiviller, and by the fact that the letter was probably never sent.

From 1789 onward, however, there were significant and unpredictable transformations in the ways in which power was distributed and legitimated in French society, and in the rules governing the linguistic modes of expression of that power. Successive factions in the Revolutionary government endeavored to obtain and retain political power by monopolizing the debate over sovereignty.[20] Particularly in the early stages of the Revolution, this debate was open to a great many participants. The archives of the various executive groups of the Assemblée Nationale overflow with memoirs, opinions, reform projects, and plans proposing all manner of legislation, from schools for the blind and electoral equality for women, to the education of the Dauphin and the implements to be given to peasant proprietors.[21] The Jardin du Roi came in for its share of reform projects. Recently enlarged and improved, it was, like the recently reformed Observatoire, a desirable property in the savant world.[22] The 1789 *Vues sur le Jardin Royal des Plantes et le Cabinet d'Histoire Naturelle,* published anonymously, suggested the union of a number of chairs of medicine and natural history from the Collège Royal, the Royal Mint, the veterinary school at Alfort, and the Jardin des Apothicaires, in the Jardin des Plantes, to form one supermedical and natural historical establishment—with the emphasis on the medical aspects.

The response to this project, by the chemist Balthazar-Georges Sage, instead proposed the transferral of the mineral collections of the Jardin to the École des Mines, which he directed. The professorial chairs were to be united with those of the Collège Royal, leaving only the botanical garden and the anatomical and botanical collections.[23] Sage portrayed the current role of the institution as being little better than that of a cabinet of curiosities, with disorder in the collections and little teaching value: two especially deep cuts against the primary functions of the Jardin according to its staff, as we shall see. On the other hand, for

20 Furet, *Interpreting the French Revolution,* 73; Outram, *The Body and the French Revolution,* 76–83.

21 E.g., AN, F/17/1310.

22 Seymour L. Chapin, "The Vicissitudes of a Scientific Institution: A Decade of Change at the Paris Observatory, 1789–1799," *Journal for the History of Astronomy* 21 (1990): 235–274.

23 Balthazar-Georges Sage, *Observations sur un écrit, qui a pour titre, "Vues sur le Jardin royal des Plantes et le Cabinet d'Histoire naturelle"* (Paris: Didot le jeune, 1790). This response also addressed proposals made by the Comité des Finances to transfer the collections of the École des Mines to the Cabinet du Roi, discussed below. See AN, D VI 26, piece 702: Sage to Comité des Finances, January 3, 1790.

Pierre-Marie-Augustin Broussonet, the perpetual secretary of the Société d'Agriculture de Paris and a former student of Daubenton, it was precisely the physical locus of the Jardin which made it a valuable asset to the Société.[24] He proposed the conversion of the Jardin into a center from which the activities of the Société could have taken place, and in which agricultural implements and the most "useful" specimens would be exhibited. Broussonet's project entailed the preservation of the posts of those Jardin naturalists who were connected with the Société—Thouin, Daubenton, and Desfontaines, in particular. Many of Broussonet's ambitions, too, reflected the agricultural interests of these men.[25]

Whether the projects produced by other savants for the reform of the Jardin appealed to the past or actual concerns of the Jardin's naturalists, however, they all had one thing in common: they served, to a greater or lesser degree, to transfer control over the Jardin into the hands of other groups of savants. Nor should the power of individual savants to intervene in matters concerning Old Regime establishments be underestimated. Condorcet in particular, the perpetual secretary of the Académie des Sciences and thus probably the most powerful man in the savant world, was at the peak of his political career in the early stages of the Revolution. For the Académie he envisaged the formation of a hierarchical system of national instruction dating from early childhood to adulthood, to be overseen by a super-Académie with no teaching role.[26] All other savant organizations were to be subsumed in the lower levels of this national organization. The Jardin was no exception: it seems clear, from correspondence between Condorcet and La Billarderie, that Condorcet had already succeeded in obtaining the survivancy of the intendant's post, and thus the prospect of academic control over the Jardin which had for so long eluded the Académie's members.[27] Thus the Jardin's fate was open to debate by other savant groups in the new regime. Even its very identity as an establishment of natural history, rather than a school of medical botany, was not particularly

24 Pierre-Marie-Augustin Broussonet, *Réflexions sur les avantages qui résulteroient de la réunion de la Société royale d'agriculture, de l'Ecole vétérinaire et de trois chaires du Collège Royal au Jardin du Roi, par P.-M.-A. Broussonet* (Paris: Imprimerie du Journal Gratuit, [1790]).

25 Several of Thouin's proposals for agricultural organizations in the mid-1780s resembled Broussonet's suggested reforms, e.g., BCMHN, MS 318: André Thouin, "Projet pour une académie d'Agriculture," 1788.

26 According to Hahn, *Anatomy of a Scientific Institution,* chapter 5, the Talleyrand proposals of 1790 for the reform of the sciences in France were based on Condorcet's project, published in 1792.

27 AN, AJ/15/507, piece 181. See Chapin, "Vicissitudes of a Scientific Institution," on similar attempts by the Académie's astronomers to annex the Observatoire.

stable. The savants attached to the Jardin responded by producing their own collective reform proposal, in which they aimed to define their own roles as naturalists, while insisting upon the natural historical function of the reformed institution.[28]

This project was written in the specific climate of a debate over sovereignty at the Jardin, but it owed its existence to a broader debate over the state of French finances. The assertion that the financial difficulties suffered by the French Crown in the 1780s were a major element in precipitating the Revolution is common in the secondary literature.[29] The primary activities of the revolutionary government in the early years were directed toward the restriction of state expenditure in all possible forms. In early 1790 Charles-François Lebrun, heading the finance committee, proposed budgets for many Old Regime establishments, including the Jardin. An individual review of each case followed in August, and created a considerable discrepancy between organizations. The budgets of the Alfort veterinary school and the Jardin were to suffer most: the former dropped from 36,000 to 14,700 livres per annum, the latter from 92,222 to 72,000 livres. The Académie's budget, on the other hand, was maintained, with an extra allotment for an annual prize. Small savings were made in the budgets for the Observatoire and the collèges. Lebrun based his proposals for cost-cutting at the Jardin on a manuscript project for reform written by J. Audrée, the bishop of Autun and a member of the Comité de Constitution, and sent to the Comité des Finances.[30] The Jardin's cuts entailed significant reforms in the Jardin's organizational structure, with the suppression of several posts, but the augmentation of the *pensions* attached to others. Four posts were to go: those of Lamarck, the botanist attached to the Cabinet; Faujas de Saint-Fond, the Cabinet's correspondent; Verniquet, the architect; and Lucas, the caretaker. The intendant's salary was to fall from 12,000 to 4,000 livres a year, since the post "ought to be the object of emulation and not of [desire for] glory."[31] Other projects,

28 *Projet de Règlements pour le Jardin des Plantes et le Cabinet d'Histoire naturelle,* September 1790, in Hamy, "Les derniers jours du Jardin du Roi," 107.

29 Bosher, *French Finances;* Florin Aftalion, *The French Revolution: An Economic Interpretation,* trans. Martin Thom (Paris: Éditions de la Maison de l'Homme; Cambridge: Cambridge University Press, 1990); Riley, *Seven Years' War;* Schama, *Citizens,* chapter 2. However, such interpretations deserve the added reflection that, as I argued in chapters 2 and 3, administrators were construing their own role during this period as that of "economic" experts on a multitude of levels; to consider the *financial* aspects of these debates over credit to the exclusion of all others is inevitably to impoverish our understanding of the political significance of proclaiming economic crisis at this time.

30 AN, F/17/1310: "Vues d'un Zélé Citoyen sur le Jardin du Roi," July 18, 1790.

31 Hamy, "Les derniers jours du Jardin du Roi," 26.

too, cut deeply into the status and salary of the intendant. Jacques-Antoine Creuzé-Latouche, a deputy with contacts at the Jardin, even suggested its abolition, tactically distancing Buffon's achievements from his post as intendant.

> You will not forget, Messieurs, the species of anathema with which your decrees and public opinion have tainted the very name "intendant" in all kinds of administration, and you will see . . . the abuses of the intendants in this school of the natural sciences, who, apart from M. de Buffon, never concerned themselves with it and never even deigned to reside in a lodging which was pompously attributed to them. You will see the present Intendant, absolutely alien to a school of natural history, receiving the emoluments of a post . . . whose utility cannot be divined.[32]

The anonymous author of the *Vues,* too, claimed that no citizen was able to succeed to Buffon's glory, and that the salary of the intendant should therefore be reduced by 6,000 livres, to be used "in improvements which the public interest dictates."[33] Needless to say, La Billarderie did not agree with the proposed removal of the intendant, or with the salary reductions. He presented his own reform project in which the Jardin's budget was significantly increased.[34] Condorcet too prepared an appeal to the Assemblée protesting against the proposed cuts.

With the new freedom available to those wishing to express, print, and distribute their own views, a conflict thus developed over who had the right to determine the fate of such establishments as the Jardin du Roi. Since everyone now had the right to *discuss* that fate, the scope of the debate widened to encompass all who had some stake in the matter; and that constituency proved to be both large and powerful. For interested participants, success in this arena was to be achieved only by the conviction of as many potential commentators as possible. The naturalists of the Jardin made up the other interested party in the debate over the Jardin's fate. Their problem was that of addressing, no

32 Jacques-Antoine Creuzé-Latouche, *Opinion de M. J.-A. Creuzé-Latouche, Membre de l'Assemblée Nationale, au sujet du Jardin des Plantes et des Académies* (Paris: Imprimerie Nationale, 1790), reproduced in Hamy, "Les derniers jours du Jardin du Roi," 94. Creuzé-Latouche corresponded with Thouin in 1790 and 1791 (see "Etât"), and his report suggests some familiarity with the naturalists' views as subsequently expressed in their own project.

33 *Vues sur le Jardin royal des Plantes et le Cabinet d'Histoire Naturelle,* quoted in Hamy, "Les derniers jours du Jardin du Roi," 78.

34 Hamy, "Les derniers jours du Jardin du Roi," 25. See also, in ibid., 129–130, La Billarderie's "Notes sur le projet de règlement pour le Jardin du Roi" of September 1790, accusing the professors of being "infinitely more occupied with their salaries than with the other interests of the establishment."

longer a single patron, but a multitude of potential allies or enemies. From the beginning, as we shall see, they set out to achieve alliances on all sides.

First Alliances

Having been informed of the date for the discussion of the Jardin's funds, probably by a contact among the deputies, the naturalists appeared together before the Assemblée to deliver a letter and an address, signed by ten of their number, to the president, Pierre-Samuel Dupont de Nemours. Following the reading of Lebrun's project, Dupont pronounced the address, henceforth referred to as the First Address, before the Assemblée.[35] According to the signatories, it was precisely because they were not merely good citizens but also good naturalists, familiar with the operation of natural laws, that they found themselves obliged to admire the work of the deputies. No distinction could be made between the operation of natural laws in the Jardin and in regenerated France.[36] In effect, the state of being a naturalist was presented as naturally leading to political engagement: Revolutionary self-presentation would underpin Revolutionary natural history.

The other striking feature of the First Address is its eulogy of Buffon—praise which contrasts quite sharply with the prevailing view in the secondary literature that Buffon's death had somehow "freed" the naturalists from despotism and the denial of credit for their work.[37] Instead, Buffon's position in reforming natural history was explicitly linked to the ideals of the Revolution itself. The naturalists emphasized the improving aims and the moral value of the great man, who "by the philosophical views and sublime images distributed throughout his works, has so well prepared spirits for the great ideas of liberty and

35 Hamy, "Les derniers jours du Jardin du Roi," 64–68. There is no reason to believe that de Jussieu was in any way opposed to the reform proposal of 1790. Consequently, it is more probable that he was in some way prevented from attending the meeting. This suggests that his absence from the list of signatories appended to the speech made by the Société d'Histoire Naturelle's members, a few days earlier, may well be due to the same cause. The standard accounts portray him as absent from the Société's list because of his opposition to Linnaeus, but this seems rather unlikely; Lamarck would have been a more obvious candidate.

36 Ibid., 97–100: "Première adresse des Officiers du Jardin des Plantes et du Cabinet d'Histoire naturelle, lue à l'Assemblée Nationale le 20 août 1790," 97.

37 Esp. Corsi, *Age of Lamarck,* 4; Letouzey, *Le Jardin des Plantes,* 247; Lepenies, *Das Ende der Naturgeschichte,* 154. In fact, such anecdotes of the naturalists' opposition to Buffon, e.g., in Laissus, "La succession de Le Monnier," and Pierre Flourens, *Des manuscrits de Buffon* (Paris: Garnier, 1860), often originate from the polemical accounts of Buffon written by the radical Linnaean Aubin-Louis Millin during the Revolution.

regeneration."[38] Passages in the First Address also mimic the style of Buffon's *Histoire naturelle,* and often the content. Such similarities reveal the extent to which the naturalists sought to exploit the fame that Buffon had acquired since the publication of the *Histoire naturelle.* The cult of Buffon, and the moral virtues ascribed to the dead naturalist in consequence, supplied material with which to bypass the existing intendant's authority, an alternative legitimation for the projects of naturalists. Buffon's image could be fashioned in a variety of different ways in order to support these competing claims to authority, both by the group of naturalists as a whole and by individuals, as in Thouin's 1788 project.[39] The naturalists were to return to it in the second of the two addresses, submitted to the Assemblée Nationale in September 1790. "Depositaries of the ideas of this man of genius, we come, in a way, only to propose to you the realization of his projects," they claimed, again going on to link the natural historical project to social reform.[40]

Buffon was, in fact, a very valuable tool with which to legitimate the desires of the group. The construction of a "great man" for the purposes of rhetorical argument was a device often used by orators in the Assemblée and later in the Convention. In historicizing Buffon as role model, then, the Jardin's naturalists were employing a characteristically Revolutionary device. The success of the *Histoire naturelle* meant that the work was probably one of the most widely owned in France in the second half of the century, and almost certainly one of the most widely known. Buffon's emphasis upon the morally educative value of the study of natural history had been supported during the 1760s and 1770s by the growth of the cult of nature sentiment and the writings of Rousseau, Bernardin de Saint-Pierre, and others.[41] It is therefore likely that many of the deputies for whom the naturalists' addresses were written would have been brought up on a diet of Buffon, and that they would have participated in the Buffon cult to some degree. Indeed, Buffon's great advantage as an element of such negotiations was that he was more highly regarded among the public than in academic circles, and

38 "Première adresse," in Hamy, "Les derniers jours du Jardin du Roi," 97–98.

39 See also Barthélémi Faujas de Saint-Fond, *Observations de M. Faujas, Adjoint à la garde du Cabinet d'Histoire Naturelle du Roi, spécialement chargé des Correspondances des Jardins et du Cabinet. Sur le rapport du Comité des Finances de l'Assemblée Nationale, article "Jardin du Roi et Cabinet d'Histoire Naturelle," page 83* (Paris: Imprimerie de Chalon, 1790).

40 "Seconde Adresse des Officiers du Jardin des Plantes à l'Assemblée Nationale, en lui présentant un projet de règlemens pour cet Etablissement," in Hamy, "Les derniers jours du Jardin du Roi," 102–107.

41 Mornet, *Le sentiment de la nature en France;* Paul van Tieghem, *Le sentiment de la nature dans le pré-romantisme européen* (Paris: A. G. Nizet, 1960).

this fame counterbalanced the negative image of him and his work that might have been presented by other interested parties such as Condorcet.[42] Even in 1793, the naturalist's reputation was politically intact to the point that he could plausibly be ranked alongside prominent Jacobin heroes. In this year the secretary of the popular and Republican society of Avre-Libre, one Dourneau-Démophile, would celebrate in verse the inauguration of a bust of Buffon along with those of other Republican "great men" in the meeting chamber: Rousseau, Voltaire, Franklin, Lepelletier, and Marat.[43] Conveniently effacing Buffon's rise to the nobility, Dourneau praised him as a philosophe who had painted nature for the glory of the French nation.

The remainder of the First Address developed two distinct themes of argument: the first concerned the relation between the Jardin and the state, the second the activities of the Jardin naturalists as a community. The interdependency of the tasks of the different naturalists was stressed, as well as their individual importance. At present, the naturalists argued, the regulations of the Jardin discriminated between savants "whom Europe places on the same level"; their effort to order their lives in conformity with revolutionary ideology was driven by the same principles as their desire to reform the Jardin. The whole passage evoked the common Revolutionary image of people struggling against the yoke of oppression. The naturalists also emphasized the transfer of sovereignty over the Jardin from the "royal" to the "national."[44] One of the most pressing concerns behind this attempt to transfer sovereignty was financial: the king's budget was severely cut by the reforms of the early Revolutionary period. "To charge the expenses of this establishment to the civil list of H.M. . . . would be to occasion the ruin of an establishment useful to the Nation and to commit an injustice toward the King," Thouin would later claim in a series of notes written at Fourcroy's request, which attempted to establish the "real" authority over the Jardin:

42 See Marie-Jean-Antoine-Nicolas Caritat, marquis de Condorcet, "Éloge de M. le Comte de Buffon," *HARS* 1788/1790, 50–84; Aubin-Louis Millin de Grandmaison, "Discours sur l'origine et les progrès de l'histoire naturelle, en France," in *Actes de la Société d'Histoire Naturelle de Paris,* ed. Aubin-Louis Millin de Grandmaison (Paris: Imprimerie de la Société d'Histoire Naturelle, 1792), xiii–xiv.

43 Dourneau-Démophile, *Couplets civiques pour l'inauguration des Bustes de Francklin, Voltaire, Buffon, Jean-Jacques Rousseau, Marat et Pelletier* (Paris: Moutard, year II/1793). The "great men" cult reflected middling attempts to define an authoritative self-image. See Outram, *The Body and the French Revolution;* R. L. Herbert, *David, Voltaire, Brutus, and the French Revolution: An Essay in Art and Politics* (London: Allen Lane, 1972).

44 "Première adresse," in Hamy, "Les derniers jours du Jardin du Roi," 97–100.

"The Jardin Royal of Paris and the Cabinet of natural history which it contains are not at all the King's property as some people are concerned to put about . . . It is the task of the representatives of the Nation to dictate the law which should govern this Establishment in the interests of its greatest utility, and that of the executive power to watch over its execution."[45] Far greater things, the naturalists argued, were to be awaited from the Jardin under the "auspices of the representatives of the Nation," providing it was properly regulated. The final sentence of the address disclaimed the naturalists' apparent authority over the prospective reforms, making over all credit to the deputies: "And to minimize the waste of the moments which you are devoting to the prosperity of France, allow us to offer you some observations concerning the internal organization of our establishment in a few days' time."[46] The address was a carefully wrought text, with echoes of Revolutionary rhetoric, expressing Revolutionary ideals, and appealing to the deputies' image of themselves while preparing them for a project which devolved control of the Jardin into the hands of the naturalists who worked there. Its strategy was successful: the deputies voted in favor of allowing the naturalists to present their own reform project within the month.[47]

In forming an association, and in addressing their views directly to those in authority, without the mediation of the intendant, the naturalists' conduct contrasted with the customary behavior within patronage relationships. The new forms of behavior that they utilized were, however, in the process of being legitimized by Revolutionary practices. Communally written documents, produced by *sociétés libres,* provincial academies, and diverse other groups of petitioners, were a common form of address to the government during the first eighteen months of the Revolution. This mode of addressing those in power borrowed the rhetoric of representation. One person became the mouthpiece for the desires of a number—the representative or spokesman. This representative did not appear before authority as an individual, but as the instrument through which the voices of many were transported before authority. The patronage system differed considerably. In addressing authority the individual would muster as many allies as possible, in the forms of friends, contacts, and their patrons, to promote his cause

45 AN, AJ/15/502, piece 42: "Nottes pour le Jardin Royal des Plantes." Creuzé-Latouche's address, similarly, rejected the claim that the Jardin was the king's property.

46 "Première adresse," in Hamy, "Les derniers jours du Jardin du Roi," 100.

47 In the same session the Société Royale de Médecine was also commanded to reform itself ("Adresse à l'Assemblée Nationale," *HSRM* 1787–1788/1790, xxxiv–xxxvj).

before the potential patron.[48] Unlike the spokesmanship situation in which, to convince, many voices spoke through one mouth, in the patronage system one voice spoke with many mouths. Numerous conflicts arising in the course of attempted reorganizations of institutions of all sorts during the Revolution contained an element of tension between these two forms of self-expression, in attempts by individuals or factions to negotiate a powerful position. In the course of the Revolution, the mustering of allies came also to depend more and more upon the mastery of the public space by rhetorical and dramatic techniques.[49] It is for this reason that the naturalists' sensitivity to the value of using rhetoric and representation in their appeals to the state is important for an understanding of the origins of the Muséum. Following the deputies' vote, the naturalists began to hold regular assemblies four times a week from August 23, for the purpose of preparing their reform project.[50] The form of these was derived from other assemblies of the Revolutionary period, notably, of course, the Assemblée Nationale itself, and also the municipal assembly of Paris, to which three of the naturalists belonged. Committees and assemblies, as well as other forms of communal behavior in which representation was the principal means of self-expression, were forms of organization which enabled the naturalists and similar groups to bypass sources of authority from the Old Regime, such as the intendant, and to construct a new basis of sovereignty in which they would have a voice, albeit a shared voice. It is thus no surprise to find that La Billarderie, who at first tried to be present at the naturalists' assemblies, was universally ignored and soon ceased to attend.[51]

The assemblies held by the naturalists contrast in an interesting way, however, with those held by the members of the Académie des Sciences during the same period and for the same purpose. Where the Académie meetings were public and full of controversy over social reform, the register of the 1790 meetings, which were held privately, re-

48 This process of "supporting," *appuyer*, was widely used throughout the eighteenth century in France and other countries. Comparable accounts of patron competitions are in Biagioli, *Galileo, Courtier;* Steven Shapin, "The Audience for Science in Eighteenth-Century Edinburgh," *History of Science* 12 (1974): 95–121; id., "Property, Patronage, and the Politics of Science: The Founding of the Royal Society of Edinburgh," *British Journal for the History of Science* 7 (1974): 1–41; Jan V. Golinski, "Utility and Audience in Eighteenth-Century Chemistry: Case Studies of William Cullen and Joseph Priestley," *British Journal for the History of Science* 21 (1988): 1–32.

49 Outram, *The Body and the French Revolution;* Schama, *Citizens;* Hunt, *Politics, Culture, and Class.*

50 AN, AJ/15/95.

51 Corsi, *Age of Lamarck,* 12; Hamy, "Les derniers jours du Jardin du Roi," 35–42.

cords the naturalists as being unanimous on all important issues to do with the Jardin's reform. Indeed, the naturalists would seem to have agreed in the Jardin assemblies over important issues on which they disagreed in the academic context.[52] It was these meetings that gave rise to the naturalists' reform project, finished on September 9, 1790. Most of the twelve naturalists of the Jardin were actively involved in the preparation of this project, in contributing to its detail and voting on the division of tasks connected with its preparation and distribution.[53] The Jardin's two honorary professors, Antoine Petit and Louis-Guillaume Le Monnier, long since retired, also asked to be allowed to contribute their signatures. The resulting document was undoubtedly one of the most radical sets of reforms proposed for any savant institution during the Revolution. It provided for the establishment of a regulated, state-dependent institution for natural history principally through a reform of the social relationships between the savants who worked there. Its implementation would equalize the working status and remuneration of each naturalist and would transfer control of events at the Jardin, to be renamed the Muséum d'Histoire Naturelle, into the hands of the naturalists themselves in their new guise as professor-administrators. These reforms involved an abandonment of the differences in social and natural historical status between the naturalists. The project claimed to express the agreement of two nobles, several qualified physicians, and a gardener, among others. In terms of salary, the Jardin employees were hardly less diverse: de Jussieu earned only 1,200 livres a year as botany demonstrator, while Thouin as head gardener received 2,400 (he had already undergone a voluntary reduction of 1,200 per annum as a donation to the revolutionary cause).[54] Fourcroy, the professor of chemistry, earned 1,500 a year, while his demonstrator Brongniart received 2,000. Under the terms of the proposed agreement each naturalist was to receive 2,500 livres per annum—an increase for some, but a substantial decrease for others, such as Daubenton, who had the highest wage at a total of 4,200 livres a year. The abolition of *cumul,* or the tenure of multiple posts and pensions, was

52 In the summer of 1790 Fourcroy and de Jussieu found themselves on opposite sides in a debate over sovereignty; de Jussieu claimed the Académie should be answerable to the king, Fourcroy that it should be answerable only to the Assemblée Nationale. When the same issue was discussed within the Jardin, however, the naturalists voted unanimously for Fourcroy's position, which had narrowly won in the Académie debates (Hahn, *Anatomy of a Scientific Institution,* 171–172; AN, AJ/15/95).

53 Although Fourcroy is given as principal author, most of the other naturalists actively participated in the meetings, in writing parts of the project, and in its distribution (AN, AJ/15/95).

54 AN, AJ/15/149.

a significant blow, for the majority of the naturalists supplemented their income from the Jardin with salaries from other chairs, such as those at the Collège Royal and the École d'Alfort. Private incomes and lecturing were also affected by Revolutionary upheavals.[55] The tax privileges which all the Jardin employees with *brevets* had obtained as members of the royal administration were also withdrawn in 1789.[56] For the naturalists the gain was to lie not in terms of salary and social status, but in the acquisition of autonomy over the institution. The most far-reaching reform of the project was to be the development of a complete administration for the establishment, run by the naturalists in the new guise of professors, with the power to regulate the museum's affairs. All decisions on such matters as choosing a director, filling vacant positions, and undertaking building works were to be made by vote; the process of voting was carefully regulated in the project. In this context there was no room for an intendant, and so it is no surprise to find that the post was to be abolished. Instead, "the Muséum d'Histoire Naturelle will be under the immediate protection of the Nation's representatives."[57] Although the duties attached to each post were briefly described, the different areas of control were left open to interpretation by the incumbents. Such a division of tasks would serve to minimize the chances of controversy between individual naturalists within the institution, while allowing them to work independently.[58]

Thus the project proposed a large-scale operation for the research and teaching of natural history. It formed the "constitution" of the new, regenerated Jardin, just as the Assemblée's constitution was to form the basis for new and regenerated France. In its language and political claims it was very similar to the Declaration of the Rights of Man: "All the professors will be equal in rights and salary."[59] In short, naturalists of the Jardin du Roi utilized a variety of strategies familiar to and fa-

55 On the salaries attached to Jardin posts, see Laissus, "Le Jardin du Roi." See Gillispie, *Science and Polity;* Taton, *Enseignement et diffusion;* and Outram, *Cuvier,* for details of the posts accumulated by different naturalists. On naturalists' medical activities see above, chapter 1; Fourcroy and Brongniart gave private chemical lectures (W. A. Smeaton, *Fourcroy: Chemist and Revolutionary, 1755–1809* [Cambridge: Heffer and Sons, 1962], 7–9; AN, AJ/15/509, pieces 271, 272, 274). Several naturalists also had independent incomes (Audelin, "Les Jussieu"; Roger Hahn, "Sur les débuts de la carrière scientifique de Lacepède," *Revue d'histoire des sciences* 27 [1974]: 347–353).

56 Rudé, *French Revolution,* 60.

57 AN, AJ/15/95: "Registre des decisions reglementaires . . . ," meeting of August 27, 1790.

58 On the Muséum as a place of overlapping spheres of authority and patronage, see Outram, *Cuvier,* chapter 8.

59 "Projet de règlements pour le Jardin des Plantes et le Cabinet d'Histoire naturelle," in Hamy, "Les derniers jours du Jardin du Roi," 107–129.

vored by the deputies themselves. Of course, other project writers concerned with the Jardin also exploited such strategies, but it is arguable that the most thoroughgoing savant deployment of Revolutionary strategy was in the naturalists' offering. As a document calculated to appeal to the temper of the deputies in 1790, the reform project was excellently designed. The naturalists portrayed themselves as facing the same sorts of problems as the deputies: they were hardworking, virtuous men, attempting to produce a moral legislation for an important purpose, attempting to achieve equality in the face of inequality, and joined by a common bond of fraternity and merit. This was the language common to both naturalists and deputies. In one respect, of course, the naturalists had an easier task than the deputies: it was not possible simply to ignore the king.

The project was never heard by the deputies of the Assemblée Nationale, however. War and pressing internal problems, such as the collapse of monarchical authority after the king's abortive flight from Paris, lessened the importance of issues like the reform of the savant institutions. The naturalists' project was not read until June 1793, during a time of considerable political turbulence which had its effects upon the naturalists as on other groups.

Political Unrest, Savant Uncertainty

The period between 1790 and 1793 has been shown to be a time of considerable disruption for individuals who were part of the savant world. In many cases the function of institutions and societies broke down to a large degree.[60] Groups drifted apart, individuals sometimes traveling to the provinces for safety or for financial reasons. Often such disintegration was a response to political divisions within the group or to the increasing number of attacks against state funding of the sciences and corporatism in the local or national assemblies. In contrast, popular societies flourished in this period.[61] In the Jardin, however, efforts were made by all the naturalists to present and maintain an image of cohesion and continuity in their teaching activities. During this period, when the fate of the Jardin was still uncertain, the naturalists made use of both patronage-style relationships and communal behavior to bolster

60 Outram, "Ordeal of Vocation"; Dhombres and Dhombres, *Naissance d'un nouveau pouvoir,* 16; Hahn, *Anatomy of a Scientific Institution;* Chapin, "Vicissitudes of a Scientific Institution"; Pierre Huard, "L'enseignement médico-chirurgical," in *Enseignement et diffusion,* ed. Taton, 171–236; Roger Hahn, "The Problems of the French Scientific Community, 1793–1795," *Actes du XIIe congrès international d'histoire des sciences* 3 (1968): 37–40.

61 Hahn, *Anatomy of a Scientific Institution,* chapters 6 and 9.

their own and others' positions. The question of sovereignty within the Jardin was not settled yet either, as will become clear.

After the submission of the project to the Assemblée Nationale on September 9, the naturalists continued their campaign. Thouin and Faujas had been charged with the arrangements for printing a brochure containing the two addresses and the project.[62] A large-scale distribution of the 2,000 copies they commissioned took place during the next weeks. The distribution list reflects the plethora of groups and individuals whom the naturalists considered to have some claim to sovereignty over the Jardin in 1790.[63] It is remarkable for the breadth of its political and social spectrum, demonstrating the naturalists' sensitivity to the multiplicity of views and potential allies available to them in promoting their project. Both the Jacobins' Club and the Société de 1789 were represented, and the Assemblée Nationale was comprehensively flooded. Among the individuals who received personal copies were the rector of the University of Paris, the permanent secretaries of the Académie des Sciences and Société d'Agriculture, the astronomers and mathematicians Charles de Borda, Joseph-Louis Lagrange, and Joseph-Jérôme de Lalande, the writers Marmontel and Jean-François de La Harpe, and the *contrôleur du bureau des finances.* Thouin also sent copies to provincial correspondents.[64]

This wide audience reflected the diverse political and institutional affiliations of the twelve naturalists themselves, a diversity frequently used by historians to emphasize the divisions within the Jardin community.[65] Thouin was involved in drawing up the *cahier des doléances* for the district of Saint-Nicolas du Chardonnet, then in electing the deputy for the district. He became a *deputé suppléant* in May 1789. De Jussieu was elected to the district assembly in September 1789, and both men served in the new municipal government until new elections in August and September 1790; along with Lacepède, they also numbered among the thirty-six administrators of the new Parisian districts elected in January 1791.[66] Lacepède rose swiftly through the ranks to become secre-

62 *Adresses et Projet de Règlemens présentés à l'Assemblée nationale par les officiers du Jardin des Plantes et du Cabinet d'Histoire Naturelle, d'Après le Décret de l'Assemblée Nationale du 20 Août 1790* (Paris: Buisson, 1790).

63 BCMHN, MS 313 (2): "2000 Exemplaires," 1790.

64 BCMHN, MSS 1971–1984.

65 Outram, "Ordeal of Vocation"; see also *Cuvier,* 171–172, where Outram claims that the Jacobin persuasions of Fourcroy must have conflicted with the moderate views of Thouin, who associated with moderates like La Revellière-Lépeaux and Roland. However, the picture must be more complex than that, for during the Terror Thouin was in the habit of writing to Fourcroy as "mon cher camarade" ('my dear comrade'; BCMHN, MS 308).

66 Hamy, "Les derniers jours du Jardin du Roi," 3–23.

tary, vice president, and president of the Assemblée Législative in the course of 1791.[67] After August 10, 1792, when the Assemblées ceased to function and were replaced by the Convention Nationale, municipal involvement was replaced by more restricted involvement with the affairs of the Section du Jardin des Plantes for de Jussieu and Thouin. By this time the Commune had superseded the municipalities as the seat of Parisian government. After several years of membership in the radical Jacobins' Club, Fourcroy became a deputy to the Convention in year II, famously succeeding the murdered Jean-Paul Marat. His colleagues Lacepède and Thouin, on the other hand, were members of the Société de 1789.[68] In addition, Thouin had connections to the minister of the interior Roland and the deputy La Revellière-Lépeaux.[69] Inevitably, with such a wide range of political affiliations, factionalism in government affected individual naturalists. But the political careers of the naturalists all had one aspect in common: they all rejected the old sovereignty in favor of the new. The one important power who is not represented on the distribution list is the king, either via his ministers or directly.

From 1790 onward, public funding of the sciences was increasingly under attack within the Assemblée Législative.[70] Other forms of organization appeared to offer hope for the continuation of natural history in France, but these risked engulfing Jardin natural history within other programs. Most of the Jardin naturalists, for example, took part in the activities of the private Société d'Histoire Naturelle during this period, and much has been written concerning its anti-Buffonian stance, based largely upon the bitter attacks against Buffon of the ardent Linnaean, Aubin-Louis Millin, perpetual secretary to the Société, in the *Actes de la Société d'Histoire naturelle* and elsewhere.[71] However, it is by no means certain that all the members of the Société equated approval of Linnaeus with rejection of Buffon. When, on July 30, 1790, a bust of Linnaeus was erected in the Jardin du Roi, the Société had originally

67 *AP,* vols. 31–36.

68 *NBU,* s.vv. "Thouin," "Lacepède," "Lamarck," "de Jussieu." For Fourcroy, see Smeaton, *Fourcroy;* Kersaint, "Fourcroy."

69 Outram, *Cuvier,* 171–172.

70 Hahn, *Anatomy of a Scientific Institution,* chapter 9.

71 Aubin-Louis Millin de Grandmaison, introduction to *Actes de la Société d'Histoire Naturelle de Paris,* ed. Millin. On the conflict between Linnaeans and others, see Gillispie, *Science and Polity,* 191–193; Duris, *Linné et la France;* Stafleu, *Linnaeus and the Linnaeans,* chapter 9. Similarly, one might qualify Corsi's claim that "those who had enjoyed Buffon's support and patronage, and those who accepted his model of natural history, underwent severe institutional setbacks and bore the brunt of a massive theoretical offensive" (Corsi, *Age of Lamarck,* 6–7), for several of those who were highly successful in institutional terms, including Daubenton, Faujas, Lamarck, and Thouin, had originally owed their entry into the Parisian scientific world to Buffon.

planned to acknowledge the Jardin naturalists' own debts by presenting the Linnaeus bust as the first of a series: the next monument, its members claimed, was to commemorate the works of Buffon.[72] In other ways, as well, the Jardin naturalists struggled to legitimate the particular natural historical tradition which their establishment represented in the face of the growing power of the Société d'Histoire Naturelle's executive. Lamarck countered Millin's printed declarations of devotion to the Swedish botanist in his editorial articles in the *Journal d'Histoire naturelle,* where much space was devoted to a critique of Linnaeus.[73] Such negotiations reflected the quiet wrestling for authority over *histoire naturelle* that was taking place within the physical site of the Jardin. By 1793, however, what had once seemed a bright future for the noninstitutional practice of natural history had dimmed. Thanks to political accusations against them, several senior figures of the Société d'Histoire Naturelle were in hiding, and, since the remaining members were predominantly younger naturalists dependent on the support of the Jardin's staff, a radical Linnaean takeover of the Jardin seemed less likely.[74] The Jardin's naturalists did not have to subscribe to all Buffon's views, but it was very important for them to preserve their own "great men" or to lose their identity (figure 12).

During the period of uncertainty between 1790 and 1793, the naturalists made use of both communal behavior and individual patronage relationships, a use which depended upon the fluctuating political scene and allowed them to capitalize on each situation in turn. So for example, individual appeals to authority, as used by other participants in the sovereignty debate, could serve as a fallback position if the communal effort appeared to be failing. Lamarck and Faujas, the two naturalists whose livelihoods were most threatened by Lebrun's proposed reforms, prepared justifications of the value of their posts at the Jardin which

72 *AP,* vol. 17, August 5, 1790: "Adresse des naturalistes, à l'Assemblée nationale."

73 See, e.g., Jean-Baptiste-Pierre-Antoine de Monet, chevalier de Lamarck, "Philosophie botanique," *Journal d'Histoire naturelle* 1 (1792): 81–91; id., "Sur les systêmes et les Méthodes en Botanique, et sur l'analyse," *Journal d'Histoire naturelle* 1 (1792): 300–306. Although Olivier and Bruguière were also editors of this periodical, Bruguière spent much time on a voyage to the Levant which would eventually claim his life, and Olivier contributed only sparsely to the journal.

74 *DSB,* s.vv. "Bosc," "Broussonet," "Delamétherie," "Bruguière," "Millin." These individuals were among the foremost Parisian naturalists unattached to the Jardin: Bosc and Millin were prominent in the Société d'Histoire Naturelle, and Broussonet was the perpetual secretary of the Société Royale d'Agriculture. On the Jardin naturalists' support for Linnaeus and membership of the Société d'Histoire Naturelle, see Duris, *Linné et la France;* Georgia R. Beale, "Early French Members of the Linnean Society of London, 1788–1802: From the Estates General to Thermidor," *Proceedings of the Annual Meeting of the Western Society for French History* 18 (1991): 272–282; Millin, *Actes de la Société d'Histoire Naturelle de Paris.*

Figure 12 Buffon as one of "the illustrious French," soon after his death. While the naturalist looks out upon the cosmos, the pedestal of his bust contains the figure of Jean-Jacques Rousseau, depicted in a famous episode when he knelt to kiss the threshold of the room in which Buffon wrote. Such accounts would reinforce the special moral role of natural history in the 1780s and 1790s, even though Buffon allegedly informed Hérault de Séchelles that Rousseau's *Confessions* had revolted him (*Correspondance,* 1:140). From Nicolas Ponce, *Les illustres Français* (Paris: Ponce, [1790–1816]). © Bibliothèque Centrale M.N.H.N. Paris 2000.

were printed and circulated to deputies during the course of 1790.[75] Even in such individual appeals, however, the naturalists were prepared to speak on behalf of each other, as Faujas's document reveals: "The economies which the committee proposes bear essentially upon the suppression of two useful positions, that of the correspondences whose creation was requested by M. de Buffon, and that of M. de la Marck, the most indefatigable and wise of Botanists, charged with ordering and directing the richest collection of plants in the world."[76] Lamarck's proposal, similarly, entailed the preservation of Faujas's post. Both authors also used their projects to set forth their views concerning the necessary reforms for the Jardin, including the employment of more savants to work on the collections. Faujas wanted the creation of a

75 Jean-Baptiste-Pierre-Antoine de Monet, chevalier de Lamarck, *Mémoire sur le Projet du Comité des Finances, relatif à la suppression de la place de Botaniste attaché au Cabinet d'Histoire naturelle* (Paris: Gueffier, 1790); Faujas de Saint-Fond, *Observations.*

76 Faujas de Saint-Fond, *Observations.*

chair of natural history and an increase in the professorial salary. The uncertainty engendered by the failure of the Assemblée Nationale to vote on the fate of the Jardin eventually led Faujas to seek other employment. On September 20, 1792, he wrote to an unknown correspondent denouncing Lacepède and Daubenton for false displays of patriotism, and soon after provisionally accepted the post of inspector of mines. Thouin's accounts for 1793 and year II recorded "Faujas, professor of geology (salary suspended pending his decision)."[77] Independently of their joint activities, the naturalists were thus prepared to take other chances available to them as individuals to express their views, as is revealed by the projects of Lamarck, Faujas, and Thouin. Naturalists found it hard to locate the site of authority in the new political system, as well as their relationship toward it, and so aimed to cover all options. Meanwhile, however, the strain of waiting caused fissure lines between old patronage groups to widen, and during 1792 the Jardin community seemed in danger of fragmenting. Looking back in July 1793, Bernardin de Saint-Pierre recalled the jealousy of different factions.[78] A bitter squabble broke out over the question of responsibility for the herbarium, kept in the Cabinet du Roi, with Antoine-Laurent de Jussieu and Desfontaines, Le Monnier's protégés, on the one hand and Lamarck, Buffon's protégé, on the other. With the threat that his post was redundant hanging over him, Lamarck asserted his claim to be curator of the herbarium and lost. When finally the Muséum was founded in 1793, Lamarck had reinvented himself as a professor of zoology of shells, worms, and microscopic animals. This was not merely, as Cuvier would later claim, because no one else wanted to study that part of natural history. Before coming to the Jardin, Lamarck had dealt in shells and other marine productions for a time; alone among the naturalists, he could therefore count as an expert in this relatively neglected area. In this guise alone could he become an essential asset to a national institution concerned with the entirety of natural history—part of a community rather than part of a patronage group.[79]

Individual naturalists also formed patronage-style relationships with powerful political figures during this period. One of the best-known of these is Thouin's patronage by Roland, twice minister of the interior,

77 AN, F/17/1221: "Devis de la depense du troisième quartier pour l'entretien du Museum National d'histoire naturelle, pendant l'année 1793 l'an 2eme. de la Republique une et indivisible." The letter is reproduced in Fayet, *La Révolution française et la science,* 81; it seems likely that it was an attempt by Faujas to demonstrate his own Revolutionary zeal.

78 Letouzey, *Le Jardin des Plantes,* 309.

79 AN, AJ/15/509, dossier "Lamarck"; Corsi, *Age of Lamarck,* 2, 48.

the results of which I have discussed at length in chapter 2. Because life as a scientific practitioner in the French Revolution was increasingly precarious, the naturalists exploited multiple strategies while their fate remained uncertain. The communal behavior characterizing the 1790 projects by all the naturalists was only one of a range of possible options of behavior open to the naturalists; however, it would prove to have several advantages over other ways of behaving. Although ministers like Roland could be valuable patrons in the short term, the fluctuating political scene made ministerial life uncertain through the early 1790s. As Hamy suggests, a financial decision concerning Brongniart's remuneration for teaching supplies was taken by "Champion de Villeneuve, the fifth titular holder of the Ministry of the Interior in a little over six months. Named one day, fired the next, these ephemeral ministers have neither the time nor the means to study the administrative questions that come within their scope."[80] Finally, in early 1794, the ministries were abolished altogether. For eighteen months after the start of 1792, the Jardin was in a state of considerable administrative confusion. La Billarderie had fled to the provinces at the very end of 1791, and was later forced to quit France when accused of forging *assignats.*[81] Meanwhile, although he had officially resigned as intendant, the Jardin continued to function without a replacement for nearly eight months. The rapidity of turnover in the ministry of the interior, responsible for the quarterly payment of the Jardin's budget, made it necessary for the naturalists to solicit their money from the ministers directly. The difficulty of legitimating these claims in the absence of any distinct authority over the establishment is clear from the draft of a letter written by some of the naturalists on January 2, 1792: "MM. Daubenton, Desfontaines, Lacepède, Thouin and Vanspaendonck officers of the Jardin [national] des Plantes, have addressed to M. the Minister of the Interior following the procedure followed by the Intendant whose post has fallen vacant by resignation, the Budget for the first quarter of the present year for the maintenance of the Botanic Garden and the Cabinets of Natural History."[82] The word "national" is crossed out in the draft, revealing the naturalists' uncertainty over the political status of the Jardin. The letter also reveals the tension between the two kinds of language with which the naturalists found it necessary to address those in authority: both the language of patronage *and* the language of commu-

80 Hamy, "Les derniers jours du Jardin du Roi," 44.

81 Ibid.; on La Billarderie's resignation, see AN, F/17/1229, dossier 1, pieces 2–4.

82 Quoted in Letouzey, *Le Jardin des Plantes,* 292: officers of Jardin to minister of the interior, January 2, 1792.

nal behavior. Despite several appeals by individual naturalists at the Jardin that Daubenton be chosen as the next intendant, it was the writer Jacques-Henri Bernardin de Saint-Pierre who was eventually chosen to fill the post in August 1792, when Terrier de Monciel was minister of the interior.[83]

Bernardin's first request for the quarterly payment due to the Jardin came in September 1792. Meantime Thouin's patron Roland had returned to his position as minister of the interior for the second time. On receipt of Bernardin's request, Roland sent Thouin a memoir on the Jardin's expenses which was critical of many aspects of the way in which the establishment was funded, and demanded his opinion. Thouin replied, sending a detailed series of notes on the accounts of the Jardin, along with a copy of the 1790 project. Despite Bernardin's undoubted value as a citizen and savant, Thouin said, "for a long time now public opinion has pronounced that this post should be suppressed, in order to attribute its functions to an administrative council composed of all the members of the establishment under the direction and the orders of the Minister of the Interior."[84] The result of this exchange was that Roland directed all his official correspondence with the Jardin thereafter through Thouin, ignoring the legitimate authority in the Jardin, Bernardin de Saint-Pierre. Between late 1792 and early 1793 Roland fed an almost continual stream of valuable plants salvaged from national and émigré properties into the Jardin, as discussed in chapter 2. Bernardin de Saint-Pierre protested:

> Permit me, to begin with, [to say] that you have at different times addressed orders to M. Thouin the elder, head gardener of the Jardin national, to travel to Versailles, Trianon, Bellevue, . . . without warning me of this. I am, however, Intendant of this garden . . . Although M. Thouin the elder has communicated your orders to me as soon as he has received them and though no one is more suitable than he to carry them out, nevertheless when I saw that I had not been given notice by your offices and that absolutely no mention of me had been made, I was almost ready to believe that a plan to overlook me [*me rendre nul*] had been formed.[85]

83 Hamy, "Les derniers jours du Jardin du Roi," mentions a letter from Thouin to the minister of the interior, Cahier de Gerville, on this subject, January 2, 1792. See also Lacepède's proposal to the other naturalists that Daubenton be elected lifelong director of the establishment (AN, AJ/15/95, meeting of August 30, 1790), a suggestion included in the naturalists' Second Address on September 9, 1790.

84 Letouzey, *Le Jardin des Plantes,* 282–283, 294–306. Original manuscripts are in AN, AJ/15/506, F/17/3903.

85 Letouzey, *Le Jardin des Plantes,* 304: Bernardin de Saint-Pierre to Roland, undated.

Roland's patronage offered Thouin the opportunity to put forward his own views on the importance of individual members of the Jardin and on its reform. For the most part, Thouin held to and elaborated upon the system of reforms established by the naturalists in the 1790 project. The important feature of this relationship, however, is that Thouin became the naturalist closest to the person having most power over the future of the Jardin, the minister of the interior. Instead of a division of power within the group, with each individual having an equal ability to comment upon the work of the others, in this situation only one person had a voice with those in power. Thouin's rejection of the intendant thus meant that the latter risked becoming "nothing" to the minister. Thouin also used this special relationship with Roland to make comments upon the utility of his colleagues' posts. Although most of these comments were favorable, he suggested, for example, that Faujas's post was useless and should be suppressed.[86] When none of the naturalists had possessed personal relationships with those in power, they had all been equally vulnerable to the danger of having their posts or even their establishment closed down by the new government. Thouin's privileged ability to speak to those in authority now meant that the tenuous balance between the individual naturalists was under threat.

However, in the turbulent political scene of 1791 and 1792, patronage relationships were highly unstable. Patrons who were liable to be deposed at any moment could not be cultivated over long periods of time, and, as I argued in chapter 1, successful settlement of claims to posts was based in part upon the historicity of patronage relationships. Roland's term of power ended in early 1793, when he resigned following the worsening of the subsistence crisis.[87] The food shortage provided ammunition for attacks by Robespierre and the Jacobin *conventionnels* on a number of moderate political groups led by Jacques-Pierre Brissot de Warville, Alexandre-Sabès Pétion, and Roland. These factions were effectively removed from the political sphere by two successive Jacobin coups on May 31 and June 2, 1793, when twenty-nine representatives and two ministers (excluding Roland, who later committed suicide) were arrested.[88] These events preceded the reform of the Jardin des Plantes by just eight days. Clearly, Roland was in no position to determine the course of events surrounding the Jardin's reform.

86 Hamy, "Les derniers jours du Jardin du Roi," 49.
87 Petersen, *Lebensmittelfrage und revolutionäre Politik.*
88 Schama, *Citizens,* 719–726.

Where historians formerly traced "party-line" allegiances within the Convention, and identified clear-cut political groups such as the Brissotins or Jacobins with modern-looking leaders, more recent studies have suggested that the locus of political debate should instead be sought in loose-knit groups congregating in, and centered around, particular meeting places, such as the salon of Mme. Roland.[89] Claims that individuals were "responsible" for the fate of any given institution should thus be treated with the utmost caution. Control of the political arena in the French Revolution was determined by the number and power of the elements that one could represent or enroll, and by the duration of such tenuous enrollments.[90] It is arguable that there were serious difficulties attached to the enlistment of an individual as patron in the Revolution. The rapidity with which the locus of power could change made such practices not only ineffective but even, in some cases, dangerous. Louis-Guillaume Bosc, a naturalist who was a close friend of both Thouin and Roland, paid for his political connections by finding it necessary to retire to the forest of Montmorency in late 1793, where he had the opportunity to observe at first hand the anatomy of various indigenous species of spiders.[91] In Thouin's case, patronage by Roland was less disruptive, resulting only in the search of the family house for hidden Rolandin supporters.[92] But patronage did not bring about the reform of the Jardin. That was left to the Comité d'Instruction Publique in the Robespierrist regime, the same period which saw the closure of many state-funded savant establishments, and the introduction of the Terror as an instrument of public control.

Étienne Geoffroy Saint-Hilaire recorded in his autobiography that Joseph Lakanal, the deputy heading the Comité d'Instruction Publique, visited the Jardin on the evening of June 9, 1793. It was Geoffroy who met the deputy and took him to Daubenton's lodgings in the Jardin. Neither of the men, Geoffroy said, had ever met Lakanal before.[93] That

89 Outram, *The Body and the French Revolution,* 130–131.

90 Outram, *The Body and the French Revolution;* Furet, *Interpreting the French Revolution,* 49–58.

91 *DSB,* s.v. "Bosc"; Outram, "Ordeal of Vocation"; Jean-Loup d'Hondt, "Louis-Auguste-Guillaume Bosc (1759–1828), conventionnel et naturaliste, premier systématicien français des bryozoaires actuels," in *Scientifiques et sociétés pendant la Révolution et l'Empire: Actes du 114e congrès national des sociétés savantes (Paris 3–9 avril 1989), section histoire des sciences et des techniques* (Paris: Éditions du CTHS, 1990), 241–258.

92 Louis-Marie La Revellière-Lépeaux, *Mémoires de Larevellière-Lépeaux, membre du Directoire exécutif de la République française et de l'Institut national; publiés par son fils sur le manuscrit autographe de l'auteur et suivies des pièces justificatives et de correspondance inédits,* 3 vols. Ed. Ossian La Revellière-Lépeaux (Paris: E. Plon, Nourrit et Cie, 1895), 1:161.

93 Fayet, *La Révolution française et la science,* 115; *CIPCN,* 1:480.

is not to say that Lakanal had previously ignored the Jardin. It was he who had put forward a motion to have the Jardin's position reviewed in the Convention in March. Slightly later he had proposed the transfer of the valuable Chantilly cabinet to the establishment.[94] However, as he did not know in person the Jardin naturalists whom he addressed, it is clear that Lakanal could not have stood in the position of patron to them as individuals. Historical accounts agree that Daubenton gave Lakanal a copy of the 1790 project, which the latter rapidly revised and abridged, possibly with some colleagues from the Comité d'Instruction Publique. On the following day, a shortened version of the Second Address and Project were read to the deputies by Lakanal, and the reforms were passed. The Jardin du Roi became the Muséum d'Histoire Naturelle.

In Lakanal's intervention, a new kind of relationship between the state and the institution was emerging: instead of a hierarchy of relationships between individuals inside and outside the establishment, through which administrative concerns were transmitted and mediated, the institution itself was being reified. Correspondence between the Muséum and the state after June 1793 thenceforth began to take the form of an exchange between different bureaucratic appendages of the state, rather than between patron and protégé. According to Hamy, the June 10 reform was successful because "minds were elsewhere," and Lakanal took the Convention by surprise.[95] The Comité d'Instruction Publique was one of the most powerful bodies of the Jacobin Republic, and since, over the next two months, Jacobin deputies would manifest considerable interest in the future of other savant establishments, it seems likely that the foundation of the Muséum was neither incidental nor accidental to Jacobin agendas for the sciences and public instruction. The post-Thermidorean mythology of the Muséum's foundation has long concealed close inspection of the special status of natural history in the period.

By the time of the Jardin's reform, many Old Regime establishments had either been closed already, or were barely functioning. Official medical and pharmaceutical teaching had ceased by decree of the Assemblée Nationale Législative after August 18, 1792, with the closure of the medical faculties and the Collège de Pharmacie in Montpellier, and the suppression of provincial pharmaceutical corporations.[96] These

94 Hamy, "Les derniers jours du Jardin du Roi," 63–66.

95 Ibid., 66.

96 Charles Bedel, "L'enseignement des sciences pharmaceutiques," in *Enseignement et diffusion*, ed. Taton, 237–257.

had been replaced by free societies of pharmacists and private practice. State-organized medical teaching did not reappear until 1794, with the establishment of three Écoles de Santé in Paris, Montpellier, and Strasbourg. Such measures were primarily concerned with the establishment of medical free trade as an ideological basis for medical training, but they were linked to the increasing polemicization of the notion of a "corporation." The many enemies of other Old Regime establishments such as the University of Paris and the academies turned this polemic to their own ends; attacks on the "corporatism" of these bodies became increasingly frequent and, as the composition of government became more politically extreme, increasingly potent. By 1793 the activity of the university had effectively ceased. It was ultimately closed by decree of the Convention Nationale on September 15.[97] The same fate had already been suffered by the academies a month before, and attempts to save at least the Académie des Sciences on the grounds of "utility" were unsuccessful.[98] The association of corporatism with inequality was too effective. On August 8, 1793, Henri Grégoire, Thouin's colleague at the Société Royale d'Agriculture, who was an active member of various committees of agriculture and public instruction over the Revolutionary period, presented a project concerning the fate of the academies to the Convention. "In this day when the sun illuminates a single people of brothers, its gaze must no longer encounter on French soil institutions which depart from the eternal principles which we have consecrated; and yet some which still bear the imprint of despotism, or where the organization clashes with equality, have escaped in the general reform; these are the academies."[99] Although Grégoire, who is today best known for his calls for racial and religious equality, went on to distance the literary academies from the Académie des Sciences, the vote taken at the end of the reading of his bill resulted in the fact that only the first two of his articles were passed by the deputies. The first suppressed "all the academies and literary societies [*tous les académies et sociétés littéraires*] patented or funded by the nation," and the second placed all the material assets of the suppressed establishments under the authority of the state until the organization of public instruction had been achieved. All the rest, clauses to maintain the public funding of courses in the sciences, and to keep alive the Académie des Sciences until after the organization of a society for the advancement of the arts and sci-

97 Hahn, *Anatomy of a Scientific Institution,* 284.

98 Ibid., chapter 8.

99 Henri Grégoire, *Rapport et projet de décret présenté au nom du Comité d'Instruction publique, à la séance du 8 août* (Paris: Imprimerie Nationale, 1793).

ences which would replace it, were ignored. Thus the closure of the most important savant establishment in France stemmed from the concern of *conventionnels* with the link between the modes of organization adopted by groups of people and the manufacture of sovereignty. The extent to which this was the case is revealed by the responses of the members of the Société Royale d'Agriculture and of the new Muséum d'Histoire Naturelle to the August 8 bill.

The Société Royale d'Agriculture was not included in the terms of the August 8 bill. In fact, the first article of Grégoire's report, proposing the closure of the academies and literary societies, is ambiguous in the French: it can be interpreted as reading "[literary] academies and literary societies," and this was undoubtedly the way in which Grégoire had written it, given the content of his address preceding the bill. On this reading the Académie des Sciences would probably not have been included in the suppression. Only a very severe reading could allow the Société Royale d'Agriculture to be interpreted as a "literary" society. Nevertheless, both the Académie and the Société ceased to hold meetings after the middle of August 1793, although the Société still featured in the 1794 budget.[100] The Muséum naturalists displayed the same concern with the interpretation of their actions. Following the closure of the Académie (of which many of the Muséum's professors were members), the naturalists wrote to the Comité d'Instruction Publique on 23 brumaire year II, asking that a clause in the decree reforming the Jardin, requiring them to hold two public meetings a year, should be suspended.[101] "We fear that it might be felt that these public assemblies resemble academic forms," wrote the secretary for the year, Desfontaines. "The Convention in passing [the decree] which suppressed [the Academies], no doubt had in view to destroy all that might smack of their regime."[102] The type of communal behavior adopted by the Jardin naturalists during the Revolution was central to their ability to sustain a particular politics of self-portrayal. It enabled them not only to achieve the reform of their institution, but to be the most successful among the Parisian naturalists at weathering the Terror: where Bosc, Broussonet, Millin, and Delamétherie all fled Paris in late 1793, the

100 Hahn, *Anatomy of a Scientific Institution;* Festy, *L'agriculture pendant la Révolution française,* vol. 1; Boulaine, "Les avatars de l'Académie d'agriculture," shows that although the Société held a couple of desultory meetings after August 8, attended by Grégoire himself, normal activity ground to a halt after the end of September.

101 Muséum naturalists who were also Académie members included Daubenton, Portal, Fourcroy, Thouin, de Jussieu, and Lamarck (Hahn, *Anatomy of a Scientific Institution*).

102 AN, F/17/3880: Desfontaines to President of Comité d'Instruction Publique, 23 brumaire year II/November 13, 1793.

administrators of the ex-Jardin were just beginning their careers in the Muséum d'Histoire Naturelle.

I have presented communal behavior so far as a means to an end: first as a new way of addressing those in authority, and second as a means of preserving the status quo in a time of political upheaval. The mere passage of the June 10 decree did not render the new institution and its professors safe from political intervention. The period 1793 to 1794 is characterized by a number of negotiations between the professors and the state as the professors struggled to maintain their autonomy over the new institution. In their behavior toward the state, the professors depended upon the communal self-portrayals which had been used in the past. A notable example is the case of the unfortunate *aide-naturaliste* Gauttier, appointed to the Muséum by the Comité de Salut Public in year III, who became a focus of the naturalists' struggle for self-government under the Terror. After Gauttier had been at the Muséum for a few months, he wrote to the Comité de Salut Public, complaining that the professors would not pay him a salary or find him any lodgings. This had followed a meeting by the professors in

Figure 13 Device of the Muséum d'Histoire Naturelle. Animals, plants, and minerals surround a beehive, symbolizing the professoriate, and a Phrygian cap, symbolizing republican liberty, in this woodcut designed by the Muséum's professor of iconography, Gérard Vanspaendonck, in 1793. Heads of corn and vines refer to the agricultural role of natural history. For contemporaries, bees were models of industry, utility, and social harmony. © Bibliothèque Centrale M.N.H.N. Paris 2000.

vendémiaire, at which a decision was effectively taken to ignore Gauttier on the grounds that he had no talent. They also objected strongly to their allies on the Comité d'Instruction Publique, who conveyed their comments to the Comité de Salut Public.

> They observe today that this citizen has been named contrary to the dispositions of a ruling which has the force of law; that the *aides-naturalistes,* being subaltern employees for whose management the professors are responsible, cannot be chosen without their intervention . . . You will sense, as we do, the necessity of maintaining good harmony between the truly estimable men who are at the head of such a precious establishment, and of employing there only men who are recommendable by their talent.[103]

From Gauttier's letter, it certainly appears more likely that he had no chance to display any "talent" which he might have possessed. The objection of the professors had more to do with the usurpation of their sovereignty than with any question of talent.

In Lacepède's case, the ties of community operated to include, rather than to exclude. According to Cuvier, Lacepède was forcibly removed from the Jardin in 1792 by concerned friends and hidden from public view at his home in Normandy.[104] To official queries, his colleagues in the Jardin responded that Lacepède was unwell, but that he attended his lectures as usual—rather a long distance to commute in 1792. Eventually Lacepède proffered his resignation in the approved style to the intendant, Bernardin, and was officially replaced by Geoffroy.[105] Thus the rhetoric of community could act as a safety net for individuals among the naturalists who were in political trouble. After the fall of Robespierre, the professors prepared a letter to the Comité d'Instruction Publique outlining the Muséum's need for a third professor of zoology; Citizen Lacepède, they suggested, might be a suitable candidate for the post.[106]

In the uncertain climate of 1793 and 1794, therefore, the relationship of the naturalists to one another was at least as important as their relationship to the state. The decree of June 10, 1793, had transferred sovereignty over the fates of individual naturalists to their colleagues. Safe-

103 AN, F/17/1233: members of Comité d'Instruction Publique to members of Comité de Salut Public, 3 brumaire year III/October 24, 1794.

104 Georges Cuvier, "Éloge historique de M. le Cte. de Lacépède, lu le 5 juin 1826," in id., *Recueil des éloges historiques des membres de l'Académie Royale des Sciences,* 3 vols. (1827; facsimile reprint, Brussels: Culture et Civilisation, 1969), 3:306.

105 AN, AJ/15/510, pieces 324, 325.

106 AN, F/17/3880. The professors' demand was granted by the Convention on 21 frimaire year III/December 11, 1794.

guarding one's post, particularly in times of prolonged absence, became a matter of enrolling allies within the group. Within the Muséum community, a discourse of fraternity served to maintain social ties. Thus, for example, when Antoine-Louis Brongniart left the Muséum just weeks after its foundation, to serve as first pharmacist to the army, he addressed his colleagues as follows in a letter of leave-taking written on July 1, 1793:

> The new Establishment formed at the Jardin National must be an imperative motive to retain all the professors who are attached to it. I regard as one of the greatest happinesses of my life to be able to count myself as one of you. May I be so fortunate as to return shortly and to take part in your labors, work together with you for public instruction, and become your model in care and zeal in fulfilling my duties.
>
> I beg you to receive, illustrious concitizens and colleagues, the homage of the fraternal sentiments of your devoted colleague,
>
> A.-L. Brongniart[107]

Brongniart had begun his letter by referring to the sovereignty of the Convention, quoting a law under which no one called up to serve in the army could be deprived of his public office. This was therefore a duplicate appeal, both to the power of the law and to the power of fraternal sentiment. Similar expressions of communal sentiment are to be found in Thouin's letters to the professors, written during his travels with Faujas as commissioners of the Republic in years II and III (1794–1795) and again in years V and VI (1796–1797), in the wake of the French army. Their task was to confiscate all the valuable specimens and objects of use to the arts and sciences in the conquered towns, and send them back to the Muséum. On 30 germinal year III (April 19, 1795), Thouin wrote from the Hague. "Without the hope of being useful to the Republic and of serving the Museum, the mounting privations that we are experiencing, far from you, would have made us cut short our voyage in order to bring us back to you. We see its end approaching with much satisfaction."[108] Another letter of 4 frimaire year V (November 24, 1796) was written from Italy near the end of the second expedition.

> It seems to me that the end of our mission is approaching and that I shall soon be able to reunite myself with you. I need it more to satisfy my heart than to sample a little rest after such a lengthy voyage. I shall seize the opportunity with much haste and pleasure.

107 AN, AJ/15/577: A.-L. Brongniart to Muséum administration, July 1, 1793.
108 AN, AJ/15/836: Thouin to Muséum administration, 30 germinal year III/April 19, 1795.

A respectful *Salut,*
A. Thouin[109]

And, on 20 pluviôse year V (February 8, 1797), from Mantua, Thouin added a postscript to his letter: "Permit me, citizens, to profit from the space which remains to me to tell citizen Geoffroy that I have fulfilled his wishes and distributed the copies of his memoir which he sent to me for the savants of Italy. I shall always hasten to do all that could be agreeable to all my colleagues and to each of them in particular."[110] Thus, in spite of the fact that the naturalists of the Jardin and Muséum were socially and politically divided, during the Revolutionary period they used communal behavior to regulate their own interactions and their exchanges with the rapidly fluctuating sites of sovereignty.[111] That this strategy was retrospectively viewed as the cause of the institution's survival is revealed in an account of a festival which took place at the Muséum in year VI to celebrate the planting of a tree of liberty.[112] The professors were accompanied by members of the Comité d'Instruction Publique and a detachment of the National Guard. While the tree was being planted by the gardeners, the report claimed, "the air *Where can one be better than in the bosom of one's family* expressed the sentiments which have always united all the members of the Museum." "National sovereignty is no longer a problem," declared the director, de Jussieu. "Reason has demonstrated it, the constitution has proclaimed it."

> Let us be united, Citizens, and we will be invincible within and without; and let us be faithful to the Constitution and the factions will extend themselves to smash against it . . . Our union, which might be cited as an example, has saved both us and this Establishment in those times of barbarism when meetings were proscribed and savants persecuted. Precious union which creates our happiness, which earns us the esteem and affection of good people, of legislators, of heads of Government, continue to inhabit in the midst of us and never distance yourself from this abode which never ceased, even in the most stormy times, to be the haven of peace.

109 AN, AJ/15/836: Thouin to Muséum administration, 4 frimaire year V/November 24, 1796.

110 AN, AJ/15/836: Thouin to Muséum administration, 20 pluviôse year V/February 8, 1797.

111 By contrast, Outram, "Ordeal of Vocation," and id., *Cuvier,* chapters 1 and 8, focuses exclusively upon individual responses to the problems of survival and power negotiation posed by the Revolution. E. R. Brygoo, "Du Jardin et du Cabinet du Roi au Muséum d'Histoire Naturelle, en 1793: La continuité par les hommes," *Histoire et nature* 28–29 (1987–1988): 47–63, suggests a more unified interpretation for the Jardin naturalists, but has little supporting material.

112 AN, AJ/15/847: "Fête civique célébrée dans le Muséum National d'histoire naturelle le Dix ventose an six pour la plantation d'un arbre de la liberté."

At the end of de Jussieu's speech, those who had attended the ceremony went on to a "fraternal banquet" in celebration of liberty.

CONCLUSION

De Jussieu's explicit association between communal behavior and self-preservation in the context of a debate over sovereignty shows that the importance of the rhetoric of community in the reform of French natural history should not be underestimated. It served to hold together different individuals in a situation where old ways of defining relationships had become untenable. As the Revolution dissolved the bonds of patronage between members of polite society, so other forms of behavior, differently legitimated, were adopted to replace them. The savants had emerged from a world in which patronage relationships controlled their degree of sovereignty over their own work and that of others, to one where the basis of sovereignty itself was continually being debated and changed. In this they differed little from any other section of the educated elite. Hence, as several historians have noted, some of the more radical revolutionaries, such as Marat, were savants who failed to obtain patrons in the old regime.[113] From denunciations of the patronage system to attacks on the locus of ultimate sovereignty, the king, was only a short step. In the case of the Jardin, Buffon's death in 1788 was an important catalyst for the changes in behavior of the Jardin naturalists, providing them as it did both with a potential "hero figure" from whom they could justly claim intellectual antecedence, and with an unwelcome new intendant whose very existence could be portrayed as undermining Revolutionary principles.

The Jardin naturalists' attempts to relocate the basis of sovereignty at the Jardin mimic in miniature the attempts of successive power-seeking groups to win sovereignty of the political world. Likewise, in attempting to revise the basis of sovereignty, the naturalists made use of the same alternatives to patronage that were being used by the deputies, and employed similar discourses. These changes in behavior were thus borrowed from, and legitimated by, Revolutionary rhetoric, and they entailed, above all, communal as opposed to individual appeals to authority in attempts to achieve autonomy and to justify autonomous action. Communal behavior also helped to provide a means for organizing individuals within the group and for communication between the group and the variable locus of political power. It served as a means of self-expression when the site of sovereignty within or outside the

113 Darnton, *Mesmerism,* chapter 3; Hahn, *Anatomy of a Scientific Institution,* chapter 5.

Jardin was not clear, as was the case in 1790 and again in 1792. It also allowed individuals within the group to appeal to a fraternal rhetoric of sentiment and social cohesion when it proved necessary to safeguard their individual positions. When power over other savants was located within the savant community, survival within the institution depended vitally upon the use of such language. Likewise, success in dealing with the new political culture necessitated the proper use of political language. In messidor year II (June 1794) Thouin requested Fourcroy to scrutinize his addresses to the Comité de Salut Public to ensure that they contained only "the most purified principles of Republicanism."[114]

I have endeavored to clothe with flesh the rather bare bones of a "Furetian" account of the language and political strategies adopted by the Jardin's naturalists, in the form of a construction of the ways in which the breakdown of patronage during the Revolution problematized personal relationships in different aspects of savant life. My treatment has implied that what was at stake in such choices of language was the proper forms which interpersonal power relations should take in a revolutionary society. Because the Jardin was virtually the only savant institution to survive the turbulent times of the Revolution with its staff intact, it provides a good setting for examining how a group of individuals coped with the renegotiation of their relationships with one another and with those in authority as personal accountability and moral probity became matters of life or death. It is perhaps because the naturalists of the Jardin were able to fashion a rhetoric of community and fraternity that they were also able to supply guarantees of good character for one another. They had exploited their position as men constantly in contact with the wonders of nature to prove their virtue. Likewise, exposure to nature could legitimate naturalists' claims to have mastered the art of living harmoniously in society. In the hands of Parisian naturalists in the late old regime, *histoire naturelle* was not a mere description of natural productions, but a system of operations for reshaping the natural world and the human observer thereof. Community, patriotism, and the national good (as manifested in the Muséum d'Histoire Naturelle) could be credibly represented as the natural outcomes of natural historical practice.[115]

Thus the self-presentation of the naturalists as united was crucial

114 Letouzey, *Le Jardin des Plantes,* 324: Thouin to Fourcroy, 4 messidor year II/June 22, 1794.

115 My argument here is radically different from Outram, "New Spaces in Natural History," who claims that the Muséum "was characterised by internal strife" (257). Such a view, however, makes it impossible to explain many crucial features of the Muséum's foundation.

to their success in reforming the Jardin du Roi and converting it into the Muséum d'Histoire Naturelle. At the time that the question of the establishment's reform arose within the Convention Nationale, the posts of all the members of the establishment were equally at risk. Only by supporting a project which, in one way or another, ensured that each of them had the same standing, could they avoid the consequences of placing too much faith in one patron or in one member of the establishment. So Lamarck was transformed into a zoologist, and Daubenton into a mineralogist, in order that those who had no alternative expertise should be able to justify their existence. The communal strategy adopted by the naturalists between 1790 and 1793 also explains why the young Geoffroy was able to become a professor at only eighteen. The formula was very successful for several years. In chapter 5 I shall explore the role played by the Muséum itself in the fashioning of Republican symbolism.

CHAPTER FIVE

The Spectacle of Nature: The Muséum d'Histoire Naturelle and the Jacobins

The greatest of all economies, because it is the economy of men, thus consists in placing them in their true position: now, it is incontestable that a good system of instruction is the primary means of achieving this.

Charles-Maurice de Talleyrand, *Rapport sur l'instruction publique fait au nom du Comité de Constitution les 10, 11 et 19 septembre 1791*

In 1893 the professors of the Muséum d'Histoire Naturelle celebrated the centenary of their institution in a commemorative volume. Here one of their number, Ernest-Théodore Hamy, described the foundation of the Muséum d'Histoire Naturelle on June 10, 1793, when the deputy Joseph Lakanal "descended from the tribune, having assured . . . the future of the natural sciences in our country."[1] This deceptively simple account, in which the moment of the Muséum's creation enabled the smooth progress of the natural sciences in France over the next century, has often reappeared in subsequent histories. But the future of the Muséum was not so secure as it appeared retrospectively. In the mid-1790s scientific organization was under heavy attack in the Convention; the Muséum was virtually the only scientific establishment to escape closure. Previous chapters of this book have suggested different reasons for its success: the perceived utility of the establishment, and the ability of the professors to manifest republican community. While the emphasis of Jardin natural history upon utility and the advancement of the arts was valued generally throughout the Revolution-

An earlier version of this chapter appeared under the title "Le spectacle de la nature: Contrôle du public et vision républicaine dans le Muséum jacobin," in *Le Muséum au premier siècle de son histoire,* ed. Claude Blanckaert, Claudine Cohen, Pietro Corsi, and Jean-Louis Fischer (Paris: Muséum National d'Histoire Naturelle, 1997), 457–479.

1 Hamy, "Les derniers jours du Jardin du Roi," 66.

ary period, the Muséum's foundation also reflected the particular concerns of Jacobin leaders.

The Revolutionary transformation of *moeurs* entailed the transformation of those practices which had been most heavily used to display social status under the Old Regime—in particular, the fine arts and natural history. Both would be closely linked in public-instruction projects of the mid-1790s. While the incorporation of the fine arts in Jacobin cultural policies has been explored at length, questions about legitimate visions of nature in the French Revolution have received far less attention.[2] From before the appearance of the *Histoire naturelle,* the power of vision had occupied an important place in natural historical epistemology and in naturalists' self-fashioning. The claims by Antoine-Laurent de Jussieu to be discovering the proper method for a natural classification were grounded, I suggest, upon just such self-presentations and upon a particular epistemological modesty constructed upon areas of agreement about what constituted the "natural" within the natural historical and chemical communities.

By drawing upon the "nature-sentiment" tradition of the later eighteenth century, exemplified in the writings of Rousseau and Bernardin de Saint-Pierre, naturalists could present themselves as transparent republicans, improved by the moral effect of seeing sublime nature. In such writings the emotional, aesthetic, and physiological senses of the term *sensibilité* became conflated in the single act of looking at Nature. For many naturalists and their publics, natural history was the science of sensibility par excellence. This status would enable natural history to be assimilated into Republican attempts to improve the public through the marshaling of emotion.[3] If naturalists were accustomed to making order within the small physical space bounded by Buffon's iron railings, deputies had the far more difficult task of ordering a new society. Their classificatory activities were aimed at generating a Rousseauist "public," no longer conceived in commercial or polite terms, but

2 On the fine arts in the Revolutionary period, see Poulot, "Le Louvre imaginaire"; Elke Harten, *Museen und Museumsprojekte der Französischen Revolution: Ein Beitrag zur Entstehungsgeschichte einer Institution* (Münster: Lit, 1989); Thomas Eugene Crow, *Emulation: Making Artists for Revolutionary France* (New Haven: Yale University Press, 1995); James A. Leith, *Media and Revolution: Moulding a New Citizenry in France during the Terror* (Toronto: University of Toronto Press, 1968); id., *Idea of Art as Propaganda;* id., *Space and Revolution: Projects for Monuments, Squares, and Public Buildings in France, 1789–1799* (Montreal: McGill–Queen's University Press, 1991). On revolutionary nature see Corvol, *La nature en Révolution;* Harten and Harten, *Die Versöhnung mit der Natur;* Richefort, "Métaphores et représentations de la nature."

3 The Revolutionary nature-sentiment cult is discussed in Pierre Trahard, *La sensibilité révolutionnaire, 1789–1794* (Paris: Boivin, 1936), chapter 6.

as a political entity, the nation. Over the course of 1793, Republican cultural reformers increasingly portrayed the task of public instruction as being the regeneration of the degraded public of the Old Regime. As one of the principal institutions for public instruction of the Jacobin period, the Muséum would be a major part of schemes to remodel public space in order to expose citizens to moral spectacles designed according to precise criteria. Such spectacular instruction became the modus operandi for a nationwide network of educational centers in which natural history's visual technology supplied a model for the spread of enlightenment, utility, and transparent self-presentation. The Muséum's fate during the Jacobin Republic thus depended upon the diversity of roles that could be played by natural history collections as sites of disciplined vision. The translation of the former royal collections into the national realm marked the point at which the establishment's spectacles of nature became visible to the largest number. In effect, the Muséum's professors offered Jacobin deputies and supporters in Paris an institution which could portray a series of possible futures, political, physical, and visual, to onlookers.

Seeing Is Believing

Naturalists at the Jardin du Roi presented themselves and natural historical praxis in general in terms of a particular sensory epistemology which privileged the visual powers. Locke's account of the formation of the understanding, in which the individual at birth was a tabula rasa which must be impressed and formed by successive sensory experiences and subsequent reflection, underpinned Buffon's epistemological concerns in the initial discourse to the *Histoire naturelle* in 1749.[4] The human learning process was one of successive accretions of sensible experience which altered and increased the faculties of the body and mind. This was also true of natural history: "once one has succeeded in gathering together specimens of everything that peoples the Universe," one initially felt defeated by the sheer immensity and diversity of nature. But there was an escape from this state of cognitive confusion and inadequacy: "in familiarizing ourselves with these same objects, in seeing them often, and, so to speak, without design, they gradually form durable impressions, which soon connect themselves in our spirit by means of fixed and invariable relations; and from there we elevate ourselves to more general views, by which we can embrace several different objects at a time; and it is then that one is in a condition to

4 Sloan, "Buffon-Linnaeus Controversy."

study with order, to reflect fruitfully, and to carve out the paths to arrive at useful discoveries."[5] Here self-development was closely tied to the faculty of vision. The abbé Étienne Bonnot de Condillac and others, including Jean-Jacques Rousseau, would remodel the claims of John Locke about the nature of the understanding. Their discussions formed the basis for what is now known as the "Locke-Condillac sensationalist psychology," but which at the time was principally understood as a theory of *éducation,* to be understood in a broader sense than the English term "education." Condillac's early writings on *éducation* were attacked as merely duplicating Buffon's arguments in the *Histoire naturelle.* However, his later publications, including the *Cours d'études pour l'instruction du Prince de Parme* of 1775, were taken as the foundation for educational reform projects in France in the second half of the eighteenth century, alongside renowned works such as Rousseau's *Émile* of 1762. Over the same period, reformist educational writers increasingly privileged sensation over reflection as the means for shaping the individual, and radical materialists such as Helvétius ended by rejecting reflection altogether.[6] The power to sense, *sensibilité,* was central to the making of natural historical knowledge at the Jardin in the second half of the century. The limits of vision, indeed, could define the limits of the naturalist's investigation. In Daubenton's view, natural historical practice terminated where human intervention became necessary for the production of knowledge: "As soon as the processes of art have destroyed the structure of minerals, or altered the organization of plants and animals, the Naturalist ceases to observe the productions of Nature: the Chemist has pulverized, dissolved, macerated, distilled, calcinated, vitrified them etc."[7] As the work of Roberts and Schaffer has shown,

5 Buffon, "Discours sur la manière d'étudier et de traiter l'histoire naturelle," *Histoire naturelle* (1749), 1:3–62.

6 François Dagognet, "L'animal selon Condillac," in Étienne Bonnot de Condillac, *Traité des animaux, où après avoir fair des observations critiques sur le sentiment de Descartes et sur celui de M. de Buffon, on entreprend d'expliquer leurs principales facultés* (1755; Paris: Vrin, 1987), 10–131; Harvey Chisick, *The Limits of Reform in the Enlightenment: Attitudes toward the Education of the Lower Classes in Eighteenth-Century France* (Princeton: Princeton University Press, 1984); Antoinette Stettler, "Sensation und Sensibilität: Zu John Lockes Einfluß auf das Konzept der Sensibilität im 18. Jahrhundert," *Gesnerus* 45 (1988): 445–460. Gilbert Py, *Rousseau et les éducateurs: Étude sur la fortune des idées pédagogiques de Jean-Jacques Rousseau en France et en Europe au XVIIIe siècle* (Oxford: Voltaire Foundation, 1997), and Jean Bloch, *Rousseauism and Education in Eighteenth-Century France* (Oxford: Voltaire Foundation, 1995), are less concerned with the model of the learning process underpinning different projects for public instruction.

7 Louis-Jean-Marie Daubenton, "Introduction à l'Histoire naturelle," in *Histoire naturelle des Animaux,* 1:i; see also BCMHN, MS 807: Daubenton, "Cours de Zoologie professé après 1786, soit à l'Ecole normale soit au Collège de France, soit à l'Ecole d'Alfort. Mammifères," f°. I.

the reforming chemist Antoine-Laurent de Lavoisier and many of his chemical supporters rejected this "sensual technology."[8] Only individuals with particular qualities of mind which enabled them to resist the seductions of the senses could make true natural knowledge. This was very distant from the older model of chemistry, represented by the Jardin's demonstrator of chemistry Hilaire-Marin de Rouelle, in which bodily experience was the very foundation for knowledge. In contradistinction to chemical reformers, for whom instrumentation and the distrust of the senses were central to the proper performance of experiments, naturalists explicitly distinguished their enterprise from the art of experimentation by insisting that the possession of sensibility was a precondition for natural historical practice. This noninterventionist sensual technology of natural history was apparent in the interest shown by the Jardin's naturalists at the end of the century in searching for *rapports,* the crucial natural relationships between beings which formed the basis for Antoine-Laurent de Jussieu's claims to be developing the *méthode naturelle* from the 1770s onward. The choice of *rapports* as the term to describe classificatory practice was significant here. It was first utilized by de Jussieu in the earliest paper that he ever presented to the Académie Royale des Sciences, in 1773.[9] The expression was, he said, comparable in one sense to the affinities "which Chemists admit in the mineral substances submitted to their examination."

What was the significance of de Jussieu's use of a specifically chemical term? *Affinités* were phenomena fundamental to pre-Lavoisierian chemistry. As the work of Metzger and Goupil has shown, chemical *affinités* or *rapports* had been given a new lease of life by Étienne-François Geoffroy in a memoir presented to the Académie Royale des Sciences in 1718.[10] Geoffroy's insistence that *affinités* and *rapports* were purely descriptive terms, referring to the relative tendency of chemicals to combine into compounds, was exploited by later chemists seeking to assert an empirical foundation for their discipline. By 1786 Guyton de Morveau, in his article "Affinité" for the *Encyclopédie méthodique,*

8 Roberts, "Death of the Sensuous Chemist"; Simon Schaffer, "Self Evidence," *Critical Inquiry* 8 (1992): 328–362.

9 Jussieu, "Examen de la famille des Renoncules."

10 Hélène Metzger, *Les doctrines chimiques en France du début du XVIIe à la fin du XVIIIe siècle* (Paris: Presses Universitaires de France, 1923), 419–420; Michelle Goupil, *Du flou au clair? Histoire de l'affinité chimique de Cardan à Prigogine* (Paris: Éditions du CTHS, 1991), chapters 4 and 5; Anderson, *Between the Library and the Laboratory,* 56, 73–74; Étienne-François Geoffroy, "Table des différents Rapports observés en Chimie entre différentes substances," *MARS* 1718/1719, 202–212.

could insist that the term was "today the expression of a purely physical action" and had no connection with occult phenomena.[11] *Affinités* or *rapports* had become part of a particular chemical ontology. De Jussieu, who had studied under Rouelle, was almost certainly exploiting this in proclaiming that his botanical *rapports* were to be compared with those of the chemists. As he explained the analogy to his Académie Royale des Sciences audience,

> Chemical affinity is the more or less powerful tendency of two bodies to contract a union; it is not the same in all [bodies]. Some unite intimately and easily; others have a weaker adherence, and can be separated by an intermediary; some do not unite at all, or only with great difficulty. Plants present pretty much the same nuances, the same gradation; they have characters in which they approach one another, and those by which they differ. The variable combination of the former with the latter has served Botanists in constituting classes, sections, orders, families, genera [and] species.[12]

In this way the classificatory relationships being developed by botanists were not creations but discoveries, prior to any theoretical intervention, transparently present to the eye of the beholder in a setting such as a botanical garden, where plants from diverse places could be brought together and compared. In a memoir presented before the Académie Royale des Sciences, de Jussieu described how the natural families were known to all botanists precisely because the relationships between the plants in them were visible to the expert observer.[13] Related plants overlapped on a spectrum of characteristics which combined in certain ways apparently determined by nature herself, rather than being forced into categories by the exclusive use of a few characters, as in artificial systems such as Linnaeus's.

For eighteenth-century naturalists, as we saw in chapter 3, plants were moving, changing, politically and chemically potent beings.[14] Natural historical *rapports*—which became the focus of the Jardin naturalists' activities during the Revolution—were what related species, but

11 Louis Guyton de Morveau, "Affinité," in *Dictionnaire de Chimie, Encyclopédique méthodique* (1786), 1:535–536, quoted in Goupil, *Du flou au clair?* 180.

12 Jussieu, "Examen de la famille des Renoncules," 214.

13 Jussieu, "Exposition d'un nouvel ordre de plantes adoptés dans les démonstrations du Jardin Royal."

14 On plants as active entities, see François Delaporte, *Nature's Second Kingdom: Explorations of Vegetality in the Eighteenth Century,* trans. Arthur Goldhammer (Cambridge: MIT Press, 1982), esp. 136–174.

also what drew them together, in a material sense. *Rapports* were pursued as an immediate goal of natural history by several naturalists in the 1780s and 1790s.[15] In publications during these decades, the term was used to describe the reality of the classificatory ties between plant or animal species. Thus, for example, while the work of Linnaeus was useful as a mnemonic aid for newcomers to natural history, the "true naturalist," according to Lamarck, sought "to form for himself an exact idea of the *rapports* . . . which the author of the universe has most decidedly placed between all the species which exist and which perpetuate themselves in Nature; in order to judge most conveniently both the whole of these natural beings, and each of them in particular."[16] Because *rapports* were not founded upon any hypothesis, they were also removed from dissent; in the most striking cases, Lamarck said, "it is not possible for any particular will to hesitate in admitting [species to a genus], or to think about rejecting them."[17] This uncompromising exclusion of subjectivity permitted naturalists to claim that "particular wills" could not intervene in the transition from observing to knowing nature. The Jardin naturalists defended their classificatory innovations on the basis of their sensory experience, both against chemists who proclaimed the authority of instruments and against competing classificatory agendas which contravened the transparency of sensory evidence. *Rapports* were true, not arbitrary, and evident to the eye of the trained observer.

It was in this respect that de Jussieu's project of natural history as a search for *rapports* became fruitful in the early 1790s. The professors covered quite diverse natural historical domains; their understanding of the ontological status of the classificatory act varied greatly, as did their theological and philosophical commitments. But, at the level of practice, a number of them found it possible to agree that the basic common denominator of natural historical inquiry should be

15 Discussions of *rapports* are in René-Louiche Desfontaines, "Histoire naturelle. Cours de botanique élémentaire et de physique végétale. Discours d'ouverture," *La Décade philosophique* 5 (1795): 449–461, 513–520; 6 (1796): 1–11, 129–143, 193–202, 321–330, 449–453; Étienne Geoffroy Saint-Hilaire, "Mémoire sur les rapports naturels des Makis Lémur," *Magazin Encyclopédique* 1 (1796): 20–36; Louis-Guillaume Bosc d'Antic, "Ripiphorus" and "Coturnix Ypsilophorus," *Journal d'Histoire naturelle* 2 (1794): 293–296, 297–298; see also Georgette Légée, "Étienne-Pierre Ventenat (1757–1806), botaniste limousin, face aux problèmes de classification et de sexualité végétales," *Comptes rendus du 102e congrès national des sociétés savantes, Limoges, 1977, Section des sciences* 3 (1977): 33–46. Ventenat, the librarian of the Bibliothèque Sainte-Geneviève, who later became a professor at the École Centrale du Panthéon, was one of the natural method's most important early publicizers.

16 Jean-Baptiste-Pierre-Antoine de Monet, chevalier de Lamarck, "Sur l'étude des rapports naturels," *Journal d'Histoire naturelle* 1 (1792): 362.

17 Ibid., 365.

the search for natural *rapports.* De Jussieu did not present himself, nor was he viewed, as having found the perfected natural classification—merely the correct foundation on which to build. During the early nineteenth century, the natural method would supersede the sexual system of Linnaeus, which had been the most widely known and used during the second half of the eighteenth century, but which was acknowledged even by its maker to be artificial. However, other naturalists, including Linnaeus and Adanson, had also generated the outlines of natural arrangements for plant species.[18] Like Antoine-Laurent de Jussieu, these two men had been the students of Bernard de Jussieu, Antoine-Laurent's uncle, described by one mid-century commentator as "the greatest botanist of the realm."[19] Bernard's claim to be developing a natural method was posthumously supported by Antoine-Laurent and others, who portrayed the bachelor botanist as simple, unambitious, and unassuming. Bernard's very failure to publish his system was thus deployed as evidence for the authenticity of his contact with the natural world. Antoine-Laurent, his uncle's natural heir, would administer Bernard's credibility in numerous ways. After his uncle's death he wrote notes for Condorcet's *éloge* at the Académie, which presented his own classificatory activities as the mere completion of a task largely achieved by Bernard; the implication was that the ideas of the greater predecessor had been implanted into the mind of his successor without alteration. This was a not uncommon strategy permitting young men to present new knowledge-claims without transgressing the limits of acceptable behavior in a patronage society.[20] Antoine-Laurent de Jussieu continued to excel in effacing himself from the scene of scientific and political labor. After the foundation of the Muséum d'Histoire Naturelle, he served as director for most of its first decade of existence. The only other two individuals elected to the post

18 Daudin, *De Linné à Jussieu;* Michel Guedès, "Jussieu's Natural Method," *Taxon* 22 (1973): 211–219; Stevens, *Development of Biological Systematics;* for Linnaeus's sketches of the natural order, see Eriksson, "Linnaeus the Botanist"; Duris, *Linné et la France;* Müller-Wille, *Botanik und weltweiter Handel,* chapters 2 and 3; James L. Larson, *Reason and Experience: The Representation of Natural Order in the Work of Carl Linnaeus* (Berkeley: University of California Press, 1971); Stafleu, *Linnaeus and the Linneans,* esp. chapter 9. For Adanson, see Michel Guedès, "La méthode taxonomique d'Adanson," *Revue d'histoire des sciences et de leurs applications* 20 (1967): 361–386; Frans A. Stafleu, "Adanson and His 'Familles des Plantes,'" in *Adanson,* ed. Lawrence, 1:163–264, esp. 178–202.

19 Tourneux, *Correspondance littéraire,* 3:512. See also Adolphe Brongniart, "Notice historique sur Antoine-Laurent de Jussieu," *Annales des sciences naturelles* 7 (1837): 5–24.

20 Marie-Jean-Antoine-Nicolas Caritat, marquis de Condorcet, "Éloge de M. de Jussieu," *HARS* 1777/1780, 94–117.

were Daubenton and Lacepède, the former for his seniority and symbolic status, the latter because of his political prominence and his mastery of sensibility as a lifestyle. De Jussieu, by contrast, played little part in political events as an individual from 1793 onward.

The first full account of de Jussieu's method was published less than a fortnight before the fall of the Bastille, in his only major work, *Genera plantarum secundum ordines naturales disposita juxta methodum in Horto Regio Parisiensi exatarum,* a text written in Latin, thus for learned male readers. Its title can be translated as "The genera of plants, arranged in natural orders, according to the method demonstrated in the King's Garden of Paris." De Jussieu adopted two different approaches to the formation of classificatory groups. First, he insisted that the principles he deployed in classification were themselves natural, "intimately inherent within plants and which present themselves easily to whomsoever observes them." Certain genera and certain orders possessed a natural existence upon which all botanists could agree. The classifier's task then consisted in studying these obviously natural groups in order to determine which characters were common to them and in which combinations, such as, for example, the relative positions of stamens, corolla, stigma, and ovaries. Thereafter, each character was to be weighted for its stability; if one character or combination of characters was stable in one genus but not in another, it must possess a lesser value for the natural economy and thus for the classifier. These natural genera and orders would provide templates which the "Botanist who is attached to following Nature" could apply to classificatory groups whose existence was not the subject of universal agreement. The characters used for generating the classes were drawn from parts judged a priori to be essential to the preservation of the species within the economy of nature—the seed, flower, and fruit. Fundamental to the natural method was the fact that the three primary divisions in the plant kingdom, the acotyledons, monocotyledons, and dicotyledons, were based upon the number of seed leaves in the germ.[21]

In fact, numerous aspects of de Jussieu's method were not dissimilar to those of his botanical peers; likewise, he was far from being the only naturalist to claim to be discovering natural plant relationships. What was different was that during and particularly after the Revolution, de Jussieu's *méthode naturelle* would *succeed* in being portrayed as

21 Henri Daudin, *Les méthodes de la classification et l'idée de série en botanique et en zoologie de Linné à Lamarck, 1740–1790* (Paris: Félix Alcan, 1926), 210–212. Quotations are translated from Daudin's French.

the most natural classification of plants. In his early lectures at the new Muséum, even the die-hard supporter of Buffon's criticisms of classifiers, Daubenton, endorsed the use of the classification of Bernard and Antoine-Laurent de Jussieu as the closest to nature.[22] The perceived "naturalness" of de Jussieu's method and his claims for the empirical transparency of plant relationships depended upon his peers' acceptance of his own authenticity as investigator of nature. More research is required to examine how, in this crucial decade, de Jussieu juggled his political and botanical status and his role as the Muséum's director. But it is evident that the epistemological status of natural historical knowledge, the self-presentation of the naturalist, and the problem of sensibility were all closely related. Moreover, de Jussieu was at pains to emphasize the grounds for agreement among investigators of nature. With the dual strategy of self-effacement and the use of epistemological resources such as *rapports* and "natural genera," de Jussieu maximized his resources for presenting and representing the natural. As we will see, such strategies were not dissimilar to those used to marshal government support for the Muséum d'Histoire Naturelle.

An emphasis upon the transparency of natural historical knowledge was widespread in writings on *histoire naturelle*. As a result, the most serious criticism that a naturalist could make of another was to accuse him of suffering from a "spirit of system." Vision filtered through "systems" was corrupt vision, which perverted the transparency of the viewer's encounter with nature. Buffon himself had qualified his discussion of the role of vision in the making of knowledge, arguing for touch as the only sense capable of rectifying the errors of the other senses.[23] But d'Angiviller's *Mémoires* of year XIII (1805) specifically pointed to Buffon's poor eyesight as a direct cause of the faults of his natural history: "If I wished to find an excuse for that spirit of system with which I believe one can justly reproach [Buffon], I might perhaps find a very natural and simple one in his organization itself. Almost entirely deprived of sight, he could only see that which he considered with attention and care; almost all the discoveries were the result of chance."[24] Buffon's descriptions of specimens relied largely upon secondhand accounts, and he lacked personal sensory access to nature. This problematized the status of his writing, which he and others described as a

22 BCMHN, MS 218: L.-J.-M. Daubenton, MS fragment; apparently a part of "Caractères distinctifs de la conformation de l'homme et des animaux" [post 1793].

23 Buffon, "Des sens en général," in *Histoire naturelle* (1749), vol. 3, chapter 8.

24 Bobé, *Mémoires*, 56.

"painting" of nature: how could a blind man make a faithful copy?[25] Critiques of Buffon's observational abilities as a naturalist dated back to the 1740s; in particular, the much-challenged experiment in which he and the English naturalist John Turberville Needham had utilized a newly designed microscope to reveal the existence of organic molecules was deployed by Buffon's opponents to question his visual powers.[26] D'Angiviller, responding to bitter attacks in the recently published memoirs of Marmontel, tried to rescue Buffon's posthumous reputation by portraying him as a prophet, rather than a savant: one who could access nature through his own inner gifts, rather than his outward sensibility. "[Buffon's] height, his fine white hair, the nobility of his physiognomy, the still greater nobility of his expressions, the majesty of his ideas and of his style which he carried even into conversation, even the feebleness of his eyesight, which did not allow him to distinguish anything, [and] prevented him from fixing his gaze, gave to his discourse an air of prophecy and inspiration which was imposing."[27] Nonetheless, such representations treated the "French Pliny" himself as an object to be seen, rather than an active viewer of nature. The reliability of observers' senses could become central to assessments of their standing as naturalists.[28]

Naturalists, like anatomists and physicians, had a particular stake in the wider negotiations over the importance of sensibility in personal,

25 As expressed in Champcenetz's satirical poem against Buffon of 1787:

> At rest after work, in my nightly repose
> The image of Buffon appeared before me
> Pompously apparelled as if at the Jardin du Roi.
> His errors had not crushed his pride at all;
> He even continued to use that affected style
> With which he was careful to paint and adorn his work
> To evade the inevitable insults of time.

(Quoted in *Correspondance,* 2:341.)

26 Sloan, "Organic Molecules Revisited"; Spary, "Codes of Passion." Such criticisms stemmed particularly from Réaumur's circle of naturalist experimenters; cf. Charles Bonnet, "Observations sur quelques auteurs d'histoire naturelle," in *Correspondance littéraire,* ed. Tourneux, 4:169; Emile Hublard, *Le naturaliste hollandois Pierre Lyonet: Sa vie et ses oeuvres (1706–1789) d'après des lettres inédites* (Brussels: J. Lebègue et Cie, 1910), 50.

27 Bobé, *Mémoires,* 54; Marmontel, *Mémoires,* 224–225, originally published in 1804 as part of his *Oeuvres posthumes.* In these years blind poets and philosophers were being rehabilitated as creative geniuses; see William R. Paulson, *Enlightenment, Romanticism, and the Blind in France* (Princeton: Princeton University Press, 1987), chapter 5.

28 Dawson, *Nature's Enigma,* cites numerous letters between Réaumur and Bonnet which suggest a similar problematization of the latter's knowledge-claims after his early loss of sight.

political, and savant experience.[29] They were the experts who could predict the relationship between conformation and moral or physical powers. Conversely, they could also relate the effects of the senses to the physical structure of the body. For example, in 1782 Mauduyt's discourse on the senses of birds distinguished between the faculties originating from the body's mechanisms and "those which emanate from the senses . . . [which] have an immediate and intimate relation with the will; they properly belong to the free animal to determine itself, and its choice for the action which it decides to take, forms its character."[30] The dominant sense in animals determined the type of character the species possessed. During the 1780s a number of Daubenton's protégés were engaged in relating conformation to moral qualities, with the notable exception of Vicq d'Azyr, who simply excluded all such discussion from his *Traité d'Anatomie* of 1786.[31]

But similar concerns also structured naturalists' self-portrayals. A couple of years after the publication of a French translation of Johann Caspar Lavater's essay on physiognomy in 1781, a new pastime reached the streets of Paris. The *physionotrace* was an instrument which, somewhat like a *camera lucida,* allowed exact profiles or full-face depictions to be produced for physiognomical readings.[32] The ancient art of physiognomy permitted inner, moral character to be read in the external features of the face and head.[33] As one reviewer of Lavater's work suggested, because of its attention to the links between external and internal properties, physiognomy was "the base of the other [sciences], or rather it is the unique science, the only one within our grasp. Everything we know [and] everything we are able to know both about ourselves and about the beings which surround us, is physiognomy."[34] In

29 Staum, *Cabanis;* Ludmilla J. Jordanova, "The Natural Philosophy of Lamarck in Its Historical Context" (Ph.D. dissertation, University of Cambridge, 1976), chapter 5.

30 Mauduyt, "Ornithologie," 1:360.

31 Vicq d'Azyr, *Traité d'Anatomie,* and see id.,"Recherches sur la structure du Cerveau, du Cervelet, de la Moelle alongée, de la Moelle épinière; & sur l'origine des Nerfs de l'Homme et des Animaux," *MARS* 1781/1784, 495–622. On sensibility, see Staum, *Cabanis.*

32 Michael Melot, "Caricature and the French Revolution: The Situation in France in 1789," in *French Caricature and the French Revolution* (Chicago: University of Chicago Press, 1988), 28.

33 On physiognomy see Graeme Tytler, *Physiognomy in the European Novel* (Princeton: Princeton University Press, 1982); Roy S. Porter, "Making Faces: Physiognomy and Fashion in Eighteenth-Century England," *Études anglaises* 38 (1985): 385–396; Martine Dumont, "Le succès mondain d'une fausse science: La physiognomonie de Johann Kaspar Lavater," *Actes de la recherche en sciences sociales* 54 (1984): 3–30; Mary Cowling, *The Artist as Anthropologist: The Representation of Type and Character in Victorian Art* (Cambridge: Cambridge University Press, 1989), chapter 1. By the end of the eighteenth century, physiognomic descriptions were standard features of *éloges.*

34 Tourneux, *Correspondance littéraire* (1782), 13:202.

Figure 14 Antoine-Laurent de Jussieu's physiognomy in the year he published his *Genera plantarum,* describing the *méthode naturelle.* By Quenedey, 1789. From Adolphe Brongniart, "Notice historique sur Antoine Laurent de Jussieu," *Annales des sciences naturelles* 7 (1837): opp. 5. By permission of the Syndics of Cambridge University Library.

1789, the year that his *Genera plantarum* was published, Antoine-Laurent de Jussieu had his portrait drawn in a *physionotrace* (figure 14). Such interests, as in the case of his experimentation with animal magnetism, should not be considered to be idle dabblings in pseudoscientific fields, for they hint at de Jussieu's wider understanding of his own classificatory enterprise. As in the case of the plants that he classified, de Jussieu apparently considered his own conformation to map onto his inner character. One of de Jussieu's most important early claims for the naturalness of his method rested on the fact that its successful discovery would be demonstrable, because naturally related plant species would share similar virtues. Physical conformation was a sure guide to the properties—medicinal, alimentary, and so on—possessed by a given plant; thus, finding the natural order was an explicitly patriotic act, of interest to botanists "who wish to unite the title of *Savant* with that of useful Citizen."[35] Familiar discourses of utility and patriotism typical of French botanical and agronomic literature from the 1760s onward were here tied to an account of the relations between conformation and natural order. While the search for plant virtues was an ancient task of botanists, a physiognomical approach to the study of plants may also have addressed the epistemological problem raised by Buffon in his introductory discourse to the very first volume of the *Histoire*

35 Jussieu, "Exposition d'un nouvel ordre de plantes," 175; id., "Mémoire sur les rapports existans entre les caractères des plantes, et leurs vertus," *HSRM* 1786/1790, 189.

naturelle, in 1749. Here the intendant controversially attacked all classifiers for claiming to found their orders upon the true inner nature of things; such essences were unknowable. Instead, he called for a new natural history based upon proper and complete descriptions and definitions of each species. Things were nothing in themselves, as they stood in relation to the observer, even when their name was known: "they begin to exist for us when we know their relations [*rapports*] and properties; it is even only by means of these relations that we can give them a definition."[36] Physiognomy may have provided a model by means of which de Jussieu sought to avoid charges of scholastic essentialism, yet still retain the botanist's privileged access to the link between external appearances and internal virtues.

All natural historical knowledge was a delicate balance between seeing and believing. The reliability of testimony depended in part upon what was to be seen in the face of the witness, and as in the case of the unfortunate Calmucks, a whole barrage of moral judgments hinged upon correct observation of conformation. Sensibility could be a way of relating outward appearance and conduct to inner nature, and in this capacity it was central to efforts at Revolutionary regeneration. On August 27, 1791, the electors and men of letters of Paris, including several naturalists, petitioned the Assemblée Nationale for the transferral of Rousseau's ashes to the Panthéon. They praised him for "the habit which he has given us of penetrating beneath the bark of false social conventions, and seeing men and things nakedly."[37] The display of inner character would be a favorite theme of Republican followers of Rousseau, who deployed a battery of oratorial, sartorial, and gestural strategies to demonstrate inner virtue.[38] The claim to be able to see things "as they are" was, however, a staple of naturalists' discourses about their own legitimacy as inquirers into the natural world. Moreover, this stance required an explicit emotional and aesthetic self-

36 Buffon, "Discours préliminaire," in *Histoire naturelle,* 1:25; see similar comments in de Jussieu, introduction to *Genera plantarum,* ii–xx. Sloan, "Buffon-Linnaeus Controversy," suggests compellingly that Buffon's concerns derived from his reading of Locke.

37 *AP,* 29:756. The petition was signed by Fourcroy and Desfontaines, as well as Broussonet, Millin, and Bosc. On Rousseau and transparency see Jean Starobinski, *Jean-Jacques Rousseau: Transparency and Obstruction,* trans. Arthur Goldhammer (Chicago: University of Chicago Press, 1988).

38 Thus, in the invention of *suspects* in the autumn of 1793, the language, dress, and behavior of one's neighbors were to be scrutinized for suspicious indications which might reveal inner corruption (Marc Bouloiseau, *The Jacobin Republic, 1792–1794,* trans. Jonathan Mandelbaum [Cambridge: Cambridge University Press; Paris: Éditions de la Maison des Sciences de l'Homme, 1987], 96–100).

presentation which was already central to natural history, as it was not for the experimental sciences.

There was a second sense in which *sensibilité* played a central role in natural history.[39] The sensibility whose seat in the animal body was being investigated by comparative anatomists in the 1770s and 1780s was also the sensibility which provoked outpourings of tears in Jean-Jacques Rousseau's readers.[40] In natural history books readers were frequently reminded of the great man's love of botany. In a world in which Buffon's and Rousseau's writings had made the study of nature and the appreciation of its beauties an essential part of a civilized upbringing, the naturalists of the generation after Buffon found a convenient form for their addresses to legislators in appeals to this dual tradition of *sensibilité*. They did not merely borrow from an external tradition of nature-sentiment literature, manifested in the writings of Rousseau, Bernardin de Saint-Pierre, and a host of lesser-known authors: sensibility was deeply embedded within natural historical writings in France during the 1780s and 1790s. Throughout the later eighteenth century, naturalists portrayed nature as the site of emotional experience in their publications, conjuring up charming scenes of verdure and pastoral peace for their readers.[41] Although some claimed that sentimental writers should be considered distinct from "scientific" naturalists, this was not a widespread view, especially among the readership of natural historical books existing outside savant institutions. The choice made by the minister of the interior to select the famous writer Bernardin de Saint-Pierre as the intendant for the Jardin des Plantes in the last six months of its existence reveals the extent to which sentimental nature could be identified with the nature of natural historians.

Similar concerns about the links between seeing and feeling appeared

39 On sensibility, see David J. Denby, *Sentimental Narrative and the Social Order in France, 1760–1820* (Cambridge: Cambridge University Press, 1994); Anne Vincent-Buffault, *The History of Tears: Sensibility and Sentimentality in France* (Basingstoke: Macmillan, 1991); Serge Moravia, "From Homme Machine to Homme Sensible: Changing Eighteenth Century Models of Man's Image," *Journal of the History of Ideas* 39 (1978): 45–60; Staum, *Cabanis;* Elizabeth L. Haigh, "Vitalism, the Soul, and Sensibility: The Physiology of Théophile Bordeu," *Journal for the History of Medicine and Allied Sciences* 31 (1976): 30–44. A larger literature exists for Britain, but predominantly concerns sensibility in literature.

40 Darnton, *Great Cat Massacre,* chapter 6; John Mullan, "Hypochondria and Hysteria: Sensibility and the Physicians," *Eighteenth Century: Theory and Interpretations* 25 (1984): 141–174; Chris Lawrence, "The Nervous System and Society in the Scottish Enlightenment," in *Natural Order: Historical Studies of Scientific Culture,* ed. Barry Barnes and Steven Shapin (Beverly Hills: Sage Publications, 1979); Staum, *Cabanis.*

41 Stafford, *Voyage into Substance;* Charlton, *New Images of the Natural;* Thomas, *Man and the Natural World;* Bourde, *Agronomie et agronomes,* vol. 2, chapter 11.

in contemporary discourses on both natural history and the fine arts. Emotional response to the arts differed only in degree, not in kind, from that induced by the sight of nature's works. Conversely, the language used to describe natural productions drew heavily upon the descriptive terminology of paintings and plays. Natural history collections, landscapes, and gardens were often referred to as spectacles or pictures (*tableaux*).[42] Polite commentators of the eighteenth century applied the criteria of taste both to the productions of art and to those of nature.[43] Cabinets reflected the owner's tastefulness, and natural history collections continued to be important in the self-construction of the educated individual up to the Revolution. Such collections were evaluated on their aesthetic merits, as well as on their scientific quality, by owners and visitors alike. Critical judgment was passed in similar ways, and often in the same paragraph, upon the productions of nature and those of art, an association made explicit in the posthumous edition of Dezallier d'Argenville's *La Conchyliologie* of 1780, which was edited by two artists: "The fine arts have as their object the imitation of nature: now, to imitate nature it is necessary to know her. Can one flatter oneself to have achieved this by always seeking her in copies which will never present her with total accuracy—is it not indispensible to go to the source, to contemplate, to study the original herself?"[44] Their proposal for creating an amateur's cabinet elided the appeal to reason and taste. The ideal collection would be located within a square room, "lighted from the north side only by large windows: the picturesque genius will decorate its other walls to the ceiling, with individuals from the three kingdoms; it will know how to make pleasing to the eyes the methodical order designed to please the spirit."[45] Such concerns with tastefulness played a far more important role than has been given credit in determining the subject matter of natural history in the polite world of eighteenth-century Paris. In 1786 Vicq d'Azyr claimed that comparative anatomy had been neglected because it was too repulsive to appeal

42 E.g., in the *Histoire naturelle,* in Pluche's *Spectacle de la Nature,* and elsewhere. In "De l'art d'écrire," written circa 1786, Buffon commented: "All the objects which nature presents to us, and in particular all the living beings, are so many objects for which the writer must not only prepare a static portrait, but [also] a moving picture, in which the forms will develop, all the traits of the portrait will appear animated, and together present all the external characters of the object" (reproduced in *Correspondance,* 1:95–96).

43 Pomian, *Collectionneurs, amateurs, et curieux;* E. C. Spary, "Rococo Readings of the Book of Nature," in *History of the Book/History of the Sciences,* ed. Marina Frasca Spada and Nicholas Jardine (Cambridge: Cambridge University Press, forthcoming).

44 Dezallier d'Argenville, *La Conchyliologie,* xiv.

45 Ibid., 195.

to "those amateurs who must be captured by the elegance and the mobility of the spectacle."[46] Forty years before, in his discussion of the arrangement of the Cabinet du Roi, Daubenton had made a radical assertion about display strategies in the collection: neither traditional methodical distributions nor purely symmetrical, tasteful orders truly managed to grasp the natural, and thus both were artificial. Nonetheless, the Cabinet was designed with areas demonstrating the most useful methods and the most beautiful arrangements; comparing the two might enable the observer to see the true natural relations which lay behind the artificial constraints of method and symmetry operating in other collections.[47] In this way Daubenton laid the foundations for discriminating between the tasteful observer, who could not look beyond the surface order of the collection and actively sought out beautiful arrangements, and the truly scientific observer for whom human arrangements were epiphenomenal to the study of nature.

But no matter how reform-minded the naturalist, in the Old Regime all were dependent upon an audience which did not fashion strict boundaries between the natural and the artistic. Rousseau's writings were particularly important in eliding the "sensory" and the "sentimental" sides of sensibility, opening the way for naturalists' self-construction as sensible individuals because they were exposed to nature. More than one naturalist constructed himself along these lines, as in the case of Lacepède. Arriving in Paris aged twenty, in 1776 the young nobleman found Parisian society open to him as a wealthy heir in a family which had been in royal service since at least the reign of Louis XIV. He attended the best salons, was presented at court, and developed his connections with leading musical and scientific figures, including Gluck and Buffon. After his initial successes in writing operas and works on electricity, Lacepède was invited by Buffon in 1784 to replace Daubenton's cousin Edmé-Louis, retiring from the post of subguard and subdemonstrator to the Cabinet owing to ill health.[48] Lacepède

46 Vicq d'Azyr, *Traité d'Anatomie,* 1:1. In "Forging Nature at the Republican Muséum" I explore in more detail the creation of a divided audience at the Muséum from 1793 onward. See also Lorraine Daston, "Nature by Design," in *Picturing Science: Producing Art,* ed. Caroline A. Jones and Peter Galison (New York: Routledge, 1998), 232–253; Lorraine Daston and Katharine Park, *Wonders and the Order of Nature, 1150–1750* (New York: Zone Books, 1998), chapter 7.

47 Louis-Jean-Marie Daubenton, "Description du Cabinet du Roy," in *Histoire naturelle* (1749), 3:1–12, esp. 5–8. This text was partially reproduced in Denis Diderot, "Cabinet d'Histoire naturelle," in *Encyclopédie* (1751), ed. Diderot and d'Alembert, 2:490–492.

48 The operas were *Omphale* (1777), at the request of Gluck; *Alcine* and *Scanderberg* (1781), at the request of the Académie Royale de Musique. Lacepède's natural philosophical works were the *Essai sur l'électricité* (1781), and *Physique générale et particulière,* 2 vols. (Paris: Imprimerie de Monsieur, 1782–1784). See Jean Théodoridès, "Le comte de Lacepède, 1756–1825: Naturaliste,

moved into the apartment over the Cabinet in the following winter to continue work on the *Histoire naturelle,* writing on oviparous quadrupeds, cetaceans, fish, and serpents.[49] Throughout his life Lacepède sought to embody the sensible public man, who turned to the study of nature as a recreation which nonetheless structured and legitimated his portrayals of himself in public and private life.[50] In his early writings Lacepède constantly resorted to nature as the source and paradigm for the beautiful in the arts. Thus, in his *Poétique de la musique* of 1785, he repeatedly demonstrated how music was grounded in nature; successful composers could create melodies which inevitably aroused feelings by using notes which were the natural expressions of the desired passions. At the same time he also sought to explain the effects of music in scientific terms.[51] His natural historical writings would be criticized by nineteenth-century naturalists for placing too much stress upon the beauty of the spectacle of nature and the emotional effect of seeing natural productions.[52] However, his emphasis upon the emotional foundation for the relationship between observer and nature was a contemporary commonplace. In the *Dictionnaire de Botanique* of 1783 Lamarck similarly linked the study of nature with the emotional life: "Has Nature an aspect more charming and gracious than this multitude of plants which form for her, as if in competition with one another, an apparel [which is] infinitely variable and always growing anew? Even the least instructed man cannot cast an attentive eye over a beautiful meadow, over a wood rich in plants, without sensing some kind of sudden joy which it would be useless to seek elsewhere. What would be [the feeling of] he who brought to bear on these objects, already so agreeable in themselves, an eye enlightened by science?"[53] The scientific gaze was here presented as a means of *enhancing* the sensible experience of nature. Other botanists such as Millin, Étienne-Pierre Ventenat, and

musicien, et homme politique," *Comptes rendus du 96e congrès national des sociétés savantes, Toulouse, 1971, section scientifique* 1 (1974): 47–61.

49 Bernard-Germain-Étienne de la Ville sur Illon, comte de Lacepède, *Histoire naturelle des Quadrupèdes Ovipares et des Serpens* (Paris: Hôtel de Thou, 1788); id., *Histoire naturelle des Serpens.*

50 Hahn, "L'autobiographie de Lacepède retrouvée."

51 Bernard-Germain-Étienne de La-Ville-sur-Illon, comte de Lacepède, *La poétique de la musique,* 2 vols. (1785; facsimile reprint, Geneva: Slatkine, 1970); id., *Physique générale et particulière.*

52 Many modern histories have seen Lacepède as a second-rate version of Buffon, as Hahn, "Sur les débuts de la carrière scientifique de Lacepède," notes. See, e.g., Robert Fox, "The Rise and Fall of Laplacian Physics," *Historical Studies of the Physical Sciences* 4 (1974): 89–136; Gillispie, *Science and Polity,* 161.

53 Jean-Baptiste-Pierre-Antoine de Monet, chevalier de Lamarck, "Botanique," in *Dictionnaire de Botanique* (1783), 1:440.

Jean-Louis-Marie Poiret, Lamarck's collaborator on the *Dictionnaire de Botanique,* similarly rhapsodized over nature in savant publications.[54] By the very end of the century, though, expressions of sensibility were being actively excised from the published materials produced by the Muséum d'Histoire Naturelle's professoriate.[55]

Buffon's emphasis upon the importance of natural history as a source for sensual experience which could irreversibly transform individuals thus gained an added significance toward the end of the Old Regime. Naturalists explored the nature and limits of sensibility in studies not just of animals, but of the sensitive plant and similar mobile vegetables.[56] De Jussieu's colleague Desfontaines published a description of plant sex in 1787, using richly erotic language to describe the movements of plant fertilization: a "sensation of love" caused plant stamens to bend toward the stigma and fertilize the flower.[57] This represented one end of a spectrum of views on plant sensibility, the other being that of Vicq d'Azyr, whose *Traité d'Anatomie et de Physiologie* of 1786 claimed that all plant movements were purely mechanical and caused by outside forces, not an internal volition.[58] Such debates had implications for the interpretation of human sensibility in the 1780s and 1790s, for they allowed the boundaries between physiological and emotional sensibility to be explored, and, with the development of the cult of sensibility, to be blurred. Moreover, they were central to naturalists' self-portrayals as sensible (thus virtuous) men. Contemporaries portrayed the study of nature and natural history as yielding an increased sensi-

54 Jean-Louis-Marie Poiret, "Plantes," in Lamarck, *Dictionnaire de Botanique* (year XII/1803), 5:394–427; Aubin-Louis Millin de Grandmaison, *Rapport fait à la Société d'Histoire Naturelle, en sa séance du 21 prairial, Par A. L. Milin [sic], l'un de ses membres, sur l'ouvrage qui a pour titre: Calendario entomologico, etc. etc., de M. Giorna le fils, membre de la même société* (Paris: Imprimerie du Magazin Encyclopédique, year III/1795).

55 This problem is further discussed in Spary, "Forging Nature at the Republican Muséum."

56 Robert M. Maniquis, "The Puzzling Mimosa: Sensitivity and Plant Symbols in Romanticism," *Studies in Romanticism* 8 (1969): 129–155. Such concerns were also prominent in the work of Linnaeus.

57 René-Louiche Desfontaines, "Observations sur l'irritabilité des organes sexuels d'un grand nombre de plantes," *MARS* 1787/1789, 468–480. Erasmus Darwin's *The Botanic Garden; a poem, in two parts. Part I. containing the Economy of Vegetation. Part II. The Loves of the Plants. With Philosophical Notes* (London: J. Johnson, 1791), similarly described plant sex in emotional and erotic terms. See Janet Browne, "Botany for Gentlemen: Erasmus Darwin and *The Loves of the Plants,*" *Isis* 80 (1989): 593–621; id., "Botany in the Boudoir and Garden: The Banksian Context," and Alan Bewell, "'On the Banks of the South Sea': Botany and Sexual Controversy in the Late Eighteenth Century," both in *Visions of Empire,* ed. Miller and Reill, 153–172, 173–193; Schiebinger, *Nature's Body,* chapter 1; id., "The Private Life of Plants: Sexual Politics in Carl Linnaeus and Erasmus Darwin," in *Science and Sensibility,* ed. Benjamin, 121–143.

58 Vicq d'Azyr, *Traité d'Anatomie,* 1–16.

bility. Naturalists themselves thus gained moral and, ultimately, political credibility from their constant exposure to the beauties of nature.

NATURAL AUTHORITIES

The writings of Rousseau, Bernardin de Saint-Pierre, and others generally emphasized nature as the setting for individual experience, and Outram has discussed the function of nature as a consolatory, Rousseauist retreat for savants during the Terror.[59] In writing of the Muséum d'Histoire Naturelle's foundation, she argues that it resulted from "the ideology, so powerful for the Jacobins, that the display of the beauties of nature in such privileged spots as the Muséum, was an essential part of human existence because it contributed to human virtue."[60] However, the reform of the Jardin in June 1793 was part of the Jacobins' plan for the setting up of institutions of *public* instruction in France, rather than solitary self-discovery. In such a context, Lamarck's claim that instruction could improve the individual's sensibility to Nature acquires a new significance.

What we know of public uses of the Jardin in the eighteenth century is principally based on two kinds of sources: the discussions of public order by the Jardin's staff, and critical commentaries by visiting experts, whether savants or connoisseurs. This dichotomy was reflected in their approach to the world of print. In making public the reform project of 1790, only 2,000 copies were printed, of which the majority went to members of the Assemblée Nationale and different Parisian political and scientific organizations. There was certainly no question of distributing the project to the Jardin's revolutionary visitors. Historians of art have viewed the opening up of the royal collections during the French Revolution as the democratization of culture, the moment at which monarchical privacy became public property. In the Revolutionary period the "public" would be more broadly construed than ever before. From being the site of *politesse* and commerce both social and financial, the public would be remodeled as the sphere of operation of political rights, the location of the national will, and the source of moral *vertu.*[61]

59 Outram, "Ordeal of Vocation"; see also Michael Fried, *Absorption and Theatricality: Painting and Beholder in the Age of Diderot* (Chicago: University of Chicago Press, 1980), chapter 3.

60 Outram, *Cuvier,* 164–165. See Gillispie, "The Encyclopédie and the Jacobin Philosophy of Science."

61 Baker, *Inventing the French Revolution,* chapter 8; Hunt, *Politics, Culture, and Class;* Roger Chartier, *The Cultural Origins of the French Revolution: Bicentennial Reflections on the French Revolution,* trans. Lydia G. Cochrane (Durham: Duke University Press, 1991), chapter 2; Mona Ozouf, "L'opinion publique," in *Political Culture of the Old Regime,* ed. Baker, 419–434; id., "Public Opinion at the End of the Old Regime," *Journal of Modern History,* supplement, 60 (1988): 1–21;

But a revalorized public clashed with the Jardin's staff over what constituted proper behavior within a newly national establishment. At the end of 1792 Bernardin appealed to the minister of the interior, Roland, for help: "Agitators are persuading the people that as the garden belongs to the Nation, all the nation has the right to go and pick the flowers. On All Saints' Day a considerable troop of men and women pillaged the flowers while threatening with violence the guards who tried to prevent them . . . I gave orders to take the delinquents to the Section in vain, the guards dared not execute them . . . There is an urgent need to oblige bad citizens to respect public properties."[62] Revolutionary violence affected the Jardin in other ways as well. The plaster bust of Linnaeus, ceremoniously placed in the Jardin by the Société d'Histoire Naturelle in August 1790, was probably smashed during the course of 1792 by *sans-culottes* who took the Latin inscription to refer to a king. Food riots over the winter of 1792 to 1793 also damaged the Jardin, which was being used as a public grain store.[63] Such infringements of public order, as defined by the Jardin's staff, reflected a much broader problem which was now a principal task of deputies, namely that of keeping order in the public realm. The Jardin was, in this sense, a political microcosm.

Just as the authority of naturalists over the Jardin could be eroded by public disorder, so too could the claims of particular factions to hold power. In August 1792 the self-proclaimed municipal authority of Paris, the Commune, suppressed the existing municipal electoral assembly. Thouin, Lacepède, and de Jussieu suffered an abrupt termination of their active political careers.[64] The Assemblée Nationale failed to obtain acknowledgment of its authority over the Commune and, its authority terminally weakened, gave way to the Convention Nationale a few days later. A little over a month later, France was declared a republic.

Lucien Jaume, "Le public et le privé chez les Jacobins, 1789–1794," *Revue française de science politique* 37 (1987): 230–248; Norman Hampson, *Will and Circumstance: Montesquieu, Rousseau, and the French Revolution* (London: Duckworth, 1983); Jürgen Habermas, *The Structural Transformation of the Public Sphere: An Inquiry into a Category of Bourgeois Society,* trans. T. Burger and F. Lawrence (Cambridge: MIT Press, 1989); Benjamin Nathans, "Habermas's 'Public Sphere' in the Era of the French Revolution," *French Historical Studies* 16 (1990): 620–644; Crow, *Painters and Public Life.*

62 Bernardin de Saint-Pierre to Roland, late 1792, quoted in Letouzey, *Le Jardin des Plantes,* 304–305.

63 BCMHN, MS 315. The smashing of the bust appears in Grégoire's self-exculpatory *Mémoires,* 61, as part of his attack against Jacobin vandalism.

64 Adolf Schmidt, *Paris pendant la Révolution d'après les rapports de la police secrète, 1789–1800,* trans. P. Viollet (Paris: Champion, 1880–1894), 1:9.

The new deputies faced the problem of potentially rebellious groups outside central government which existed in uneasy alliance with different governmental factions. An enduring yet genuinely republican government needed simultaneously to appear to manifest the public will and to be able to direct public violence away from itself. Public instruction would be invoked for this purpose. On March 20, 1793, the deputy Lanthenas, a member of the Comité d'Instruction Publique, foresaw a new role for public instruction. "No one," he argued "has considered it as a revolutionary power, before envisaging it as a means of moral and physical perfection for the species."[65] A republican instruction would change the public's "moral and political opinions." In fact, Lanthenas was suggesting that a principal task of public instruction was to justify the Convention's grip on sovereignty. During the early 1790s, public-instruction projects presented to the Assemblée and Convention gradually abandoned proposals for the formation of a largely autonomous body of savants to supervise public instruction. In a debate of July 1790, most deputies supported a project which praised the social role of savants and advocated substantial financial support by the state. The deputy Martineau was a lone voice in 1790, but not in 1794: "Are you afraid of running short of savants? It's cultivators you need, save your rewards for them."[66] Savant claims to expertise over the natural world were increasingly treated by radicals as evidence of a despotism of knowledge. Moreover, interventionist programs for universal education were increasingly criticized. Government must appear to be a manifestation of the general will, coming from within the heart of the citizen, rather than an order imposed from above.[67] Over the course of 1793, reform proposals came to favor the "self-educating nation," in which virtue could be engendered and corruption undone through moral display and patriotic example.[68]

Successive political factions in the Convention and Directoire struggled to create a form of sovereignty which would enable effective politi-

65 François-Xavier Lanthenas, "Bases fondamentales de l'instruction publique et de toute constitution libre ou moyens de lier l'opinion publique, la morale, l'education, l'enseignement, l'instruction, les fêtes, la propagation des lumières et le progrès de toutes les connoissances au gouvernement national républicain . . . ," *AP*, 64:458.

66 *AP*, 17:445. Martineau's comments were most likely an allusion to Jean-Jacques Rousseau's association of savants with corruption in the *Discours sur les sciences et sur les arts* (1751; New York: Modern Language Association of America; London: Oxford University Press, 1946), esp. 150.

67 Roland's *bureau pour l'esprit public*, an organization concerned with the management of the public, was founded in 1792. See Schmidt, *Paris pendant la Révolution*, 1:129–132; Rudé, *French Revolution*, 89–106.

68 Dominique Julia, *Les trois couleurs du tableau noir: La Révolution* (Paris: Belin, 1981); Michel Grenon, "Science ou vertu? L'idée de progrès dans le débat sur l'instruction publique, 1789–1795," *Études françaises* 25 (1989): 177–190.

cal action and recruit other powerful groups to their cause. This was to be achieved by the mastery of Revolutionary symbols, such as Liberty, Nature, and Justice. Nature, one of the most important revolutionary symbols, was often invoked in Revolutionary political discourse to legitimate the rejection of Old Regime customs.[69] However, the meaning of nature was contested. In 1792 and 1793 France was gripped by a serious grain shortage. Some *sans-culottes* viewed the Revolution as the harbinger of a time of plenty for all, when France, as the setting for the only natural society, would become fertile. Famine was thus evidence of governmental error, and signified the moderate deputies' failure to capture the meaning of the symbol Nature adequately. The Jacobin coups of May 31 and June 2, 1793, ending the power of Roland, Brissot de Warville, and their supporters, were legitimated in part on this basis.

The Muséum's naturalists, along with governmental supporters of the new agriculture, offered a different version of Nature. In a speech of brumaire year II (November 1793), Grégoire insisted that political reform would not automatically bring about plenty. "You propose, I have been told, to acclimatize foreign plants and cultures among us: but our soil has everything; nature has placed among us that which is necessary to us. I begin by denying that assertion: nature, in truth, has given us a fertile soil, and that is pretty much all."[70] The Muséum naturalists' version of Nature would be implemented within the public teachings offered by their institution, as a central part of the Jacobin public-instruction project. Their contribution to public instruction was thus twofold. First, in lectures they taught citizens to interpret Nature's laws as evidence of the hegemony of the new government in general and the ruling faction in particular. The surviving lectures from this period reveal that the Muséum's audience received a heavy dose of pro-government rhetoric from professors who numbered among their ranks renowned orators such as Fourcroy, known also from the Convention and the Jacobins' Club, Lacepède, Desfontaines, and Brongniart.[71] Sec-

69 Schama, *Citizens,* 799–800; Hunt, *Family Romance,* 154; Herbert, *David, Voltaire, Brutus;* Maurice Agulhon, *Marianne into Battle: Republican Imagery and Symbolism in France, 1789–1880* (Paris: Éditions de la Maison des Sciences de l'Homme; Cambridge: Cambridge University Press, 1979), 16–19; Ronald Paulson, *Representations of Revolution, 1789–1820* (New Haven: Yale University Press, 1983), chapter 1.

70 Henri Grégoire, *Nouveaux développemens sur l'amélioration de l'Agriculture, par l'Etablissement de Maisons d'Economie rurale; présentés par le citoyen Grégoire à la Séance du 16 brumaire, l'an deuxième de la République une et indivisible* (Paris: Imprimerie Nationale, year II/1794).

71 Kersaint, "Fourcroy"; Laissus, "La succession de Le Monnier." Brongniart's courses at the Jardin were "très brillante" (AN, AJ/15/509, piece 274); Lacepède, as a deputy to the Assemblée Nationale during 1791, was accustomed to public oratory.

ond, the political credit which the naturalists accrued through these practices also enabled them, uniquely among the savant institutions in 1793–1794, to become an official voice for Nature within the Republic.

There were important similarities between the problems confronted by naturalists in the making of natural historical knowledge, and by deputies in the maintenance of political authority. Republican sovereignty and natural historical credibility were both assured only by the capture of sources of authority which were both invisible and intangible. Successful representations of Nature would enable deputies to embody the public will while effacing the operations of their personal wills from the legislative process.[72] Political action could thus be legitimated by the claim that laws were indeed drawn from Nature. Throughout 1793 and 1794 the making visible of the true sovereignty of the legislators via public instruction relied upon spectacle, both natural and artificial, to create its most potent effect. In May 1793 Dutard, one of a network of spies fielded by the minister of the interior, Dominique-Joseph Garat, explained the problem of public order as a problem of the social art. "I have heard it said to a painter that to judge a painting, to see all the general effects within it, it is necessary to place oneself exactly in the vantage point which is proper to it. He who lacks the visual ray will only see a confusing mass of diverse colors, which will no longer present anything more than half-images to him . . . It is the same in our Revolutionary politics, few men have seen this whole [which is] so necessary and the true relationships [*rapports*] which exist between the diverse classes which form the social body."[73] While the legislators of the Republic scrambled to capture the "vantage point" for the spectacle of society and to present it to citizens, naturalists were claiming the ability to see, directly and without interfering "systems," the laws of nature in operation, the order in the apparent chaos into which the Revolution had tumbled. As Thouin's Angers correspondent Merlet de la Boulaye put it: "Civil war is ravaging our department. The Greek and the Trojan are pillaging us . . . You may be surprised that I am busying myself with decorating and with planting a botanical garden in the middle of the dissensions of civil war, but I am at my post and I see that men may burn, pillage, massacre, but that they cannot act against the immutable order of nature. When they have tired of

72 Hampson, *Will and Circumstance*, 57–58, sees this as a central feature of Robespierre's political doctrine.

73 Dutard to Garat, May 9, 1793, quoted in Adolf Schmidt, *Tableaux de la Révolution française publiés sur les papiers inédits du département et de la police secrète de Paris*, 3 vols. (Leipzig: Veit, 1867–1870), 1:198–199.

throttling each other, I am happy to think that my compatriots will be well pleased to find a useful and instructive garden ready-made."[74] Naturalists thus claimed to be able to capture Nature within natural history collections, whether in a cabinet or garden. The synopsis of nature presented here enabled the observer to visualize the entirety of the natural order, the whole beauty of the Universe, the natural relationships between individuals and species, through his or her memory of a few specimens. This function is perhaps most clearly expressed in Lacepède's manuscript lectures of year III (1794): "Citizens, it is in this first room of the Cabinet of Natural History that you will recapitulate in your memory everything which has been the subject of our lectures so far . . . There will grow your ideas on that which one can never study, Nature! What if, after our Course on the plant Kingdom, and that on the mineral Kingdom, I show you in detail in the other two rooms . . . of the Cabinet of Natural History, the other riches which present immense utilities and real charms? Ah! Then you will bless the Republic and the national Convention."[75] The claim of several of the Muséum's naturalists to be searching for *rapports* during the early 1790s, relations which were natural and evident to the senses and indicated underlying natural laws, was thus of considerable importance for natural history as a form of Republican instruction. Collections were scientific instruments, but, crucially, their artificiality was transparent to contemporary viewers. What they appeared to display was in fact invisible: Nature herself.[76]

The same principles can be seen at work in the Revolutionary festivals designed by Robespierre's disciple, the artist Jacques-Louis David. In his hands the use of managed spectacle as a means to control the public assumed its most elaborate form.[77] Robespierre's speech of floréal

74 Merlet de la Boulaye to Jean Thouin, Angers, April 3, 1793, quoted in Letouzey, *Le Jardin des Plantes,* 319.

75 BCMHN, MS 1529: Lacepède, year III/1794.

76 Here I rely on Pomian's argument that collections perform a semiotic function by denoting through a series of visible signs the capture of the invisible. See Pomian, "Entre l'invisible et le visible: La collection," in *Collectionneurs, amateurs, et curieux,* 15–59; Baudrillard, "System of Collecting"; Hooper-Greenhill, *Museums and the Shaping of Knowledge;* but also Latour, *Science in Action,* chapter 2; Simon Schaffer, "Natural Philosophy and Public Spectacle in the Eighteenth Century," *History of Science* 21 (1983): 1–43.

77 Ozouf, *Festivals;* Marie-Louise Biver, *Fêtes révolutionnaires à Paris,* preface by Jean Tulard (Paris: Presses Universitaires de France, 1979); Jean Ehrard and Paul Viallaneix, eds., *Les fêtes de la Révolution: Colloque de Clermont-Ferrand, juin 1974* (Paris: Société des Études Robespierristes, 1977); Béatrice de Andia and Valérie Noëlle Jouffre, eds., *Fêtes et Révolution* (Dijon: Musée des Beaux-Arts, 1989–1990); A. Lamadon, "Fêtes en Révolution, 1789–1794," *Revue d'Auvergne* 103 (1989): 59–82; Roy Strong, *Art and Power: Renaissance Festivals, 1450–1650* (Woodbridge, Suffolk: Boydell, 1984); Parker, *Portrayals of Revolution.* For the involvement of

year II (May 1794), soon after the institution of the cult of the Supreme Being, pointed to the dual nature of a Revolutionary spectacle.

> There is . . . a sort of institution which must be considered an essential part of public education, and which necessarily belongs to the subject of this report. I am talking about public festivals.
>
> Gather men together, and you will make them better; for men assembled try to please themselves, and they can only please themselves with things that render them estimable; give their reunion a big moral and political motive, and the love of honest things will enter with pleasure into all hearts; for men cannot see each other without pleasure.
>
> Man is the greatest object there can be in nature; and the most magnificent of all spectacles, is that of a great people assembled.[78]

The organization of the public in the rituals of moral festivals was a representation in miniature of the way in which, for Robespierre and his supporters, society as a whole should function. The public should be controlled by art (that is, the social art), but at the same time, by its emotional response to Nature.[79] Visualization was the principal means of forming and directing patriotic feelings.

David's design of *fêtes* was an extension of Republican efforts to harness the authority of Nature as a legitimating strategy.[80] In these festivals the public itself became the subject of a new form of art, designed to contain and depict emotion on a scale hitherto unknown. Festivals were living paintings, a spectacle which blurred the line between the natural and the artificial. In this respect, indeed, they resembled the natural history collection itself, as seen by contemporaries. Festivals were composed of natural bodies, staged like theatrical pieces, and represented the symbols of the Republic. Like other forms of Revolutionary bodily display, they aimed to render citizens transparent, to peel off that "bark" of corruption. Robespierre and his acolytes endeavored to maintain strict control over the authenticity of spectacle.[81] On 11 messidor year II (June 29, 1794) Joseph Payan, a member of the Commission Exécutive d'Instruction Public, attacked attempts to convert

---

David, see David Lloyd Dowd, *Pageant-Master of the Republic: Jacques-Louis David and the French Revolution* (Lincoln: University of Nebraska Studies, 1948); Herbert, *David, Voltaire, Brutus.* Daniel Rabreau, "Architecture et fêtes dans la nouvelle Rome," in *Les Fêtes de la Révolution,* ed. Ehrard and Viallaneix, 355–376, note that nature was an omnipresent theme of the festivals.

78 Maximilien Robespierre, "Rapport fait au nom du comité de Salut publique, sur les rapports des idées religieuses et morales avec les principes républicains et sur les fêtes nationales," *La Décade philosophique* 1 (1794): 243–244.

79 Baker, "Politics and Social Science."

80 Dowd, *Pageant-Master of the Republic;* Leith, *Idea of Art as Propaganda.*

81 Parker, *Portrayals of Revolution,* chapter 2.

the great festivals into theater pieces, claiming that this would reduce the power of the great spectacle of nature to the level of a mere magic-lantern show—in other words, to falsehood.[82] Similar metaphors were utilized by naturalists, as in Lacepède's ichthyology course of floréal year III (May–June 1795), where the wrong approach to nature was portrayed as "substituting the imperfect images and empty specters of a magic lantern, for the spectacle of nature."[83]

Eyewitness accounts of the Festival of the Supreme Being bear out the success of David in evoking mass emotions.[84] The emotional power of the festivals depended upon their success in representing the spectacle of nature that lay behind the new social order. At the Festival of the Supreme Being on 20 prairial year II (June 6, 1794), Robespierre declared: "The true priest of the supreme Being is Nature, its temple the Universe, its Cult truth, its festivals, the joy of a great people assembled to tie the sweet knots of Fraternity, and Vow death to Tyrants."[85] For the festival makers Nature served as the legitimator of the construction of the proper social order. But the "true priest" of the festival was Robespierre, who appeared on a raised plinth at center stage, dressed in a special ceremonial garb of blue coat and gold-colored trousers.[86] The plan for the festival detailed the direction of the public gaze: "The National Convention, preceded by a stirring music, shows itself to the people: the president appears at the raised tribune in the center of the amphitheater; he makes felt the motives which determined this solemn festival; he invites the people to honor the author of nature." And at the end of the ceremony, "the piercing tone of the trumpet rings out . . . The people arranges itself; it is in order: it leaves."[87] The people was to be the observer of itself in this spectacle, but that gaze was then to be translated toward those who represented what could not be seen: Nature. Having been addressed in the name of Nature and of the Supreme Being, the people was then to order itself.

Like the naturalists, Republican deputies endeavored to make them-

82 Joseph Payan, "Rapport et arrêté de la Commission de l'instruction publique," 11 messidor year II/June 29, 1794, quoted in Julia, *Les trois couleurs,* 352.

83 "Introduction au Cours d'Ichthyologie, donné dans les galeries du Muséum d'Histoire naturelle, par le citoyen Lacepede, et commencé le 13 floreal de l'an 3e," *Magazin Encyclopédique* 1 (year III/1795): 449.

84 Ozouf, *Festivals*; Parker, *Portrayals of Revolution,* chapter 2; Schama, *Citizens,* 748–750.

85 A contemporary paraphrase of a sentence from Robespierre, "Rapport," 243, inscribed at the base of a colored image of Nature published to celebrate the Festival of the Supreme Being in year II.

86 *NBU,* s.v. "Robespierre."

87 Robespierre, "Rapport," 250–251.

selves transparent to their audience, to remove the intrusive evidence of individual labor in the making of social knowledge.[88] This particular aspect of public-instruction reforms has been viewed askance by twentieth-century historians, who find in it echoes of "a compulsive ideology and foretaste of the mass manipulation used more recently by totalitarian regimes," in the words of Robert Palmer.[89] Perhaps in consequence, the Muséum's status as an institution offering Jacobin public instruction has been much underplayed in the secondary literature. Yet throughout the Revolution, the naturalists at the Muséum explicitly offered the legislators the kinds of resources which the festivals were designed to supply. The invisibility of both naturalists and deputies in the legislative process was emphasized in the naturalists' Assemblée presentation of August 1790: "accustomed to consider the great and magnificent spectacle of the power of nature, and of the unity of her laws, they believed that their admiration had not changed in directing itself toward the immortal work that the national power is constructing with your hands."[90] The naturalists continued to translate their audience's gaze from the spectacle of Nature to that of the government after the Jardin's reform. In Mertrud's closing lecture for the course on animal anatomy, given in the Muséum's new amphitheater on 11 germinal year II (March 31, 1794), he indicated his role in revealing natural spectacle to his students at the start of the lecture: "Nothing is small in nature to the eyes of the enlightened observer; our duties are to admire its greatness, to know its laws, to research all the properties which have been attached to it by the creator's hand, to extract from it the useful things and to apply them to the needs of men. Citizens, I have only been able to sketch this picture feebly; it is a part of the task which has been assigned to me. Other professors will enlarge this spectacle and will present to you all its majesty." By the end of the lecture, however, he had made the transition to the spectacle of legislative power: "Is there in the Universe a greater spectacle than that which the legislators present, who crush traitors with one hand, and with the other sustain and protect this immense people which has committed itself to their care?"[91]

88 Shapin, "Pump and Circumstance"; Schaffer and Shapin, *Leviathan and the Airpump;* Latour, *Science in Action,* discusses the making invisible of traces produced by scientific practitioners.

89 Robert R. Palmer, *The Improvement of Humanity: Education and the French Revolution* (Princeton: Princeton University Press, 1985), 190–191.

90 "Première adresse," in Hamy, "Les derniers jours du Jardin du Roi," 97–100.

91 Antoine-Louis-François Mertrud, *Discours prononcé à l'Amphitéâtre du Museum National d'Histoire Naturelle, pour la clôture du cours de l'Anatomie des animaux, le Primidi 11 Germinal l'an deuxième de la République Française une et indivisible* (Paris: n.p., year II/1794). As Harten and

The "Museological Moment"

The Muséum d'Histoire Naturelle's success as an institution of public instruction under a Jacobin regime depended upon the naturalists' skills in demonstrating natural laws through visual displays which would exemplify order but also enable the moral and physical re-creation of the public. The visitor to the Muséum d'Histoire Naturelle was to be transformed emotionally, aesthetically, and rationally through the encounter. I have already discussed how the menagerie's foundation in floréal year II (May 1794) was justified on the basis that it could reveal to the visiting public the consequences of partaking in Revolutionary society, by directing the public gaze toward the operation of Nature's laws within a Republican garden. The moral effect of seeing nature at the new establishment was not restricted to the sight of living animals. The "Essais de Botanique morale," published in the journal *La Décade philosophique,* established a close relationship between botany as moral education and as visual spectacle, explicitly appealing to Rousseau and Bernardin de Saint-Pierre.[92] "Botany does not refine *moeurs* merely by [engendering] healthy and moderate inclinations; it attaches [man] more strongly to the fatherland by the image of its productions, by pastoral tableaux, by the memories which it engraves in the heart, by the monuments which it can embellish in such an interesting manner."[93] The Muséum d'Histoire Naturelle's reification was inseparable from its ability to provide a site for exhibition, and thus from its image as a dispenser of patriotic knowledge. It was not the only establishment to have such a role. Revolutionary programs for the institutional reform of the fine arts were supported by identical concerns. On July 27, 1793, some seven weeks after the foundation of the Muséum d'Histoire Naturelle, the Convention decreed the foundation of a Muséum National des Arts at the Louvre.[94] The development of museological proj-

---

Harten suggest, "Das Naturgeschichtemuseum verstand sich als ein Kraft der Revolution" ('the natural history museum understood itself as a power of the Revolution'; *Die Versöhnung mit der Natur,* 59).

92 Bernardin de Saint-Pierre, the intendant of the Jardin des Plantes until its reform, went on to become the first professor of *morale* at the École Normale in year III/1794, confirming the link between moral issues and natural history.

93 *La Décade philosophique* 1 (20 messidor year II/July 8, 1794): 455. See Jean-Marc Drouin, "L'histoire naturelle à travers un périodique: *La Décade philosophique,*" in *La nature en Révolution,* ed. Corvol, 175–181; id., "Le Jardin des Plantes à travers la *Décade philosophique,*" in *Les jardins entre science et représentation,* ed. Jean-Louis Fischer (Paris: CTHS, forthcoming).

94 Germain Bazin, *The Louvre* (London: Thames and Hudson, 1971); McClellan, *Inventing the Louvre.* The Muséum National des Arts opened in early August 1793, but soon closed again for repairs, and functioned only intermittently for the next decade.

ects with the aim of suppressing factionalism and creating a virtuous public which supported the existing order had begun with d'Angiviller's 1770s project for a public Louvre.[95] But they had their most visible effect, in terms of institutional reform, in years II and III under Jacobin rule. In this "museological moment," as Pickstone has entitled it, an epidemic of museum foundation in every sphere broke out; although such institutions were short-lived, for the most part, they installed the museum as the principal manifestation of Jacobin cultural reform.[96] Addressing the Convention on 13 messidor year II (July 1, 1794), Bertrand Barère proposed the conversion of all Paris into a Revolutionary museum, "a new monument for public instruction."[97] Nature, art, and architecture were to be simultaneously deployed to construct a total Republican environment in which every public space would contribute to Barère's "vast plan for regeneration."[98] While natural history played a smaller part in this great transformation, it is nonetheless in this context that the Muséum's function in years II and III can be understood. The common institutional policy regarding natural history and the arts was reflected in their similar aesthetic function. Both natural and artificial objects of display generated similar effects upon the observer, and both, accordingly, received similar treatment by the Convention's deputies in their reforms of public instruction. Thus a *Décade philosophique* article of messidor year II (June–July 1794) noted that both natural history and the arts relied upon the presentation of continual spectacle for their ability to perfect the mind.

95 Dominique Poulot, "Le musée entre l'histoire et ses légendes," *Débat* 49 (1988): 69–83, sees the Revolutionary museum as legitimated by appeals to a virtuous, spontaneous public "mobilisable pour le Bien par la simple vue du Beau et du Vrai" ('mobilisable for Good through the mere sight of the Beautiful and True'; 72). See also Miriam R. Levin, "'Ideology' and Neoclassicism: The Problem of Creating a Natural Society through Artificial Means," *Consortium on Revolutionary Europe, 1750–1850: Proceedings* (1981): 177–187; Harten, *Museen und Museumsprojekte;* Poulot, "Le Louvre imaginaire," 183, 196; id., "Musée et société dans l'Europe moderne," *Mélanges de l'école française de Rome, moyen âge–temps modernes* 98 (1986): 991–1096; Anthony Vidler, *The Writing on the Walls: Architectural Theory in the Late Enlightenment* (London: Butterworth Architecture, 1989); Hooper-Greenhill, *Museums and the Shaping of Knowledge;* id., "Museum in the Disciplinary Society"; Vergo, *New Museology.* The "democratizing" of display described in some of the above somewhat glosses over the extremely contested nature of the sovereignty that such displays were designed to evoke.

96 John V. Pickstone, "Museological Science? The Place of the Analytical/Comparative in Nineteenth-Century Science, Technology, and Medicine," *History of Science* 32 (1994): 111–138.

97 Quoted in Harten, *Museen und Museumsprojekte,* 28. Barère had been a member of the Comité d'Instruction Publique at the time of the foundation of the Muséum d'Histoire Naturelle.

98 Bertrand Barère, in *Le Moniteur,* 20 (year II/1794): 192 (12 germinal/April 1), quoted in Leith, *Space and Revolution,* 149, 151. The refashioning of public space is discussed in ibid., and Ozouf, *Festivals,* chapter 6.

> The *fine arts* perfect our intelligence; they help powerfully in the development of our moral faculties. What is the goal of almost all the *liberal arts?* that of imitating nature, of offering to us the closest image to her. Well! it is this continuous spectacle of nature, both real and imitated, which increases our sensations, offers us continual occasions to compare, forces us to make a larger number of judgments, in consequence enriches our memory with a larger number of ideas, and causes our mind to grow. And what is the mind? . . . We know that it is strong or weak, shallow or profound by virtue of the number of sensations that we have experienced, and of the number of comparisons which we have had occasion and interest to make based on these same sensations.[99]

The Muséum National d'Histoire Naturelle and its alter ego, the Muséum National des Arts, shared the power to present moral spectacles to the public. In both cases, the institutions took on a quasi-religious function.

> This great monument elevated to Nature, is such at this moment, by the care which the Convention has taken to enrich it with a multitude of precious objects dispersed in various cabinets, that one can rest assured that there is no other nation in the entire world, which can glorify itself with possessing an equally vast and equally important collection . . . It is well worthy of a free people preparing the bases of an education founded upon Nature, that is to say upon the study of facts . . . This central and necessary point of the sciences can be considered as a temple where everyone can come to consult Nature, who will always respond to them herself, by displaying her collected riches.[100]

Similarly, the Muséum National des Arts was described by the members of the Commission des Arts who presided over its foundation as a "temple of nature and of genius."[101] Both museums provided the first

99 *La Décade philosophique* 1 (1794): 401 (10 messidor/June 28). This periodical would become the principal mouthpiece for the group of philosophers, writers, and savants now known as the *idéologues* after the name of the new science of ideas, *idéologie,* that they sought to invent. During the second half of the 1790s, the Muséum professoriate would possess numerous connections with *idéologues* in government and in the newly founded Institut National des Sciences et Arts. See Georges Gusdorf, *La conscience révolutionnaire: Les idéologues* (Paris: Payot, 1979); Ludmilla J. Jordanova, *Lamarck* (Oxford: Oxford University Press, 1984); Hahn, *Anatomy of a Scientific Institution;* Staum, *Cabanis;* Léon Szyfman, *Jean-Baptiste Lamarck et son époque* (Paris: Masson, 1982), chapters 23–29.

100 *La Décade philosophique* 1 (1794): 519.

101 André Michel and Gaston Migeon, *Les grandes institutions de France. Le musée du Louvre: Sculptures et objets d'art du moyen age, de la renaissance, et des temps modernes* (Paris: Renouard, 1912); see also the *Journal d'instruction publique* of August 1793, quoted in Harten, *Museen und Museumsprojekte,* 185–187. In his *Observations,* Faujas described Buffon as the founder of the "Temple of Nature" at the Jardin. Other references to the Muséum of natural history as a temple are in Georges Cuvier, *Notice historique sur Daubenton, lue à la séance publique de l'Institut national*

institutional context for the Jacobins' attempts to control the public spirit by the redirection of the public gaze in an explicitly educational setting.

During the summer of 1794, after Robespierre had temporarily suppressed opposition to his regime by guillotining Hébert and Danton, the Comité de Salut Public passed numerous decrees favoring the two museums and establishing similar establishments in the departments, along with another highly important site of display, the Palais National.[102] Fourcroy, who replaced the assassinated Marat in the Convention on July 22, 1793, was heavily involved in these projects, as an active member of the Jacobins. Along with David, Boisset, and Granet, he was one of the Comité de Salut Public's most effective implementers of the Republican program of *instruction publique*.[103] Thouin too was involved by Fourcroy in the adornment of some of the most public places in Paris.[104]

In floréal year II Fourcroy announced a lightning visit by the Comité de Salut Public to his Muséum colleagues.[105] Shortly thereafter, the Commission des Travaux Publics, headed by Robespierre's protégé, the architect Fleuriot-Lescot, was ordered by the Comité to report on the progress of the new menagerie.[106] The presence of several of the

---

de France du 15 germinal an 8 (Paris: Institut National, year IX/1800); Bernard-Germain-Étienne de La-Ville-sur-Illon, comte de Lacepède, *Discours d'ouverture du cours d'histoire naturelle de l'homme, des quadrupèdes, des cétacées, des oiseaux, des quadrupèdes ovipares, des serpens et des poissons* (Paris: n.p., year VII/1799), 2, 3; id., *Discours d'ouverture et de clôture du cours d'histoire naturelle donné dans le Muséum national d'Histoire naturelle, l'an VIII de la République* (Paris: Plassan, year VIII/1800), 5; id., *Discours de clôture du cours d'histoire naturelle de l'an IX* (Paris: n.p., year IX/1801), 20. This rendering was far more common than the description of the Muséum as an Elysium referred to by Outram ("New Spaces in Natural History," 255–256), which seems only to have been utilized on the occasion of Daubenton's burial in the Jardin.

102 Harten, *Museen und Museumsprojekte*, part 2; Leith, *Space and Revolution*, 154–155; Daniel Sherman, *Worthy Monuments: Art Museums and the Politics of Culture in Nineteenth-Century France* (Cambridge: Harvard University Press, 1989), chapter 3. For example, the purchase of the Joubert collection on 5 prairial year II/May 24, 1794 (*Actes CSP*, 13:717; AN, F/14/187[B]).

103 Kersaint, "Fourcroy"; Margaret Bradley, "The Financial Basis of French Scientific Education and the Scientific Institutions of Paris, 1700–1815," *Annals of Science* 36 (1979): 451–492; *Actes CSP*, vols. 10–17; AN, F/14/189[B]. On 25 floréal year II/May 14, 1794, Fourcroy, Granet, and David were chosen to supervise the improvement of the Palais National and its grounds. On 27 floréal year II/May 16, 1794, and 8 messidor year II/June 26, 1794, Fourcroy and Boisset were ordered to supervise the improvement of the Muséum d'Histoire Naturelle. On 19 messidor year II/July 7, 1794, Fourcroy, Granet, and David were selected to supervise improvements at the Muséum National des Arts (*Actes CSP*, 13:512, 546, 14:538, 773).

104 *Actes CSP*, 13:719.

105 Fourcroy to administration, 27 floréal year II/May 16, 1794, quoted in Letouzey, *Le Jardin des Plantes*, 323. See BCMHN, MS 457; *Actes CSP*, 13:544–545, 538.

106 Fleuriot-Lescot was appointed to the commission by Robespierre and the Comité de Salut Public some two weeks after his selection, with Joseph Payan, to head the Paris Commune after

professors to give the commission a guided tour of the establishment may have contributed to its decision to submit a project which "considered at the same time what was to be done provisionally in this garden, and what the Comité de Salut Public could order definitively for the construction of the most beautiful garden of natural history possible."[107] This construction never took place. It was to involve the extension of the Muséum's grounds to more than five times their existing size, from approximately 50 arpents to approximately 270. The projected cost of this to the nation was estimated at four million livres—a vast sum, inflation notwithstanding. The area into which the Muséum was to spread included the entirety of the lands formerly occupied by the monks of Saint-Victor, some unused docklands along the Seine, and the lands and properties of certain émigrés. It would cover the land between the Pont Bernard in the east and the hospital in the west, fronting the river all along this stretch. Plans for this undertaking were under active consideration for approximately eighteen months, and were made public in the periodical press.[108]

The commission's second report painted the imaginary Muséum in glowing colors: "The eyes would not be able to see anything more beautiful, the imagination could not figure for itself a more varied or more enchanting spectacle. In this vast garden, agreeable to all ages, useful to all Estates, the Physician would come to study nature, the young man would learn to love it, the philosophe, the Old Man would find the source of the most profound meditations here, the most beautiful texts for thought. The public already feels its charms, it has never been so frequented." They portrayed "the beauty of such a place, . . . the living whole of so many diverse objects, . . . the enchanting effect of all the marvels of nature, of all the productions united and offered to man by his kind, to admire them, to study them and to understand them." The formation of the new establishment offered, too, unparalleled opportunities for the Republic's artists: "Nothing will be lacking to them, the finances of a great Republic, a vast Space where imagination will be untrammeled, the marvels of the three Kingdoms, the living force of nature and all the Elements united to work together with Them

---

the execution of Chaumette and the former mayor of Paris, Pache. The Commission des Travaux Publics was one of a number of commissions set up by the Comité de Salut Public to replace the ministries after 12 germinal year II/April 1, 1794 (*Actes CSP,* 12:664–665).

107 BCMHN, MS 457, f°. 1: "Premier Raport au Comité de Salut public fait par la Commission des travaux publics relativement au Jardin des plants" (27 floréal year II/May 16, 1794); f°. 2: "Second Raport fait par la Commission des Travaux publics au Comité de Salut public relativement au Jardin des Plantes."

108 AN, F/14/187B; *La Décade philosophique* 1 (floréal year II/May 1794): 445–446.

Figure 15 "View of the School of Plants." Jean-Baptiste Hilair's ten watercolors, painted shortly before Robespierre's fall, depicted the Muséum as a peaceful place of recreation and leisure, free of Revolutionary violence. As a gardener in *sans-culotte* dress presents a woman and children with a rose, others are observing de Jussieu's version of the natural order, as presented by the labeled plants in the beds of the École de Botanique. Natural history thus offered Republican viewers both order and sentiment. Bibliothèque Nationale, France: Collection Destailleur, tom. 6 (Cabinet des Estampes). Cliché Bibliothèque Nationale de France.

in the execution of the most beautiful garden in the universe."[109] *Artiste* opposition to the Old Regime's patronage and exclusive corporations, headed by David, now a member of the Comité d'Instruction Publique, had already led to the closure of the Académie des Sciences and Société d'Agriculture. This report conferred upon natural history an aesthetic role which was unique among the sciences.[110] The professors themselves envisaged the new space as offering them the opportunity to present yet more, and bigger, tableaux of nature to their public: Thouin suggested that the teaching of botany could be completed by "the most complete Picture of the plants which cover the surface of the globe, arranged in a methodical order, based upon natural *rapports*."[111]

But above all, the commissioners drew the attention of the Comité

109 BCMHN, MS 457.

110 Hahn, *Anatomy of a Scientific Institution*, chapter 5.

111 BCMHN, MS 308: "Programme pour la Distribution du Terrain affecté au Museum National d'Histoire Naturelle," post floréal year II/May 1794; Harten and Harten, *Die Versöhnung mit der Natur*, 57–63.

de Salut Public to the power of the Muséum to display the future: the enlarged establishment "would present in miniature complete examples of the whole of nature, *in the end it would be the abridgement of the physical world as regenerated France will be that of the moral world.*"[112] The Comité approved the Commission's *projet de décret* at the end of floréal, and a series of decrees concerning the purchase and protection of lands adjoining the Muséum were passed in the summer of 1794.[113] Thus the Muséum became a central point for the representation of France's Republican future to citizens.[114] Its existence as a Republican institution was inseparable from its ability to dispense images of the future. This function involved physical objects as well; the Jardin supplied trees of liberty from year II onward, eventually becoming the official supplier of the trees of liberty required by law to be planted outside public buildings.[115]

Centers of Enlightenment

The decrees passed by the Convention and Comité de Salut Public between autumn 1793 and summer 1794 reveal that the Muséum was involved in a new kind of public instruction. It was to be a center of enlightenment, reproducing natural historical instruction across France. On 13 brumaire year II (November 3, 1793), based upon a plan commissioned from Thouin by the deputy Grégoire, the committees of public instruction and alienation approved the setting up of a system of botanical gardens and natural history collections, with a primarily agricultural emphasis, in every department in France. The purpose of these collections was to enable the acclimatization of exotic plants and animals in the different climates that France offered, for their utility and beauty. The collections were to imitate on a much reduced scale

112 BCMHN, MS 457: "Second Raport fait par la Commission des Travaux publics au Comité de Salut public relativement au Jardin des Plantes," original emphasis.

113 "Projet d'arreté présenté par la Commission des Travaux Publics au Comité de Salut Public relativement à un Jardin national des Plantes," year II/1794; *Actes CSP,* 13:544–546, 27 floréal year II/May 16, 1794; 14:538, 8 messidor year II/June 26, 1794.

114 On the Muséum as utopia, see Harten and Harten, *Die Versöhnung mit der Natur,* 9–19, 57–63; Outram, *Cuvier,* 183–185.

115 BCMHN, MS 1905: [Jean Thouin], "Registre des Dons faits par le Jardin National à Ses Correspondants, à des Cultivateurs ou Amateurs répandus dans les Départemens de la République dans Ses Colonies et dans les différentes parties du Monde," [year II/1794]; [A. Thouin], "Notes sur la plantation et le choix des arbres qui doivent être consacrés à la Liberté," undated; Franz Joseph Hell, *Suite des Notes sur les Arbres de la Liberté,* undated; AN, AJ/15/847: "Jardins. Arbre de la Liberté, an 6–an 8"; see also Ozouf, *Festivals,* 250–256; Andrée Corvol, "The Transformation of a Political Symbol: Tree Festivals in France from the Eighteenth to the Twentieth Centuries," *French History* 4 (1990): 455–486.

the sorts of activities that went on at the Muséum in Paris. "We do not present a fantastic picture," enthused Grégoire, "in saying that by [passing the decree] you will make France into a vast garden."[116]

The Convention voted to preserve foreign plants for public utility and instruction on 16 nivôse year II (January 5, 1794), and again on 8 pluviôse (January 27), and continued to favor the establishment of regional gardens with the surplus from the Muséum d'Histoire Naturelle throughout the spring and summer of year II.[117] The committees of alienation, public instruction, and finance, the Paris Commune, the Ministry of the Interior (until its abolition), the Robespierrist commissions, and numerous other ephemeral commissions were involved in lengthy negotiations over the issue. Decrees were passed between brumaire and messidor year II (October 1793–June 1794) which partially subordinated the provincial gardens to the Muséum d'Histoire Naturelle, and promoted the dismantling of national collections in émigré houses and gardens. For the next five years the Muséum staff in Paris was heavily occupied in making order from the chaos of specimens arriving from the pilfered collections in the rest of France and Europe, and in compiling collections of superfluous specimens for the departmental gardens. The decrees founding the École Normale on 9 brumaire year III (October 30, 1794) and the Écoles Centrales on 7 ventôse year III (February 25, 1795), both proposed during the Robespierre regime, underlined the Muséum's role as a center of centers.[118] It was to be a source of political, agricultural, and moral "lumières," which would, according to Thouin, "instruct active Republicans, who, having come from all the departments, would return there to report the knowledge necessary to the perfection of the Sciences and the Arts which watch so closely over the happiness of mankind."[119] In the opening lecture of his year II course, Mertrud, the professor of animal anatomy, offered his students, on behalf of the national museum of natural history, "rich knowledge of the three kingdoms of Nature; gather it on

116 Henri Grégoire, *Rapport et projet de decret, sur les moyens d'améliorer l'agriculture en France, par l'établissement d'une maison d'économie rurale dans chaque département, présentés à la Séance du 13 du premier mois de l'an deuxième de la République Française, au nom des Comités d'Aliénation et d'Instruction publique, par le Citoyen Grégoire* (Paris: Imprimerie Nationale, year II/1793), 24. Also mentioned in Harten and Harten, *Die Versöhnung mit der Natur,* 19.

117 See chapter 2; also Harten, *Museen und Museumsprojekte,* 73. Musées des beaux-arts founded in 1793–1794 still exist in Nancy, Poitiers, Reims, Toulouse, and Tours, to name but a few. Thus the policy of establishing provincial institutions also operated in tandem to favor the fine arts and natural history.

118 Julia, *Les trois couleurs,* 154–156, 264–267.

119 BCMHN, MS 308: Thouin, "Programme pour la Distribution du Terrain affecté au Museum National d'Histoire Naturelle," undated (post floréal, year II/May 1794).

all sides, to spread it thereafter in the bosom of the Republic, to the profit of humanity; your successes will not go unrewarded."[120]

This model of public instruction was based upon the courses in the manufacture of gunpowder and saltpeter offered to citizens from several districts during ventôse year II (February 1794), by order of the Comité de Salut Public, in the Muséum's chemistry laboratory.[121] Fourcroy's lectures in this Revolutionary cause were so popular that the experiment was repeated at the École de Mars during the summer of year II.[122] Again, the aim was that instructed students should return to their departments, and pass on the light of knowledge to their countrymen. During the spring and summer of year II, natural history teaching was portrayed as operating in the same way. On 6 floréal year II (April 25, 1794) the text of the decree allotting funds to refurbish botanical gardens expressed the same view: "The Muséum is, so to speak, a common reservoir, which will furnish the other gardens and receive exchanges from them; these gardens will spread enlightenment in their vicinity, by the example of an enlightened culture."[123] The museums and the different écoles established under the Jacobins revealed a common approach to public instruction, the spreading of enlightenment from center to periphery through the sheer force of Republican zeal. Enlightened students would travel back to the provinces to shine on their compatriots, so to speak. The creation of provincial gardens was portrayed explicitly as a solution to the public-order problem in a draft commentary, possibly written by Jean Thouin, André's brother. "The Gardens destined for the instruction of the young pupils of the Nation must be *simple agreeable* and *instructive* . . . It is in placing the productions of nature before the eyes of young Citizens early on, [and in] inspiring them with love for [those productions,] that one succeeds in making them known to them and that one manages resources of several kinds against false tastes and against boredom and idleness source of all troubles."[124]

120 Antoine-Louis-François Mertrud, *Discours prononcé dans l'amphithéatre du Muséum Nationale d'histoire naturelle, à l'ouverture du cours de l'anatomie des animaux; le primidi 21 nivos, l'an deuxième de la République françoise, une et indivisible* ([Paris]: Imprimerie de la rue des Droits de l'Homme, [year II/1794]).

121 Julia, *Les trois couleurs,* 293–296.

122 Ibid., 292; Dhombres, *Les savants en révolution,* chapter 6; Martin S. Staum, "Science and Government in the French Revolution," in *Science, Technology, and Culture in Historical Perspective,* ed. Louis A. Knafla, Martin S. Staum, and T. H. E. Travers (Calgary: University of Calgary Press, 1976); Charles Coulston Gillispie, "Science and Secret Weapons Development in Revolutionary France, 1792–1804: A Documentary History," *Historical Studies in the Physical and Biological Sciences* 23 (1992): 35–152.

123 *CIPCN,* 1:510.

124 BCMHN, MS 315: Jean or André Thouin, untitled, undated fragment.

The new Muséum, the epitome of middling management of an economy of specimens, had now become a site at which specimens were used to manage people.

Since the nature, object, and purpose of display had changed with the changing political and moral order, so too disciplines which had relied before the Revolution upon display as their principal attraction for an audience had to be altered. Fourcroy portrayed the Convention as having "torn from luxury and idleness" the lands of the nobility, which had been converted to grow "useful plants": "Collections of minerals and of animals are no longer, as before, heaped up without taste and without method, resembling treasuries rather than the museums of studious men."[125] The reform of both natural history and the arts involved the physical dismantling of old collections of luxury and tyranny, and the refashioning of their component parts into objects of public instruction and for public view. It was in this way that both art and nature were to be re-created as instruments of public control via individual experience. Fourcroy's speech to the Convention, suggesting ways of encouraging the sciences and arts, referred to the translated specimens as "treasures for public instruction," designed to create engineers, astronomers, orators, historians, painters, physicians, naturalists, and chemists—all those categories of people of which "the nation has need."[126] The value assigned to specimens, whether productions of art or nature, had changed dramatically with the new requirements for representing moral order in society.

Likewise, the Muséum's new menagerie was partly populated with animals perforce removed from the former royal menagerie at Versailles. A caricature concerning the transfer of the royal family to the Temple prison in 1792 (see figure 16) made its point by characterizing the king, the queen, and the royal children as rare and dangerous beasts. But the eventual transfer of the royal animals to the Muséum in 1794 symbolized, as the Muséum's naturalists and other supporters continued to emphasize, their conversion from monuments of despotic luxury

125 Antoine-François de Fourcroy, *Discours sur l'Etat actuel des Sciences et des Arts dans la République Française. Prononcé à l'ouverture du Lycée des arts le dimanche 7 Avril 1793, l'an second de la République* ([Paris]: n.p., [1793]). On this body see W. A. Smeaton, "The Early Years of the Lycée and the Lycée des Arts: A Chapter in the Lives of A. L. Lavoisier and A. F. de Fourcroy," *Annals of Science* 11 (1955): 257–267, 309–319.

126 Antoine-François de Fourcroy, "Rapport et projet de décret sur l'enseignement libre des sciences et arts," 19 frimaire year II/December 9, 1793, quoted in Julia, *Les trois couleurs,* 75–76; see also Langins, *La République avait besoin de savants.* On the power relations implicit in museum displays, see, among others, George W. Stocking Jr., "Essays on Museums and Material Culture," in Stocking, *Objects and Others;* Bennett, introduction to *Birth of the Museum.*

Figure 16 "Rare Animals. Or the transfer of the Royal Menagerie to the Temple, August 20, 1792." Menagerie animals needed to be adapted from their associations with monarchical degeneracy and luxury to become objects of virtue, as revealed by this caricature of the royal family's incarceration in the Temple. The first efforts by the Jardin's naturalists to create a menagerie at the establishment dated from this period. By de Vinck, 1792. Bibliothèque Nationale, France: Cabinet des Estampes. Cliché Bibliothèque Nationale de France.

to representations of republican virtue.[127] Like the rest of the Muséum's displays, the menagerie permitted the opening up of formerly royal or private property to the sight of a new public of citizens, and also a conversion of the "currency" of natural history into the coinage of republicanism.

Successive Revolutionary authorities appointed naturalists of the Jardin/Muséum to undertake the natural historical part of the dismantling, cataloguing, and transport of confiscated collections. The minister Roland's part in Thouin's stripping of *émigré* and national gardens in and around Paris has been discussed in chapter 2. Roland also obtained a decree from the Convention Nationale on March 27, 1793, commanding the Chantilly natural history collection, discovered walled

127 Bernard-Germain-Étienne de La-Ville-sur-Illon, comte de Lacepède, and Georges Cuvier, *La ménagerie du Muséum national d'Histoire naturelle, ou les animaux vivants, peints d'après nature . . .* (Paris: Miger, year X/1801), 1–6; Osborne, "Applied Natural History," 132–133.

up in a tower, to be transferred to the "national collection" at the Jardin, along with 15,000 livres to enable the necessary alterations to be made to receive it.[128] The process was supervised by the Commission des Monuments, established in 1790 to safeguard valuable objects of display in the arts and the sciences.[129] The transfer of the Chantilly specimens to the new Muséum took place only after Roland's ousting from office, with Geoffroy, Vanspaendonck, and Lamarck traveling to Senlis to oversee the process and catalogue the many specimens, under the auspices of Daubenton and the new minister of the interior, Garat. These two events marked the first major transfers of specimens to the new national museum of natural history. However, the redistribution of "monuments of the arts and sciences" reached its height under the Jacobins. On August 24, 1793, the Convention set up a Commission Temporaire des Arts with the task of making inventories of all of the confiscated, "national" collections. Five days later, its thirty-eight members, drawn from all areas of the sciences and arts, began the first of many meetings at the offices of the Comité d'Instruction Public.[130] Thouin and Desfontaines represented the botanical section of the commission; Lamarck later took a place on the zoological section after the departure of the *adjoint*, Alexandre Brongniart, for Bayonne. Throughout the existence of the commission, there were at least twice as many members for natural history as for any other subject.

During the next eleven months, these commissioners traveled in their allotted area—a circle of thirty *lieues* around Paris—making inventories of valuable collections belonging to émigrés, *condamnés*, and the royal family. Thouin and Desfontaines finally deposited their inventories with Robespierre's executive commission of public instruction on 3 thermidor year II (July 21, 1794).[131] They had catalogued twenty-

128 AN, AJ/15/836.

129 Dominique Poulot, "Naissance du monument historique," *Revue d'histoire moderne et contemporaine* 32 (1985): 418–450; Rhoda Rappaport, "Revolutions, Accidents, and 'Bouleversements,'" *Histoire et nature* 19–20 (1981–1982): 57–58. The term "monument" was still widely used in the sense of "markers" or "indicators."

130 BCMHN, MS 315. A list of the members appointed to the Commission Temporaire des Arts, which first met in late August 1793, is in *CIPCN,* 2:508–510.

131 The executive commission had been set up by the Comité de Salut Public on 12 germinal year II/April 1, 1794, to oversee "the conservation of national monuments, public libraries, museums, cabinets of natural history and precious collections; . . . schools and the mode of teaching; all that concerns scientific inventions and research; the fixing of weights and measures; spectacles and national fêtes; the formation of tables of population and political economy" (*Actes CSP,* 12: 327, art. 5). For the finished reports, see AN, F/17/1050, F/17/1219, F/17/1223, F/17/1224, and F/17/1238.

five different collections or depots, with a total of 1,061,675 trees.[132] Thirteen remained to be done. In each transfer, valuable and rare specimens were directed to the central Muséum d'Histoire Naturelle, or, in the case of objects of the arts, to the Muséum National des Arts. Less valuable objects and copies were placed in storage in specially reserved depots such as the former Maison des Petits-Augustins. Living plants also had depots; the Trianon garden formed the chief of these, where Antoine Richard, Thouin's and Desfontaines's correspondent, cared for them until they could be moved elsewhere. Often, if a terrain was particularly valuable for the purposes of natural history, cultivation, or rural economy, Thouin made special appeal to the Convention, via Fourcroy, for it to be set aside from the sale of *biens nationaux*.[133]

The productions of art and nature which were confiscated, listed, and sent back to Paris came from many places around France and later from other parts of Europe conquered by the Revolutionary army, from Belgium and Holland in years III and IV, and from Italy during year VI. Once the best specimens had been skimmed off for the two national collections, the remaining specimens were sorted for transfer to departmental collections, packaged, and sent out by the members of the commission.[134] A detailed study on the extent to which the national museums' consignments reached their targets and were used as the basis

132 BCMHN, MS 315: A. Thouin, "Projet d'Organisation de la Section d'Histoire naturelle arreté la 30 Aoust 1793 dans l'ass. des 6 Commissaires"; id., "Exposé Succincte des Travaux de la Division de Botanique Section de l'Histoire Naturelle de la Commission temporaire des Arts," 30 messidor year II/July 18, 1794; id., "Apperçu des Travaux qui restent à faire pour la division de Botanique, Section de l'Histoire Naturelle de la Commission temporaire des Arts," circa August 1793; BCMHN, MS 312; Guillaumin, "André Thouin"; Letouzey, *Le Jardin des Plantes,* 359.

133 BCMHN, MS 308: "Métairie centrale de la République," "Nottes sur les moyens d'Employer utilement les Domaines nationaux réservés par le decrêt du floréal 2de.," "Commission d'Economie rurale adjointe au Comité de Salut Public," "Programme pour la Distribution du Terrain affecté au Museum National d'Histoire Naturelle."

134 Studies on this subject have concentrated almost exclusively on objects of the arts sent back by the commissions between 1793 and 1799; see Edouard Pommier, "La fête de thermidor an VI," in *Fêtes et révolution,* ed. Béatrice de Andia and Valérie Noëlle Jouffre (Dijon: Musée des Beaux-Arts, 1989–1990), 178–215; id., "La Révolution française et l'origine des musées," paper presented at Department of History of Art and Architecture, Cambridge, April 1992; id., *Le problème du musée à la veille de la Révolution* (Montargis: Musée Girodet, 1989); Eugène Müntz, "Les annexations de collections d'art et de bibliothèques et leur rôle dans les rélations internationales, principalement pendant la Révolution française," part 2, *Revue d'histoire diplomatique* 9 (1895): 375–393; Ferdinand Boyer, *Le monde des arts en Italie et la France de la Révolution et de l'Empire: Études et recherches* (Torino: Società Editrice Internazionale, 1969); Cecil Gould, *Trophy of Conquest: The Musée Napoléon and the Creation of the Louvre* (London: Faber and Faber, 1965), chapters 2 and 3.

Figure 17 "Triumphal entry of the monuments of the sciences and arts into France." A parade of the objects of the sciences and arts collected by Thouin and the other commissioners in Belgium and Holland, and celebrating the anniversary of Robespierre's fall from power. Living plants and animals are clearly visible. By Berthault, after Girardet, 1798. © Bibliothèque Centrale M.N.H.N. Paris 2000.

for collections is lacking. However, it seems likely that the Terror, the problems of centralized bureaucracy, and widespread administrative chaos hampered these optimistic consignments from Paris. Nonetheless, Thouin's documents suggest at least a partial success for the new network.[135]

The activities of the commission, which resulted in an almost continual influx of new objects into the new museums, were partly stimulated by episodes of what the deputy Grégoire would later term *vandalisme* in national properties.[136] Rare and valuable specimens could best be surveyed in a few official depots run by trained men. Among Geoffroy's

135 BCMHN, MS 313.

136 Henri Grégoire, "Rapport sur les destructions opérées par le Vandalisme, et sur les moyens de le réprimer. Par Grégoire, Séance du 14 Fructidor, l'an second de la République une et indivisible, suivi du décret de la Convention Nationale," in *Oeuvres,* 2:258–278; id., "Second rapport sur le vandalisme . . . Séance du 8 brumaire l'an III," in *Oeuvres,* 2:321–334; id., "Troisième rapport sur le vandalisme . . . Séance du 24 frimaire l'an III," in *Oeuvres,* 2:335–358.

and Lamarck's duties in Senlis was the total valuation of the collection, which they performed with Gaillard, a merchant of natural history specimens, and two administrators from the district of Senlis. The Jardin's savants were regarded as experts in matters of natural historical value by the deputies, playing a part in conserving the wealth of the nation.[137] However, drawing up the inventory was also part of the Jacobins' goal to make the Muséum d'Histoire Naturelle and its sister establishment, the Muséum des Arts, the richest and most astonishing collections in the world.

The objects from the confiscated collections were not merely moved from one place to another, but translated in a far more fundamental way by being renamed and philosophically and physically reordered by the commissioners.[138] No matter what the original order in the collection had been, Thouin and Desfontaines chose the order preferred at the Muséum to make their national inventory: Linnaean nomenclature and de Jussieu's method. De Jussieu's method also entered the syllabus for the Écoles Centrales.[139] In this sense the naturalists of the former Jardin began to acquire truly national control over the natural order, by applying it to national specimens with which the collections of the departments were to be stocked, and by controlling the moral interpretation of natural productions.

137 AN, AJ/15/836: Desfontaines to Garat, August 3, 1793, Paris. However, the professors also kept a safe distance from involvement in the commercial valuation of specimens, as revealed by the presence of Gaillard among the commissioners.

138 BCMHN, MS 315: André Thouin, "Notes relatives à la confection des Inventaires, à la conservation et au transport des objets du Regne vegetale qui se trouvent dans les dépots de la République."

139 For early teaching texts see Éloi Johanneau, *Tableau synoptique de la méthode botanique de B. et A. L. Jussieu* (Paris: Imprimerie de la République, year V/1797); [Jean-Marie Morel], *Tableau de l'École de Botanique du Jardin des Plantes de Paris, ou Catalogue général des Plantes qui y sont cultivées et rangées par classes, ordres, genres et espèces, d'après les principes de la Méthode naturelle de A. L. Jussieu. Suivi d'une Table alphabétique des Noms vulgaires des Plantes le plus fréquemment employées en Médecine, dans les Arts, la décoration des Jardins, etc.; avec les Noms des Genres et des Espèces auxquels elles se rapportent* (1st ed., Paris: Didot le jeune, year VIII/1800; 2d ed., Paris: Méquignon l'aîné, year IX/1801). Many of the naturalists teaching at the Écoles Centrales were protégés of the professors: for example, Alexandre Brongniart, Georges Cuvier, Sébastien Gerardin, and Dominique Villars. The implementation of natural history teaching at the Écoles Centrales, which taught teenage boys, was left to the professors; initially an important though elementary part of the curriculum, natural history lost ground in successive reforms and vanished at the closure of the Écoles Centrales, playing no role in the curricula of the new Lycée system that replaced them (Julia, *Les trois couleurs;* Pierre Lamande, "La mutation de l'enseignement scientifique en France [1750–1810] et le rôle des écoles centrales: L'exemple de Nantes," special issue of *Sciences et Techniques en Perspective* 15 [1988–1989]; Williams, "Science, Education and the French Revolution").

Conclusion

Governmental support for the Muséum, and for natural historical and fine arts establishments in general during years II and III, stemmed from a particular program for public instruction which sought to generate a new public order while simultaneously distancing legislators from the scene of taxonomic production. The Muséum served as a site of instruction in public behavior, rather than private conduct. During the early years of the Republic, virtually all aspects of the individual became public issues. Not only dress and language were to conform with Revolutionary existence, but even (or perhaps especially) the expression of the emotions.[140] At the Muséum even the feelings appropriate to the good citizen confronted by the moral spectacle of Nature were being defined. In effect, as public space grew, private space shrank until even the individual's inner moral state was subject to intervention which aimed at overcoming corruption; but those rejections of the Old Regime and its ways had to come from the heart, from within every citizen, since the foundation of the Republic was nature itself, the nature ineradicably located within humanity.

Natural history at the Muséum was valuable for Jacobin deputies because it exemplified the invisible mastery of spectacle. In the small area of the Muséum, the naturalists were empowered to display the links between the operation of Nature's laws within the universe as a whole, and the operation of those same laws within the new Republic. The goodness and perfection of those laws was made evident by the professors in directing their audience how to look at the collections. Citizens experienced an emotional response to the sight of nature within the carefully ordered precincts of a natural history collection; through their innate *sensibilité,* they would be both morally and physically regenerated at the Muséum. The translation of the objects of art and nature and the Revolutionary festivals were thus aspects of a single enterprise, in which public spectacle was being recast so as to address the sensibility of the citizen.

During the Revolution the establishment's function as the center for a patronage-based correspondence declined. No longer exclusively reliant upon networks of protégés and patrons for its income, it was now a center from which Revolutionary propaganda streamed, both via the professors' teachings and via the distribution of specimens which

140 Outram, *The Body and the French Revolution,* chapter 5; Vincent-Buffault, *History of Tears,* chapter 5.

would permit the display of the moral spectacle of nature in other settings. The replication of the natural historical concerns of the Muséum naturalists throughout Republican France also denoted the expansion of the Muséum's boundaries to encompass the nation; through emulation, a garden utopia would gradually spread across the face of Republican France. Even the Muséum's role in the distribution of agricultural knowledge and food plants was a response both to the need of the *conventionnels* to master the symbols of virtuous republican government and to the subsistence crisis. The experiment, however, was too brief to have any lasting effect upon French agricultural practices, as historians have frequently pointed out.[141]

In short, the Muséum d'Histoire Naturelle was the location at which the future perfection of France could be displayed to all citizens. Although these projected futures for natural history were not fulfilled, their scope indicates the considerable importance of the Muséum and the twelve professors for the new political system. The Jardin and the naturalists who worked there benefited greatly from this translation of the needs of the Jacobins via their own practices. As a site for the invisible manufacture of natural order, the Muséum received the powerful support of the Convention, including a continual influx of specimens and the promise of yet more spatial and financial resources. Thus, to take contemporary claims about Jacobin scientific and artistic iconoclasms at face value is to read the Robespierre regime through the eyes of the Directoire.[142] Little has been written concerning the Robespierrists and the sciences and arts which does not draw upon these polemical portraits of cultural barbarism. Such accounts, however, were often generated by individuals who had served under Robespierre and the Comité de Salut Public, and who sought to cover their tracks after 9 thermidor by portraying themselves as moderates forced to obey the tyrant. Thus Fourcroy and Grégoire both labored to portray the Robespierre regime as espousing policies of "vandalism" after 9 thermidor, as they sought to distance themselves from the discredited dictatorship; nonetheless, as we have seen, both were heavily involved in the execution of its public instruction projects. For example, in April 1793—just when he was later to claim that Jacobin "*vandalisme*" had flourished—

141 Festy, *L'agriculture pendant la Révolution française,* 1:91–95; Roger Price, *Economic Modernisation of France,* chapter 2; but compare Rosenthal, *Fruits of Revolution.* Boulaine, *Histoire de l'agronomie en France,* argues for slow transformation.

142 E.g., Fayet, *La Révolution française et la science;* Gillispie, "The Encyclopédie and the Jacobin Philosophy of Science"; Roger Hahn, "Elite scientifique et démocratie politique dans la France révolutionnaire," *Dix-huitième siècle* 1 (1969): 229–235.

Fourcroy had attacked those who regarded the Republic as a wantonly destructive political system.[143]

The spread of natural history teaching across France can thus also be seen as the growth of a regime of public management. The future of natural history and the boundaries of the Muséum d'Histoire Naturelle depended upon the success of that program. It is appropriate, then, that the failure of the project to enlarge the Muséum should have been linked to its failure to fulfill its role in controlling its audience. After 9 thermidor, former members of Robespierre's Comité d'Instruction Publique and the Commission Temporaire des Arts, including Fourcroy, Thibaudeau, Louis-François-Antoine Arbogast, and Noël-Gabriel-Luce Villar, continued to press for the enlargement plan to be carried out. However, if the Old Regime Jardin had catered to an elite audience, the new public education system of which the Muséum was a part was explicitly democratic in character. It was to prove very difficult to control how viewers not inducted into the expert visual culture of polite society would interpret savant claims. The École Normale which opened in late 1794 in the Muséum's amphitheater closed after less than a year, as students, unable to follow the professors' lectures, frittered away departmental resources in Paris cafés.[144] Within months of the fall of Robespierre, the fivefold increase proposed in the Muséum's surface area had been reduced to a mere doubling.[145] The Muséum still had its Jacobin supporters, but a *projet de décret* presented in 1795 by Thibaudeau, at the end of a long and impassioned speech describing the Muséum's power as natural spectacle, succeeded only in obtaining Lacepède's reappointment as third professor of zoology; other demands, including an increase in the professors' salary, the payment of outstanding debts for building works, and the purchase of lands to be

143 Compare Fourcroy, *Discours sur l'Etat actuel des Sciences et des Arts,* and Grégoire, "Rapport sur les destructions opérées par le Vandalisme." The latter report dates from the month after Robespierre was executed. See also Anthony Vidler, "Grégoire, Lenoir, et les 'monuments parlants,'" in *La carmagnole des muses,* ed. Jean-Claude Bornet (Paris: Armand Colin, 1986), 131–159; Gabriele Sprigath, "Sur le vandalisme révolutionnaire, 1792–1794," *Annales historiques de la Révolution française* 52 (1980): 510–535; Hahn, *Anatomy of a Scientific Institution,* 289–290; Outram, "Ordeal of Vocation." Other legislators with similar concerns included Freron, Lanthenas, Boissy d'Anglas, Daunou, and La Harpe.

144 Julia, *Les trois couleurs,* 168–171.

145 Bradley, "Financial Basis of French Scientific Education"; Kersaint, "Fourcroy"; AN, F/14/187^B. On 21 frimaire year III/December 11, 1794, the Convention passed a decree approving the Muséum's reduced enlargement, and on 17 prairial year IV/June 5, 1796, the Directoire authorized the purchase or exchange of all lands concerned in the enlargement project. See Guillaume-Jean Favard de l'Anglade, *Rapport fait par Favard au nom d'une commission spéciale, sur des bâtimens et terreins réunis au Muséum d'histoire naturelle. Séance du 15 Brumaire an 6* (Paris: Imprimerie Nationale, year VI/1797).

attached to the establishment, were ignored.[146] Meanwhile the foundation of the new Institut National des Sciences et des Arts in the same year added an overwhelming rival to the scientific scene.[147] Outside the Muséum, too, the exercise of centralized authority gave way before a tangle of competing territorial claims, as owners of adjoining properties protested to the Convention about the manner in which they were recompensed for their land.[148] After a year of taking no action, the Directoire appointed a special commission on 9 fructidor year V (August 26, 1796) to report on the Muséum. In the interests of economy and propriety, but probably above all because of the severe weakening of their political connections, in 1797 the professors renounced their claims to have the expansion project implemented.[149] The last vestiges of their ambitious vision had disappeared.

146 Thibaudeau, *Rapport . . . sur le Muséum national d'histoire naturelle.*

147 Hahn, *Anatomy of a Scientific Institution,* chapter 10.

148 *Pétition des Propriétaires des Maisons et Terreins environnans le Jardin des Plantes, à la Convention Nationale* (Paris: Démonville, year III/1794–1795). Proprietors objected to being "recompensed" with the speedily devaluing *assignat;* see Dieudonné Dubois, "Rapport fait . . . au nom d'une commission, sur un Message du Directoire exécutif du 15 nivôse dernier, tendant au rapport de la loi du 9 du même mois, interprétative de celle du 17 prairial de l'an 4, concernant les terreins destinés à l'agrandissement du muséum d'histoire naturelle" (Paris: n.p., year V/1797).

149 Favard de l'Anglade, *Rapport;* AN, F/17/1229, dossier 9. As correspondence between local proprietors and the Comité d'Instruction Publique reveals, the Muséum's value as a place of public instruction, particularly of agriculture, was not in question.

CONCLUSION

Possible Futures

Institutions appear to be things of bricks and mortar, of paper and records, rules and regulations. Human beings associated with them are often described as *belonging to* institutions, so that the establishment itself seems to possess a life, an individuality which outlasts the confines of human existence. Such perceptions of solidity and permanence on our part are belied by the history of the Jardin des Plantes in the second half of the eighteenth century. For the question needs to be addressed from the other side in the first instance: how was the establishment's meaning fashioned through the interplay of social relations between different individuals? Even the physical structure of the establishment was dependent for its construction and maintenance upon such negotiations, let alone its existence as an institutional whole rather than as a collection of scattered buildings by the river Seine. Rights of access, readings of collections, networks of correspondence, titles of employment, all manifested the culture within which the establishment existed, as well as the particular history and constitution of the place itself. In a similar way, particular forms of knowledge may survive their conditions of production. They may be rejected, but they may also be taken up and used for new purposes within new scientific cultures. Sometimes their continued force is legitimated by an explicit acknowledgment of their historical origins, but even in such cases, appeal to the past is a proactive move, serving to construct a particular version of events which will support current claims. Like the meaning of natural history itself, therefore, the institution required continual reinvention, particularly after 1789, when a succession of new regimes ensured that its role was continually open to challenge.

The move to fashion the Jardin/Muséum as an entity essentially independent of human agency suggests some interesting historical avenues. While objects and abstractions, such as books, institutions, or ideals, must be continually refashioned at every moment and by every user in order to persist, they may simultaneously be ascribed powers in their own right by those users.[1] If the Jardin/Muséum was a site within which the powers of nature were invoked in order to recruit scientific authority, it remains to be understood how those powers can be handled by the historian. While recent histories of the French Revolution, and a good deal of this volume, have emphasized the symbolic and the discursive as manifestations of power, one should also accept the caution of those who ask what changed as a result of these maneuverings around the display of sovereignty. Sociologists of scientific knowledge have argued that power is not something that is possessed, but something that is exercised: "When you simply *have* power . . . nothing happens and you are powerless; when you *exert* power . . . others are performing the action and not you."[2] If one considers the Muséum's history up until the fall of Robespierre on 9 thermidor year II, naturalists did appear to gain in political power. But the Muséum's subsequent fate suggests that the successful capture of Nature, a major symbol of sovereignty, by the naturalists did *not* give them more than local power, or rather, that the tenuous links between scientific and political power could be maintained only for a short period.

The Jardin/Muséum expressed the multiplicity of futures opening up at the end of the century. Its moral and material boundaries grew or shrank under different political regimes, just like the boundaries of France itself. Its space was ordered and divided according to intricate patterns shaped by past, present, and future usage, in an arena where convenience, tradition, and innovation jostled for position. Of the many possible futures being envisaged for the institution, however, few were actually implemented as reforms of the material fabric of the Muséum. My aim is not to offset the utopian visions of the professorial administration with a picture of a "real" Muséum, where infant trees were tied to stakes, captured specimens moldered in the amphitheater, and pro-

1 Nicholas Jardine, "The Laboratory Revolution in Medicine as Rhetorical and Aesthetic Accomplishment," in *Romanticism and the Sciences,* ed. Andrew Cunningham and Nicholas Jardine (Cambridge: Cambridge University Press, 1992), 304–323.

2 Bruno Latour, "The Powers of Association," in *Power, Action, and Belief,* ed. Law, 264–265; Barry Barnes, *The Nature of Power* (Cambridge: Polity Press, 1989).

fessors struggled to resolve ongoing difficulties produced by the lack of space. It is to suggest that the Muséum's actuality, both as material object and as symbol, was the outcome of constant negotiations between different groups: the professors, the deputies, but also hungry laborers and Revolutionary visitors. The total control of the visitor's body and gaze identified by historians of architecture as the material construction of power relations within institutions is arguably never fully achieved. It is always balanced by the visitor's power to escape such enforcements by the creation of different, conflicting meanings for displayed objects and the bodily transgression of institutional boundaries, as demonstrated in Thouin's problems with the Jardin's various publics. One should perhaps write, therefore, of multifarious powers, perpetually making and unmaking futures.[3] Cuvier might castigate the Muséum's disastrous condition in 1795, the year of his arrival in Paris, but for the professors the packaged mountains of newly acquired specimens, awaiting their intercalation into the national galleries, were evidence both of the bounty of nature and the military might of the Republic, the promise of things to come.[4] For many Jacobin deputies, the Muséum and its professors did indeed represent a miniature of the regenerated earth, a display of fraternal love and social harmony within a microcosm of Nature's order. In a plan of messidor year II concerning the enlargement of the Muséum's terrain, the architect Molinos proposed a large building of circular shape, to harbor the "great book of Nature" (figure 18). Inside, the floor area would display the earth's topography, the ceiling the constellations, and a subterranean gallery the mineral kingdom and rock formations. Animals and plants would adorn the inner space, surrounding a giant statue of the goddess Nature, and the walls would be decorated with the names and characters of animals, plants, and philosophers.[5] The collection as microcosm was not new, by any means; but it had not previously been used to abolish the symbols of church and crown in favor of a Republican

3 Michel de Certeau explores some implications of extending Foucault's "technologies of power" to a society of multiple powers. See Michel de Certeau, "Micro-Techniques and Panoptic Discourse: A Quid Pro Quo," in *Heterologies: Discourse on the Other,* trans. Brian Massumi (Minneapolis: University of Minnesota Press, 1986), 185–192; id., *The Practice of Everyday Life,* vol. 1, trans. Steven Rendall (Berkeley: University of California Press, 1988).

4 Corsi, *Age of Lamarck,* chapter 1.

5 AN, F/17/1229, dossier 9, piece 185: Molinos, "Disposition générale pour le Muséum d'Histoire Naturelle et son jardin," 3 messidor year III/June 21, 1795, discussed in Harten and Harten, *Die Versöhnung mit der Natur,* 57–63. For earlier plans to remodel the Jardin's terrain, see Yves François, "Notes pour l'histoire du Jardin des Plantes: Sur quelques projets d'amenagement du Jardin au temps de Buffon," *Bulletin du Muséum d'Histoire Naturelle,* series 2, 22 (1950): 675–693; Falls, "Buffon et l'agrandissement du Jardin du Roi."

Figure 18 Molinos's ink and wash drawing of year II shows the projected enlargement of the Muséum's terrain, accepted by the Comité de Salut Public prior to 9 thermidor, but never implemented. Document conservé au Centre historique des Archives nationales, France: réf. F/17/1229, dossier 9, piece 186.

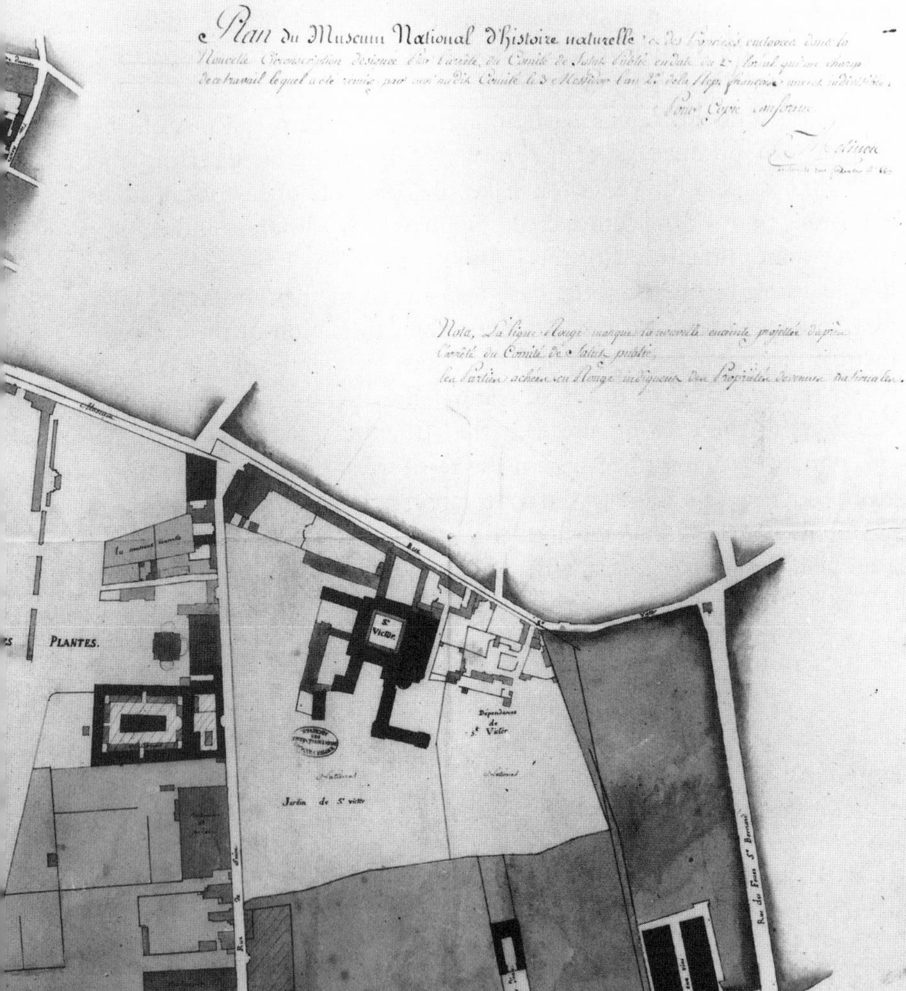

Plan du Museum National d'Histoire naturelle et des Propriétés enclavées dans la Nouvelle Circonscription designée par l'arrêté du Comité de Salut Public en date du 2e floréal qui m'a chargé de ce travail lequel a été remis par moi audit Comité le 3 Messidor l'an 2e de la Rep. française une et indivisible.

Pour Copie conforme

Molinos

Nota, La ligne Rouge marque la nouvelle enceinte projettée d'après l'arrêté du Comité de Salut public,
les Parties achées en Rouge indiquent des Propriétés devenues nationales.

PLANTES

St Victor

Dependances de St Victor

National

Jardin de St victor

Rue des Fossés St Bernard

Quay St Bernard

pantheism. In the event, however, even the power of display was denied the naturalists with the weakening of Jacobinism. Molinos's temple exists only on paper; nonetheless, the very possibility of Republican approval for such an ambitious project for natural history indicates the importance of that discipline and of the Muséum for a particular, ephemeral politics of knowledge.

Architectural historians have argued that museum space could be designed to produce particular forms of behavior by directing the movement of the future visitor. Likewise, patterns of historical usage and projections of the future could be invoked by different individuals to constrain present uses of the Jardin's terrain.[6] As late as 1823, André Thouin's brother Gabriel designed a plan of the Muséum which was probably a final attempt to obtain the completion of the enlargement project first mooted in year II (figure 19). Here the Muséum is barely recognizable, its principal features dwarfed by the monumental additions to the existing site. But the structures of the Old Regime are, nonetheless, inscribed within the territory. The labyrinth and the greenhouses are pointers to the contemporary boundaries of the establishment. Even their presence in the image is significant, for in such a utopian design there was, after all, no need to preserve any features of the past. The future, like the past, could thus be evoked as grounds for action in the present. Gabriel was not portraying the future as it would be, for he and his brothers could never have lived to see the completion of this grandiose scheme. Instead, he offered the future as it could be were it to be realized at that moment—the fulfillment of the ambitions of the existing staff for the Muséum's role. Such prescriptive claims were absolutely necessary for the negotiations between naturalists, patrons, laborers, and visitors which continually refined the multiplicity of projected futures into the solid actuality of the institution's day-to-day existence.[7]

The Muséum was, in effect, a heterotopia as well as a utopia.[8] Perhaps it is more true to say that it could become a utopia *because* it was a heterotopia. I should like to suggest that the Jardin's transformation into the Muséum was possible only because naturalists were *bricoleurs,*

6 Forgan, "The Architecture of Display"; see also Bennett, *Birth of the Museum,* chapters 1 and 2.

7 Geoffrey Hawthorn, *Plausible Worlds* (Princeton: Princeton University Press, 1992); Nicholas Jardine, *The Scenes of Enquiry* (Oxford: Clarendon, 1991).

8 Dominique Poulot has suggested that, "prior to the implementation of a new culture of the object, a 'carnivalesque' disorder temporarily ruled" in Revolutionary attempts to create museums ("Le Louvre imaginaire," 172).

generating a variety of meanings and practices within a space which was very different from the homogeneous world of the laboratory—indeed, which itself enclosed more than one laboratory. Its multiplicity as physical site was admired by contemporaries, who actively espoused landscaping in botanical gardens in order to provide simulacra of different climatic and topographical conditions for introduced plants and animals.[9] In this respect the Jardin/Muséum differs considerably from the kind of settings for the production of scientific knowledge that have been the focus of recent studies of scientific practice.[10] Other naturalists, looking to the Muséum d'Histoire Naturelle as a model for their own enterprises, could thus tailor its meaning to their own purposes.

Throughout the early nineteenth century no one version of natural history would reign supreme at the Muséum. This may reflect the fact that natural history offered less scope for the kinds of strategies for engineering decisive proofs which have been explored by historians of experimental practice.[11] In the axis between the Old Regime past and the Republican future, it was possible to transform the practice of natural history in France. There were changes in the status of the institution and its practitioners, changes in the political relations and meaning of natural history, changes in the function of display, and changes in the economy of collecting. There were also changes in naturalists' ability to account for the natural world—to present a "true" picture of nature. That these changes took place, that they were considerable, that they

9 The directors of the Montpellier botanical garden responded to the year II/1794 circulars by praising the landscaping activities of the garden's founder on these grounds (Harten and Harten, *Die Versöhnung mit der Natur,* 46). See also chapter 3.

10 Shapin, *Social History of Truth;* id., "The House of Experiment in Seventeenth-Century England," *Isis* 79 (1988): 373–404; Latour and Woolgar, *Laboratory Life;* Michael Lynch, *Art and Artifact in Laboratory Science: A Study of Shop Work and Shop Talk in a Research Laboratory* (London: Routledge and Kegan Paul, 1985). On place-specific knowledges see Ophir and Shapin, "Place of Knowledge"; John A. Agnew and James S. Duncan, eds., *The Power of Place: Bringing Together Geographical and Sociological Imaginations* (Boston: Unwin Hyman, 1989); Pratt, *Imperial Eyes;* Driver and Rose, introduction to *Nature and Science.*

11 My reflection on these issues has been shaped against my reading of works such as Gooding, *Experiment and the Making of Meaning;* Latour, *Science in Action;* David Gooding, Trevor Pinch, and Simon Schaffer, eds., *The Uses of Experiment: Studies in the Natural Sciences* (Cambridge: Cambridge University Press, 1989); Pickering, *Science as Practice and Culture;* and work by Schaffer and Shapin mentioned in the introduction.

Figure 19 The Thouin family's utopia. Gabriel Thouin's "Project of enlargement for the jardin des plantes of Paris" had its origins in the Jacobin enlargement project of year II. The circular building at *C* was probably Molinos's microcosm, which, in the original version, contained a giant statue of the goddess Nature. From G. Thouin, *Dessins pour tous sortes de jardins* (Paris, 1823). © Bibliothèque Centrale M.N.H.N. Paris 2000.

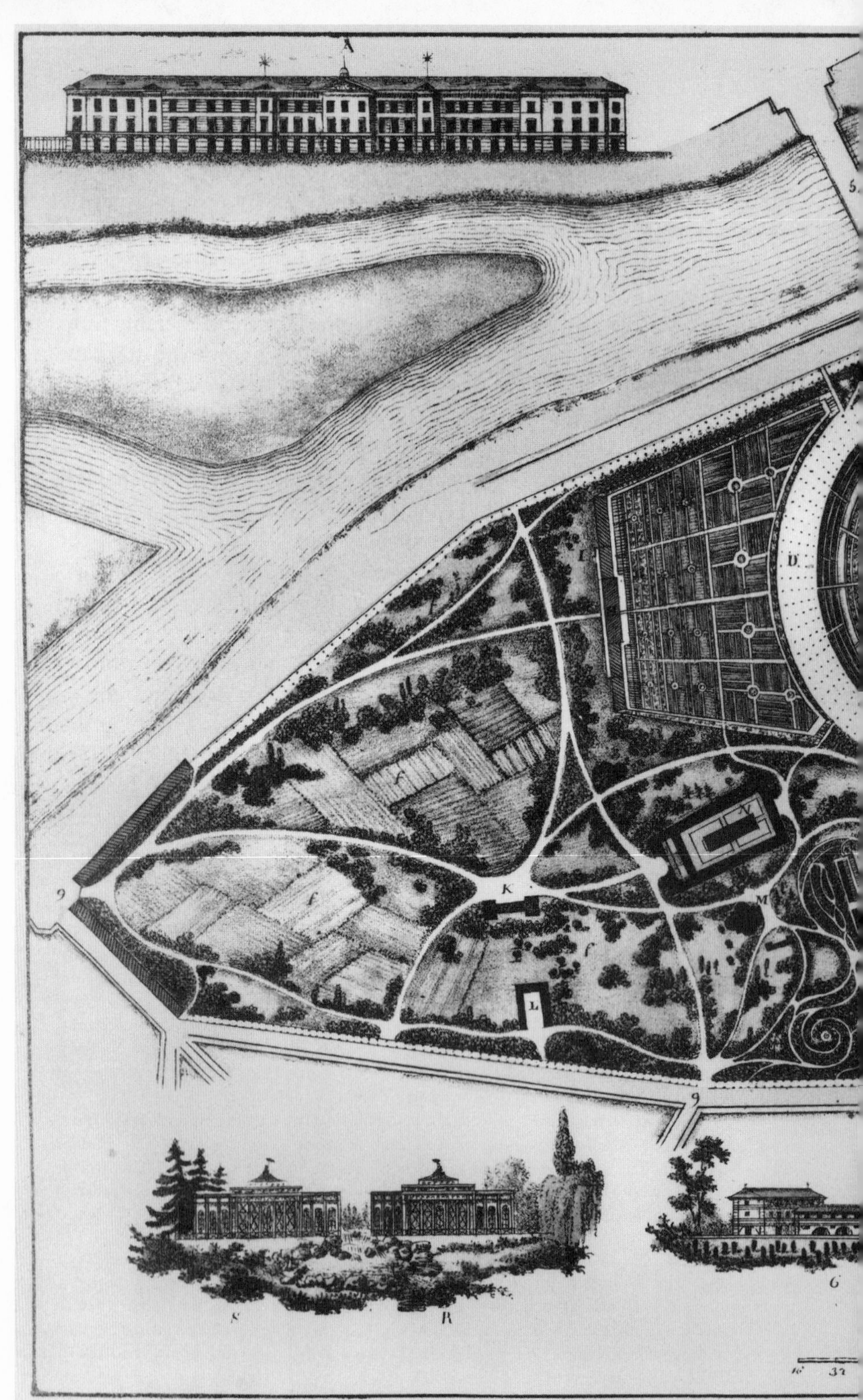

Thouin Del.

N° 13 - Projet d'agrandissemen

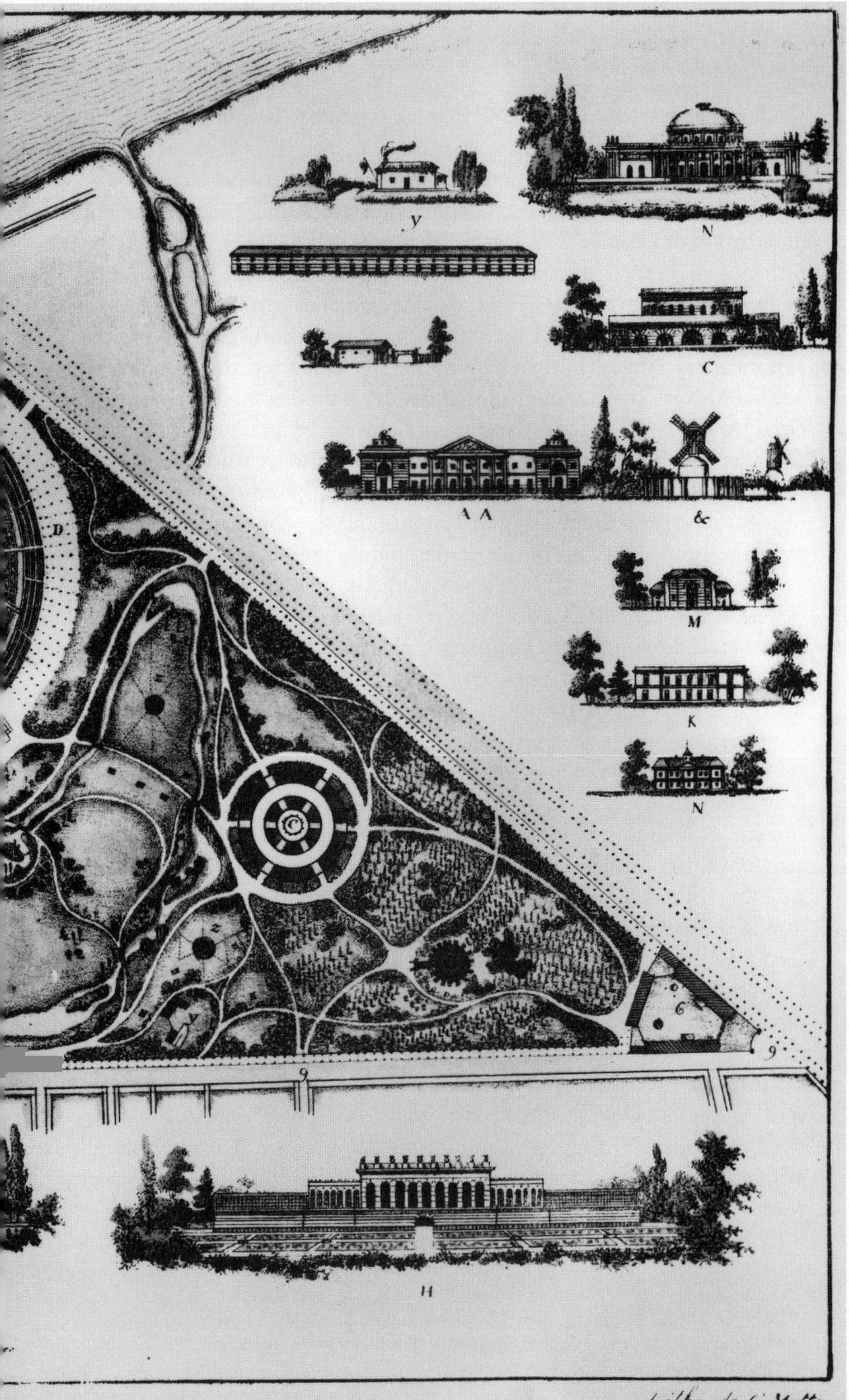

Litho. de C. Motte

rdin des plantes de Paris.

produced a new world of natural historical practice within a new political world is not denied by any historian in the field. However, there are still sweeping disagreements on the nature of these changes and the appropriate resources to be deployed in explaining them. Since the publication of Daudin's work at the beginning of the twentieth century, histories of French natural history have predominantly concentrated upon the problems of classification and nomenclature.[12] Classificatory issues were indeed part of the currency of eighteenth-century natural historical exchanges, but they by no means exhausted the meaning of natural history. In attending so exclusively to these issues, such historiography not only makes many natural historical practices and practitioners invisible, but it also replicates eighteenth-century categorizations of natural historical practice in which the "philosophical" aspects of natural history, the making of systems and the conferring of names, were the proper activity of gentlemen, while their social subordinates occupied their time with other activities.[13]

The Jardin du Roi can certainly be portrayed as a site riven with—or structured by—diverse patronage, political, and scientific commitments, and social distinctions. But to take these as the only categories for historical analysis is to perpetuate a history in which the Jardin's transformation into revolutionary, national Muséum through Jacobin agency is impossible to explain, since there is no way of speaking about how it might have been possible to construct a common ground for agreement. This last was a particularly dangerous problem for scientific activity during the Revolution. Rather than study scientific controversies, therefore, I have considered the rules operating in a particular culture of scientific (and political) agreement. In order to give an adequate account of the Jardin's transformation, I have had to shift the scene of my inquiry away from preexisting literature in two ways. In the first place, my study has paid particular attention to those practices specific to natural history which have been ignored, marginalized, or devalued since the eighteenth century. This opens the possibility of placing diverse activities, including ordering, collecting, preserving, cultivating, observing, representing, describing, and naming, on a par, although evidently I have not placed equal weight upon all of these in the present work.

The second, and more fundamental, translation in my move away

12 Daudin, *De Linné à Jussieu;* id., *Cuvier et Lamarck.*

13 Marie-Noëlle Bourguet, in "Voyage, statistique, histoire naturelle: L'inventaire du monde au XVIIIe siècle," rapport de synthèse, université de Paris I, Panthéon, Sorbonne, 1993, discussed in Pickstone, "Museological Science?" 127.

from older histories and their preoccupation with classification as the principal, characteristic mode of natural history reflects my borrowings from a range of recent models of knowledge as practice. To suggest that a study of natural historical practices does not address the problem of natural historical knowledge is to miss the implications of this book: practices did not merely "relate to" knowledge, they *constituted* knowledge in a fundamental way. Attention to the practices of natural history reveals a plethora of shared concerns among naturalists, not just at the Jardin du Roi or in France, but in Europe and the colonies. These concerns have not been addressed in most of the secondary literature on eighteenth-century French natural history, probably because they address activities which no longer appear natural historical: agricultural and horticultural improvement, animal breeding, and naturalization, common interests which tied naturalists in many places, including the Jardin, closely to the centers of governmental power. Classification, in this light, appears to be an epiphenomenon of these economic and improving activities, and, as the case of Banks in England also shows, one did not have to be a famous classifier to be a famous naturalist. The emphasis upon classification has been the consequence of a methodology which treated naturalists as isolated great thinkers, their published output the principal index of their concerns. If, on the contrary, one explores naturalists' associations, their institutional and patronage affiliations, a very different account of the meaning of their natural historical enterprise becomes possible. It should be emphasized, however, that the naturalists at the Jardin operated in many overlapping spheres. This study itself should be regarded as exploratory, and no doubt fashions many problematic boundaries by ignoring chemistry and the theory of the earth, and by focusing on naturalists' relationships within the Jardin to the exclusion of other social settings, such as the Académie Royale des Sciences and the Société d'Histoire Naturelle.

A little over a decade ago, Steven Shapin suggested a categorization of practices involved in making natural knowledge, which has since been refined upon by others.[14] Following his lead, I want to suggest that natural historical practice might be viewed as being composed, not of ideas, but of technologies. The term implies organized methods of making things happen, and Shapin used it to suggest how competing knowledge-claims can be validated in the eyes of members of a particular community. These structuring technologies, he argued, are not

14 Shapin, "Pump and Circumstance"; Roberts, "Death of the Sensuous Chemist"; Nicholas Jardine and E. C. Spary, "Introduction: The Natures of Cultural History," in *Cultures of Natural History*, ed. Jardine, Secord, and Spary, 3–13.

merely incidental to the success of a particular form of knowledge (natural philosophy, natural history) in a given culture; in a most important way, they define what that knowledge "is," making meaning and belief possible. Whether such technologies are peculiar to scientific practice or more general resources of conviction deployed in other spheres of life, they produce knowledge-claims which are by definition social, since their meaning derives not from the individual practitioner but from its acceptance or rejection by a competent community.

In chapters 1, 2, and 4 I considered the linguistic technology of natural history. Both writerly style and forms of address were put at risk by the Revolution; naturalists moved from a culture of *politesse,* gift exchange, and honorific language to a discourse of community and patriotic sentiment which was also a purified Revolutionary form of expression. Such shifts were temporary, but they were also deeply political, affecting the social form which natural historical practice took, as well as naturalists' self-presentation before other naturalists, the basis for their credibility and status, and the location of sovereignty within natural historical exchange. Honorific language implied a standard of gentlemanly conduct which was intrinsically antidemocratic even while its shared use professed to level the gardener and the count. That leveling also depended upon the perpetuation of an economy of exchange which itself was a social technology, fashioning social inequity through the deployment of material objects and depending for its ultimate success upon the perfection of a technique of subordination and domination emergent within the patron-protégé relationship.

A whole material technology was required to maintain the currency reserve upon which that economy rested, so that concerns about preservation and physical manipulation of specimens and of space were implicated within natural historical practice. But concerns to present the material world as acting in particular ways were also fundamental to classificatory enterprises such as de Jussieu's, as well as to the massive Republican enterprise of moving scientific and artistic objects around France. They were evident too in the utopian plans of naturalists and *agronomes* discussed in chapter 3, where naturalization was viewed as a means of transforming France and the French at a physical, as well as moral, level. Here the physical remodeling of the natural world was to be taken to extremes; every successfully acclimatized natural production could present material evidence of the truth of the naturalists' knowledge-claims.

But the changed world which naturalists had helped to create would also alter the role and practice of natural history itself. With the collapse of the polite world of the Old Regime, the Jardin's currency of credit and control was also destroyed. Revolutionary upheavals probably affected the majority of individuals with the money, leisure, interest, and lands to practice and patronize natural history. Personal wealth and leisure were virtual necessities for individuals seeking to acquire scientific instruction and to participate in the competitive game of specimen collecting. The Jardin du Roi's natural historical enterprise in the 1780s had required the existence of many networking individuals with access to different sources for specimens. In particular, naturalizing enterprises were under threat. Long portrayed as generalized acts of patriotism, in practice naturalization projects had been heavily promoted primarily to an audience of enlightened improving landowners, who were mostly nobles or the very wealthy: precisely those whom the Revolution damaged most. Thus the Jardin's naturalists needed to adapt natural history in their establishment, and to some extent in the rest of France, to the new conditions and opportunities with which the Revolution faced them.

Such transitions are illustrated by Thouin's seed-distribution lists. Whereas in 1792 his consignments were still overwhelmingly to botanical correspondents at other gardens, in 1794, after the decrees concerning departmental gardens, consignments to cultivators and deputies predominated.[15] The correspondence network which had formed the center of his natural historical enterprise in the Old Regime had dwindled away. But in exchange the Paris institution had become ever less dependent upon its provincial counterparts. The fame of the Muséum would allow it to stand as an emblem of French cultural superiority. By attaching their concerns to state interests, naturalists achieved greater control over the order of specimens than ever before. Even the enforced return of confiscated natural history collections to their erstwhile owners in 1815 did not entail the collapse of their enterprise, since the professors claimed it was impossible to separate out the original specimens from the national series, and substituted their own ordered duplicate collections. In place of *politesse* came a situation which held both promises and dangers for those who, their subsequent self-portrayals notwithstanding, owed their posts to the historical peculiarities of the patronage society: a handful of Burgundians, some sensible nature worshippers, several respectable and well-to-do physicians, a couple of

15 BCMHN, MS 313: "Etat de la Distribution des Graines en 1792," "Distribution des Graines en l'an 3eme de la Repub." (actually year II). Thouin's first record of a consignment for the purposes of stocking a departmental garden was on 20 pluviôse year II/February 8, 1794, to Marin, the deputy for Montblanc.

pharmacists, and an artist. They were there because of their past successes and those of their patrons within the royal world of patronage and politicking; but they could only remain in place by adroit appropriation of the symbols, language, and opportunities of the bewildering succession of new regimes which followed the storming of the Bastille.

Perhaps the greatest threat to the Jardin and to its naturalists was the Revolutionary crisis over the value of scientific expertise in France. The naturalists of the Muséum d'Histoire Naturelle needed to defend their institution and their position within it against the abandonment of state protection for the organized practice of the sciences that was being advocated by democratic extremists. The social structure of natural history had thus also to be refashioned to suit Republican life. Daubenton, acutely sensitive to the politics of scientific knowledge in the 1790s, was attentive to such issues, as the record of his lecture to the École Normale of year III reveals: "If a long life span . . . and the resource of a large collection of objects of Natural History, have made me acquire knowledge that you lack, let us attempt to reestablish equality between us."[16]

Attempting to establish certain kinds of knowledge as authoritative in a radically democratic society was very problematic. In chapter 5 I argued that the natural relations which bound families in de Jussieu's system were to be evident to the senses of habituated observers. This kind of education of the senses, which was potentially open to all with the proper bodily organization, was appropriate to a democratic public instruction. But it also focused attention on the naturalizing of mental function, since, in the wake of the democratic Republic, a hierarchy of knowledge was to be reestablished only through the claim that individuals would naturally respond differently to *éducation.* Epistemological innovations by the *idéologues,* a group of political theorists and philosophers with whom the Muséum's professors would have strong connections in the second half of the 1790s, enabled the reinvention of a natural hierarchy of knowledge which mapped onto a hierarchy of social status and authority.[17] If the categories of such taxonomies could be shown to be drawn from nature, then he who used them escaped accusations of being an aristocrat of knowledge. The conditions of practice of natural history in the First Republic thus did much to encourage agreement among naturalists about which classificatory categories were genuinely natural.

16 *La Décade philosophique* 4 (1795): 356.

17 Wilda A. Anderson, "Scientific Nomenclature and Revolutionary Rhetoric," *Rhetorica* 7 (1989): 45–53.

The *histoire naturelle* project at the Jardin/Muséum could be tailored to suit a diversity of different regimes while its practitioners continued to proclaim their search for a *unitary* Nature and for *universal* human happiness. Naturalists and legislators, elite collectors, sensible readers, radical reformers, all appealed to nature as the source of authority and of true order in society. But that which was being invoked—nature—was asked to perform differently in each case. The naturalists themselves expressed common views, albeit for a brief time, concerning the Muséum's role in a republic and their own conduct within that institution, not to mention the fact of their cooperation on a broadly defined classificatory project and their shared interest in naturalization and regeneration. A substantial secondary literature has explored the divisions arising among the Muséum's professoriate in the more stable conditions of the first decades of the new century.[18] But if the consensus of the 1790s was fleeting, it should not be interpreted as contrived or assumed. To do so is to insist that intentions are accessible to the historian, a problematic claim. In the same way, historians have often suggested that someone who afterward repented of his Jacobin ties had always been a covert moderate. This is to assert the transparency of biographical accounts written after 9 thermidor, thus under particular requirements for self-justification. It seems more fruitful to view eighteenth-century savants as having multiple allegiances and multiple roles. This practice was advantageous in the patronage society of the old regime, but was potentially deadly in the Revolution. One of the most coherent "programs" in the sciences of Old Regime France, Lavoisier's new chemistry, failed to enforce a single identity for chemical practitioners which, in the eyes of republican deputies, superseded other ways of measuring status. Lavoisier was executed—as a tax farmer, not a chemist—in 1794. While Lavoisier was guillotined, his chemical supporters Fourcroy, Hassenfratz, and Guyton de Morveau were heavily involved in the Comité de Salut Public and Comité d'Instruction Publique's efforts to legislate on the sciences and public instruction.[19] In other words, political categories were not being generated by savants. For these reasons I have tried to adopt a slightly different approach to discussing the relation between scientific practice and political allegiance in the Revolutionary period from that hitherto espoused in most

18 Outram, *Cuvier;* Corsi, *Age of Lamarck;* Appel, *Cuvier-Geoffroy Debate;* A. W. Brown, "Some Political and Scientific Attitudes to Literature and the Arts in the Years Following the French Revolution," *Forum for Modern Language Studies* 2 (1966): 230–252; and others already mentioned.

19 Dhombres, *Les savants en révolution,* 105.

secondary accounts, often written by historians seeking to defend particular scientific practitioners from charges of "left-wing" or "right-wing" allegiances. Precisely because of the transient and fluctuating nature of the political world in the period between 1789 and 1795, it seemed essential to treat political categories as equally ephemeral. I am not, therefore, asserting that all the Muséum's naturalists were Jacobins. Nor am I following a commonly held view that several of them were "moderates" forced into positions of political prominence by the pressures of the Terror. The content of particular political positions was being invented in this very period; political allegiances were not universal, timeless givens that we can unproblematically impose upon participants in the debates. Political categories, after all, emerged out of these negotiations, rather than determining them. To treat them otherwise is effectively to deny the open-endedness of such negotiations.

The mingling of savant roles with political status was particularly problematic after the fall of Robespierre. Almost immediately it began to be necessary for functionaries who had formerly helped to create or implement Jacobin policies to renounce their earlier political involvement. Savants thus worked to unravel their affinities and affiliations as scientific practitioners from their public existence. Accordingly, many later historians have repeated the Revolutionary savants' accounts of themselves as politically singular (when in fact their political affiliations were usually plural) and scientifically universal (when in fact their savant circles were very restricted and entry into them carefully policed). The sciences themselves could be protected against the dangers of political turbulence only by subsequent portrayals of them as politically neutral activities.[20] Savants were continually engaged in rewriting their political roles after 1789, but such reworkings were not always very successful. For example, Fourcroy was haunted by claims that he had allowed personal motives to drive him to use his political power to intervene in the savant world, by failing to save Lavoisier in 1794—in other words, he had mixed three incompatible categories of being, the personal, savant, and political.[21] Among the Muséum's major supporters were deputies who themselves tailored their identity to successive regimes, particularly Joseph Lakanal. But it also won support from the

20 Outram, "Ordeal of Vocation."

21 Outram, *Cuvier,* chapters 4 and 5; *NBU,* s.v. "Fourcroy"; Kersaint, "Fourcroy"; Smeaton, *Fourcroy;* Alexis Eymery, *Dictionnaire des Girouettes, ou Nos contemporains peints d'après eux-mêmes: ouvrage dans lequel sont rapportés des discours . . . écrits sous les gouvernemens qui ont eu lieu en France depuis vingt-cinq ans . . .* (Paris: A. Eymery, 1815).

radical reformers of public space during the Comité de Salut Public's period of power.

Through the Republic and First Empire the Muséum d'Histoire Naturelle never ceased to perform as a cultural instrument of the state, although the professors would lose their scientific supremacy by the foundation of the Institut National des Sciences et des Arts in 1795, and their right of self-determination by an imperial decree promulgated in year X.[22] Thus one major shift in the institution's function, which arguably predated the Revolution, was its re-creation as a body answerable to a modern abstraction, the nation, rather than as a mirror for an early modern manifestation of power, the absolute monarch. Although described in much museological theory, little attention has been paid to the mechanics of this transition as manifested within collections.[23] The Muséum d'Histoire Naturelle offers a particularly useful setting for such an inquiry, since it was judged to be a model scientific institution by numerous nineteenth-century writers. Its value as an innovation was evident to contemporaries even while they rejected its political origins.

It was crucial to the Muséum's future after 1795 that its institutional existence and meaning should be separable from the particular Republican context of its creation and the particularly Republican content of its practitioners' claims. If naturalists' claims concerning nature were to retain credibility in other political settings, both their expert status and their freedom from political dependency urgently needed to be re-created. Foremost among those involved in that process was a newcomer to the Muséum, Georges Cuvier, about whom much has been written.[24] But Cuvier was notoriously *un*successful in creating a space for scientific practice which was universally agreed to be politically neutral. In fashioning a new and conservative version of natural history, he explicitly countered the republican implications of the accounts of the natural world successively advanced by his colleagues Lamarck and Geoffroy Saint-Hilaire. In so doing he involved himself in three de-

22 AN, AJ/15/916. The professors' power to appoint their own colleagues was abolished by a law of year X/1802, imposing ministerial control over the choice of new candidates. AJ/15/539 reveals the abrupt alteration of tone in official communications to the Muséum over the two years that elapsed between Dolomieu's and Haüy's appointments as professors of geology.

23 One fruitful line of approach is exemplified by Crow, *Painters and Public Life.* But see also Poulot, "Musée et société"; id., "Le Louvre imaginaire"; Harten, *Museen und Museumsprojekte.* The conflicts created by the rapidly changing uses of the Muséum immediately after its foundation are explored in Spary, "Forging Nature at the Republican Muséum."

24 Outram, *Cuvier,* chapters 4 and 5.

cades of very public and highly politicized conflict within the Muséum's professoriate. The famous disputes among naturalists at the Muséum in the 1820s and 1830s revolved around two principal issues: the extent to which living beings could transform themselves through self-exertion within a measurable time span, and the possibility that great upheavals or revolutions of the globe could effect the total extinction of particular species. It is only in learning how, in previous decades, the Muséum d'Histoire Naturelle had already served as a principal site for constructing the Revolution as a *natural* phenomenon, that we are able to comprehend the stakes involved in natural history, long before the outbreak of open controversy. Discussions of the nature of the earth, of living bodies, and of their interrelations, possessed political significance from the word go. The naturalists' concerns to naturalize perfectibility and revolution in the nineteenth century, however, reflected their profoundly different modes of appropriation of the political Revolution which they had recently experienced.[25]

25 Appel, *Cuvier-Geoffroy Debate;* Outram, *Cuvier;* Desmond, *Politics of Evolution,* 1–100; Richard W. Burkhardt Jr., "Lamarck, Evolution, and the Politics of Science," *Journal for the History of Biology* 3 (1970): 275–298; id.,"The Inspiration of Lamarck's Belief in Evolution," *Journal for the History of Biology* 5 (1972): 413–438; William Coleman, *Georges Cuvier, Zoologist: A Study in the History of Evolution Theory* (Cambridge: Harvard University Press, 1964); Laurent, *Paléontologie et évolution en France,* chapters 1 and 2. What are widely referred to in the secondary literature as "geological catastrophes" were, of course, known as "revolutions of the globe" to French naturalists writing in the early 1800s.

APPENDIX

Holders of Scientific Posts at the Jardin du Roi/Muséum d'Histoire Naturelle, 1750–1793

Georges-Louis Leclerc de BUFFON (1707–1788), ennobled to comte de Buffon, 1771; member, Académie Royale des Sciences and Société d'Agriculture de Paris
1739–1788, intendant

Auguste-Charles-César de flahaut de LA BILLARDERIE, marquis d'Angiviller (1724–1793)
1788–January 1792, intendant

Jacques-Henri BERNARDIN DE SAINT-PIERRE (1737–1814)
1792–1793, intendant

Antoine DE JUSSIEU (1686–1758), D.med.; member, Académie Royale des Sciences
1711–1758, démonstrateur de l'intérieur des plantes sous le titre de professeur en botanique

Louis-Guillaume LE MONNIER (1717–1799), premier médecin du roi, 1770; D.med.; member, Académie Royale des Sciences
1759–1786, professeur de botanique

René-Louiche DESFONTAINES (1750–1833), D.med.; member, Académie Royale des Sciences and Société d'Agriculture de Paris
1786–1833, professeur de botanique

Louis-Claude BOURDELIN (1696–1777), D.med.; member, Académie Royale des Sciences
1743–1777, professeur de chimie; frequently seconded by Paul-Jacques MALOUIN (1701–1777)

Based on Laissus, "Le Jardin du Roi."

Pierre-Joseph MACQUER (1718–1784), D.med.; member, Académie Royale des Sciences
1777–1784, professeur de chimie
Antoine-François de FOURCROY (1755–1809), D.med.; member, Académie Royale des Sciences and Société d'Agriculture de Paris
1784–1793, professeur de chimie aux écoles du Jardin Royal des Plantes
1793–1809, professeur de chimie générale
Antoine FERREIN (1693–1769), D.med.; member, Académie Royale des Sciences
1751–1769, démonstrateur et opérateur des opérations pharmaceutiques
Antoine PETIT (1722–1794), D.med.; member, Académie Royale des Sciences
1769–1778, professeur royal d'anatomie au jardin royal
Antoine PORTAL (1742–1832), D.med.; member, Académie Royale des Sciences
1778–1793, professeur d'anatomie aux écoles du Jardin Royal des Plantes
1793–1832, professeur d'anatomie humaine
Bernard DE JUSSIEU (1699–1777), D.med.; member, Académie Royale des Sciences
1722–1777, sous-démonstrateur de l'extérieur des plantes
Antoine-Laurent DE JUSSIEU (1748–1836), D.med.; member, Académie Royale des Sciences
1778–1793, sous-démonstrateur de l'extérieur des plantes
1793–1826, professeur de botanique dans les champs
Jean-André THOUIN (d. 1764)
1745–1764, jardinier en chef
André THOUIN (1747–1824), member, Académie Royale des Sciences and Société d'Agriculture
1764–1793, jardinier en chef 1793–1824, professeur de culture
Guillaume-François de ROUELLE (1703–1770), maître apothicaire; member, Académie Royale des Sciences
1743–1768, démonstrateur en chimie sous le titre de professeur en chimie
Hilaire-Marin de ROUELLE (1718–1779), apothecary to duc d'Orléans
1768–1779, démonstrateur de chimie
Antoine-Louis BRONGNIART (1742–1804), premier apothicaire to Louis XVI
1779–1793, démonstrateur de chimie aux écoles du jardin royal
1793–1804, professeur des arts chimiques

Louis-Jean-Marie DAUBENTON (1716–1799), D.med.; member, Académie Royale des Sciences and Société d'Agriculture de Paris
1745–1793, garde et démonstrateur du cabinet d'histoire naturelle et du jardin royal
1793–1799, professeur de minéralogie
Antoine MERTRUD (d. 1764)
1748–1764, démonstrateur en anatomie et chirurgie
Jean-Claude MERTRUD (d. 1787)
1764–1787, démonstrateur en anatomie et chirurgie
Antoine-Louis-François MERTRUD (d. 1802), D.med.
1787–1793, démonstrateur en anatomie et chirurgie
1793–1802, professeur d'anatomie animale
Edmé-Louis DAUBENTON (1732–1785)
1766–1784, garde et sous-démonstrateur du cabinet d'histoire naturelle du roi
Bernard-Germain-Étienne de La-Ville-sur-Illon, comte de LACEPÈDE (1756–1825)
1784–March 1793, garde et sous-démonstrateur du cabinet d'histoire naturelle
1794–1825, professeur de zoologie (quadrupèdes ovipares, cétacés et serpens)
Étienne GEOFFROY SAINT-HILAIRE (1772–1844)
1792–March 1793, adjoint à la garde du cabinet d'histoire naturelle
March–June 1793, garde et sous-démonstrateur du cabinet d'histoire naturelle
1793–1841, professeur de zoologie (quadrupèdes et oiseaux)
Barthélémi FAUJAS DE SAINT-FOND (1741–1819)
1787–1793, adjoint à la garde du cabinet, chargé de la correspondance
1793–1819, professeur de géologie
Jean-Baptiste-Pierre-Antoine de Monet, chevalier de LAMARCK (1744–1829), member, Académie Royale des Sciences
1778–1789, correspondent of Jardin du Roi
1789–1793, botaniste du roi attaché au cabinet d'histoire naturelle
1793–1829, professeur de zoologie des insectes, vers et animaux microscopiques
Magdeleine-Françoise BASSEPORTE (1701–1780)
1735–1780, peintre en miniature
Gerard VANSPAENDONCK (1746–1822)
1780–1793, peintre en miniature, having possessed letters of succession since 1774
1793–1822, professeur d'iconographie (after 1785, his painting duties largely carried out by Pierre-Joseph REDOUTÉ)

BIBLIOGRAPHY

Primary Sources

Manuscript

Archives Nationales, Paris. AJ/15/95–103, AJ/15/149, AJ/15/502–503, AJ/15/505–507, AJ/15/509–512, AJ/15/514, AJ/15/521, AJ/15/539, AJ/15/565, AJ/15/577, AJ/15/836, AJ/15/847, AJ/15/879, AJ/15/916, D VI 26, F/10/201, F/10/210, F/10/366, F/12/1223–1224, F/14/187B, F/14/189B, F/17/1050, F/17/1219, F/17/1221, F/17/1223, F/17/1224, F/17/1229, F/17/1233, F/17/1238, F/17/1310, F/17/3880, F/17/3903.

Bibliothèque Centrale du Muséum d'Histoire Naturelle, Paris. MSS 47, 56, 216, 218, 219, 284, 308, 312–315, 318, 352, 357, 457, 471, 691, 807, 870, 882, 1197, 1321, 1384, 1459, 1529, 1905, 1971–1985, 2323.

Bibliothèque du Port de Brest. MS 1507.

Huntington Library, San Marino, CA. HM 9017.

Printed

Adanson, Michel. *Familles des Plantes.* 2 vols. Paris: Vincent, 1763.

———. "Mémoire sur un mouvement particulier découvert dans une plante appellé *Tremella.*" *MARS* 1767/1770, 564–589.

———. "Examen de la question: si les espèces changent parmi les plantes; nouvelles expériences tentées à ce sujet." *MARS* 1769/1772, 31–48.

Adresse des Naturalistes, à l'Assemblée Nationale, du 5 Août 1790, soir. Paris: Imprimerie Nationale, 1790.

Adresses et Projet de Règlemens présentés à l'Assemblée Nationale par les officiers du Jardin des Plantes et du Cabinet d'Histoire Naturelle, d'Après le Décret de l'Assemblée Nationale, du 20 Août 1790. Paris: Buisson, 1790.

Andry, Charles-Louis-François. "Recherches sur la mélancholie." *HSRM* 1783/1787, 89–159.

Andry, Charles-Louis-François, and Michel-Augustin Thouret. "Observations et Recherches sur l'usage de l'aimant en médecine, ou Mémoire sur le Magnetisme animal." *HSRM* 1779/1782, 531–688.

[Ane (possibly Charles-Joseph Panckoucke)]. *De l'homme, et de la reproduction des differens individus. Ouvrage qui peut servir d'introduction & de défense à l'Histoire naturelle des animaux par M. de Buffon.* Paris: n.p., 1761.

Aulard, François-Victor-Alphonse, ed. *Recueil des actes du Comité de Salut Public, avec la correspondance officielle des représentants en mission et le registre du Conseil Exécutif Provisoire.* 28 vols. Paris: Imprimerie Nationale, 1889–1951.

Bailly, Jean-Sylvain, Benjamin Franklin, Jean-Baptiste Le Roy, Gabriel de Bory, and Antoine-Laurent de Lavoisier. "Exposé des Expériences qui ont été faits pour l'examen du Magnétisme animal." *HARS* 1784/1787, 6–15.

Baras, Marc-Antoine. *L'état actuel des établissemens destinés à l'instruction publique.* Paris: Imprimerie de la Convention Nationale, 1793.

Beauharnais, Marie-Anne-Françoise, comtesse de. "Aux incrédules. Épitre envoyée à M. le Comte de Buffon." *Journal de Paris,* November 7, 1778.

Bernardin de Saint-Pierre, Jacques-Henri. *Études de la nature.* 4th ed. Paris: Didot, 1792.

———. *Paul and Virginia. Elizabeth. The Indian Cottage. The Whole Newly Translated.* London: J. F. Dove, 1828.

———. *Voyage à l'Île de France: un officier du roi à l'Ile Maurice, 1768–1770.* Introduction by Yves Bénot. Paris: La Découverte/Maspero, 1983.

Bertholon, Pierre-Joseph. *De l'electricité du corps humain dans l'état de santé et de maladie; ouvrage couronné par l'Académie de Lyon, dans lequel on traite de l'électricité de l'Atmosphere, de son influence & de ses effets sur l'économie animale, des vertus médicales de l'électricité, des découvertes modernes & des différentes méthodes d'électrisation.* Paris: P. F. Didot, 1780.

———. *De l'électricité des végétaux. Ouvrage dans lequel on traite des effets de l'électricité de l'athmosphère sur les plantes, de ses effets sur l'économie des végétaux, de leurs vertus médico- & nutritivo-électriques, & principalement des moyens de pratique de l'appliquer utilement à l'agriculture, avec l'invention d'un électro-végétometre.* 1784. Lyon: Bernuset, 1788.

———. *De la salubrité de l'air des villes, et en particulier des moyens de la procurer.* Montpellier: J. Marcel Aîné, 1786.

Bertin, Exupère-Joseph. *Des moyens de conserver la santé des blancs et des nègres, aux Antilles ou climats chauds et humides de l'Amérique.* Paris: Méquignon, 1786.

Bobé, Louis, ed. *Mémoires de Charles Claude Flahaut, comte de la Billarderie d'Angiviller: Notes sur les mémoires de Marmontel publiés d'aprés le manuscrit.* Copenhagen: Levin and Munksgaard, 1933.

Boisset, Joseph-Antoine. *Rapport et projet de décret relatifs à l'établissement des jardins des plantes dans les departemens par Boisset, membre du Comité de commerce.* Paris: Imprimerie Nationale, [1794].

———. *Notes sur la necessité d'établir un jardin des plantes dans chaque département, faisant suite d'un rapport sur le même sujet . . . Apperçu de la dépense annuelle des Jardins de Botanique, projetés dans les départemens de la République.* [Paris]: Imprimerie Nationale, [1794].

Bordeu, Théophile de, Antoine de Bordeu, and François de Bordeu. *Recherches sur les maladies chroniques, leurs rapports avec les maladies aiguës, leurs périodes, leur nature: & sur la manière dont on les traite aux eaux minérales de Bareges, & des autres sources de l'Aquitaine.* Paris: A. Ghio, 1775.

Bosc d'Antic, Louis-Guillaume. "Coturnix Ypsilophorus." *Journal d'Histoire naturelle* 2 (1794): 297–298.

———. "Ripiphorus." *Journal d'Histoire naturelle* 2 (1794): 293–296.

Bougainville, Louis-Antoine de. *Voyage autour du monde par la frégate du roi La Boudeuse et la flûte l'Etoile, en 1766, 1767, 1768, & 1769.* 2d ed. 2 vols. Paris: Saillant et Nyon, 1772.

Brieude, Jean-Joseph de. "Topographie médicale de la haute Auvergne." *HSRM* 1783/1787, 257–340.

Broussonet, Pierre-Marie-Augustin. "Mémoire pour servir à l'histoire de la respiration des Poissons." *MARS* 1785/1788, 174–196.

———. "Observations sur la régénération de quelques parties du corps des Poissons." *MARS* 1786/1788, 684–688.

———. *Réflexions sur les avantages qui résulteroient de la réunion de la Société royale d'agriculture, de l'Ecole vétérinaire et de trois chaires du Collège Royal au Jardin du Roi, par P.-M.-A. Broussonet.* Paris: Imprimerie du Journal Gratuit, [1790].

Bucquet, Jean-Baptiste-Michel. "Mémoire sur la manière dont les animaux sont affectés par différens fluides aériformes méphitiques, et sur les moyens de remédier aux effets de ces fluides." *HSRM* 1776/1779, 177–192.

Buffon, Georges-Louis Leclerc, comte de. *Oeuvres philosophiques de Buffon.* Ed. Jean Piveteau. Paris: Presses Universitaires de France, 1954.

———. *Buffon: Les Époques de la nature: Édition critique.* Ed. Jacques Roger. Mémoires du Muséum National d'Histoire Naturelle, series C, Sciences de la terre, 10. Paris: Muséum National d'Histoire Naturelle, 1988.

Buffon, Georges-Louis Leclerc, comte de, Louis-Jean-Marie Daubenton, Philibert Guéneau de Montbeillard, and Gabriel-Léopold-Charles-Aimé Bexon. *Histoire naturelle, générale et particulière, avec la description du Cabinet du Roi.* 36 vols. Paris: Imprimerie Royale, 1749–1789.

Buffon, Henri Nadault de, ed. *Buffon: Correspondance générale recueillie et annotée par H. Nadault de Buffon.* 2 vols. Geneva: Slatkine, 1971.

Catalogue des livres de la Bibliothèque de feu M. A. F. de Fourcroy. Paris: Tilliard frères, 1810.

Catesby, Mark. *The Natural History of Carolina, Florida, and the Bahama Islands/Histoire Naturelle de la Caroline, la Floride, & les Isles Bahamas.* London: Catesby, 1731–1743.

Condillac, Étienne Bonnot de. *Essai sur l'origine des connaissances humaines.* 2 vols. Amsterdam: Pierre Mortier, 1746.

———. *Traité des sensations, à Madame la comtesse de Vassé.* 2 vols. London and Paris: De Bure l'aîné, 1754.

———. *Traité des animaux, où après avoir fair des observations critiques sur le sentiment de Descartes et sur celui de M. de Buffon, on entreprend d'expliquer leurs principales facultés.* 1755. Ed. François Dagognet. Paris: Vrin, 1987.

———. *Cours d'étude pour l'instruction du Prince de Parme, aujourd'hui S. A. R. l'infant D. Ferdinand; Duc de Parme.* 16 vols. Parma: Imprimerie Royale, 1775.

Condorcet, Marie-Jean-Antoine-Nicolas Caritat, marquis de. *Du commerce des bleds, pour servir à la réfutation de l'ouvrage sur la législation et le commerce des grains.* Paris: Grangé, 1775.

———. *Réflèxions sur le commerce des bleds.* London: n.p., 1776.

———. "Éloge de M. de Jussieu." *HARS* 1777/1780, 94–117.

———. "Éloge de M. de Linné." *HARS* 1778/1781, 66–82.

———. "Éloge de M. Duhamel." *HARS* 1782/1785, 131–155.

———. "Éloge de M. Macquer." *HARS* 1784/1787, 20–30.

———. "Éloge de M. le Comte de Buffon." *HARS* 1788/1790, 50–84.

———. "Éloge de Camper." *HARS* 1789/1793, 45–52.

———. "Éloge de Fougeroux." *HARS* 1789/1793, 39–44.

———. "Éloge de Turgot." *HARS* 1789/1793, 31–38.

———. *Esquisse d'un tableau historique des progrès de l'esprit humain.* Paris: Agasse, year III/1794.

———. *Oeuvres complètes de Condorcet.* 21 vols. Brunswick: Vieweg, year XIII/1801.

———. *Condorcet: Écrits sur l'instruction publique.* 2 vols. Annotated and introduced by Charles Coutel and Catherine Kintzler. Paris: Edilig, 1991.

Coupé, Jacques-Michel. *De l'amélioration générale du sol français dans ses parties négligées ou dégradées.* Paris: n.p., year III/1794.

———. *De la tenue des bois.* Paris: Imprimerie Nationale, year III/1794.

———. *Des animaux de travail et de leur tenue.* [Paris]: Imprimerie Nationale, year III/1794.

Cranston, Maurice, ed. *Rousseau Selections.* New York: Macmillan, 1988.

Creuzé-Latouche, Jacques-Antoine. *Opinion de M. J.-A. Creuzé-Latouche, Membre de l'Assemblée Nationale, au sujet du Jardin des Plantes et des Académies.* Paris: Imprimerie Nationale, 1790.

Crosland, Maurice P., ed. *Science in France in the Revolutionary Era Described by Thomas Bugge, Danish Astronomer Royal and Member of the International Commission on the Metric System, 1798–1799.* Cambridge, MA: Society for the History of Technology and MIT Press, 1969.

Cuvier, Georges. "Mémoire sur les cloportes terrestres." *Journal d'Histoire naturelle* 2 (1794): 18–30.

———. "Mémoire sur la structure interne et externe, et sur les affinités des animaux auxquels on a donné le nom de vers." *La Décade philosophique* 5 (1795): 385–424.

———. *Tableau élémentaire de l'histoire naturelle des animaux.* Paris: Baudoin, year VI/1797.

———. *Notice historique sur Daubenton, lue à la séance publique de l'Institut national de France du 15 germinal an 8.* Paris: Institut National, year IX/1800.

———. *Recueil des éloges historiques des membres de l'Académie Royale des Sciences.* 3 vols. Paris: Levrault, 1827. Facsimile reprint, Brussels: Culture et Civilisation, 1969.

———, ed. *Histoire naturelle de Buffon, mise dans un nouvel ordre, précédée d'un notice sur la vie et les ouvrages de cet auteur.* 36 vols. Paris: Ménard et Desenne, 1825–1826.

———. *Oeuvres complètes de Buffon.* Paris: Baudoin frères, 1826–1836.

Cuvier, Georges, and Étienne Geoffroy Saint-Hilaire. "Mémoire sur les rapports naturels du Tarsier (*Didelphis Macrotarsus* Gm.), lu à la société d'Histoire naturelle, le 21 messidor, an troisième, par les citoyens Cuvier et Geoffroy." *Magazin Encyclopédique* 2 (1795): 147–154.

Daire, Eugène, ed. *Physiocrates: Quesnay, Dupont de Nemours, Mercier de La Rivière, l'abbé Baudeau, Le Trosne, avec une introduction sur la doctrine des physiocrates, des commentaires et des notices historiques.* 2 vols. Paris: Guillaumin, 1846.

———. *Collection des principaux économistes: Mélanges d'économie politique,* 2 vols. Paris: Guillaumin, 1847.

D'Alembert, Jean Le Rond. "Climat (Géog.)." In *Encyclopédie, ou Dictionnaire raisonné des sciences, des arts et des métiers,* ed. Denis Diderot and Jean Le Rond d'Alembert. Paris: Briasson, 1753, 3:532–534.

———. "Éloge de M. le baron de Montesquieu." In *Encyclopédie, ou Dictionnaire raisonné des sciences, des arts et des métiers,* ed. Denis Diderot and Jean Le Rond d'Alembert. Paris: Briasson, 1755, 5:iii–xviii.

Darwin, Erasmus. *The Botanic Garden; a poem, in two parts. Part I. containing the Economy of Vegetation. Part II. The Loves of the Plants. With Philosophical Notes.* London: J. Johnson, 1791.

———. *Zoonomia, or The Laws of Organic Life.* 2 vols. London: Johnson, 1794.

Daubenton, Louis-Jean-Marie. "Botanique." In *Encyclopédie, ou Dictionnaire raisonné des sciences, des arts et des métiers,* ed. Denis Diderot and Jean Le Rond d'Alembert. Paris: Briasson, 1751, 2:340–345.

———. "Description (Hist. nat.)." In *Encyclopédie, ou Dictionnaire raisonné des sciences, des arts et des métiers,* ed. Denis Diderot and Jean Le Rond d'Alembert. Paris: Briasson, 1754, 4:878.

———. "Méthode." In *Encyclopédie, ou Dictionnaire raisonné des sciences, des arts et des métiers,* ed. Denis Diderot and Jean Le Rond d'Alembert. Paris: Briasson, 1765, 10: 458–460.

———. "Mémoire sur les os et les dents remarquables par leur grandeur." *MARS* 1762/1766, 206–229.

———. "Mémoire sur les différences de la situation du grand trou occipital dans l'homme et dans les animaux." *MARS* 1764/1767, 568–602.

———. "Mémoire sur le mécanisme de la rumination, et sur le tempérament des bêtes à laine." *MARS* 1768/1770, 389–398.

———. "Observations sur les bêtes à laine parquées pendant toute l'année." *MARS* 1772/1775 (1), 436–444.

———. "Observations sur l'animal qui porte le musc, et sur ses rapports avec les autres Animaux." *MARS* 1772/1775 (2), 215–220.

———. "Mémoire sur les remèdes les plus nécessaires aux troupeaux." *HSRM* 1776/1779, 312–320.

———. "Mémoire sur l'amélioration des bêtes à laine." *MARS* 1777/1780, 79–87.

———. "Mémoire sur le régime le plus nécessaire aux troupeaux, dans lequel l'auteur détermine par des expériences ce qui est relatif à leurs alimens et à leur boisson." *HSRM* 1777–1778/1780 (2), 570–578.

———. "Mémoire sur les laines de France, comparées aux laines étrangères." *MARS* 1779/1782, 1–11.

———. "Observation sur la disposition de la trachée-artère de différentes espèces d'Oiseaux, & sur-tout de l'oiseau appelé *Pierre.*" *MARS* 1781/1784, 369–376.

———. *Histoire naturelle des Animaux.* Paris: Panckoucke, 1782. Part of *Encyclopédie méthodique, ou par ordre de matières. Par une Société de gens-de-lettres, de savans, et d'artistes.* 190 vols. Paris and Liège, 1782–1830.

———. *Instruction pour les Bergers et pour les Propriétaires de Troupeaux.* Paris: Ph.-D. Pierres, 1782.

———. *Tableau méthodique des minéraux, suivant leurs différentes natures, et avec des caractères distinctifs, apparens, ou faciles à reconnoître.* Paris: n.p., 1784.

———. "Mémoire sur le premier Drap de Laine superfine du crû de la France." *MARS* 1784/1787, 76–84.

———. "Observations sur la comparaison de la nouvelle Laine superfine de France, avec la plus belle Laine d'Espagne, dans la fabrication du Drap." *MARS* 1785/1788, 454–464.

———. "Mémoire sur la pierre de poix, *Pechstein* des Allemands." *MARS* 1787/1789, 86–91.

———. "Mémoire sur les pierres figurées et principalement sur la pierre de Florence." *Magazin Encyclopédique* 1 (1794): 38–45.

———. *Instruction pour les Bergers et pour les Propriétaires de Troupeaux; Avec d'autres Ouvrages sur les Moutons et sur les Laines.* 3d ed. Annotated by J. B. Huzard. Paris: Imprimerie de la République, year X/1802.

Daubenton, Louis-Jean-Marie, Mathieu Tillet, Jean-Sylvain Bailly, Antoine-Laurent de Lavoisier, Pierre-Simon Laplace, Charles-Augustin de Coulomb, and Jean D'Arcet. "Rapport des Mémoires et Projets pour éloigner les Tueries de l'Intérieur de Paris." *MARS* 1787/1789, 19–43.

Daubenton, Louis-Jean-Marie, Félix Vicq d'Azyr, and Henri-Alexandre Tessier. "Sur les maladies des moutons." *HSRM* 1777–1778/1780, 157–161.

Daubenton, Marguérite. *Zélie dans le désert; par Madame D. . . . Nouvelle Edition, avec un supplément.* 4 vols. in 2. Geneva: n.p., 1793.

Dazille, Jean Barthélémy. *Observations générales sur les maladies des climats chauds, leur cause, leur traitement, et les moyens de les prévenir.* Paris: P. F. Didot le jeune, 1785.

La Décade Politique, Philosophique, et Littéraire; par une Société de Républicains. Vols. 1–6. Paris Bureau de la Décade, years II–IV/1793–1796.

Décret sur le Jardin National des plantes, le Cabinet d'histoire naturelle de Paris, du 10 juin 1793, l'an deuxième de la République; Précédé du Rapport du Citoyen Lakanal, Député de l'Arriège à la Convention, Membre du Comité d'instruction publique; imprimé par ordre de la Convention Nationale. Paris: Imprimerie de la Convention Nationale, 1793.

Dehorne, N., Jean-Noël Hallé, Antoine-François de Fourcroy, and Michel-Augustin Thouret. "Rapport sur la voierie de Montfaucon." *HSRM* 1786/1790, 198–226.

Delamétherie, Jean-Claude. *Essai analytique sur l'air pur, et les différentes espèces d'air.* 2d ed. 2 vols. Paris: Cuchet, 1788.

Deleuze, Joseph-Philippe-François. *Histoire et description du Muséum Royal d'Histoire Naturelle.* 2 vols. Paris: Levrault, 1823.

Description anatomique d'un caméléon, d'un castor, d'un dromadaire, deux ours et une gazelle. Paris: Léonard, 1669.

Description sommaire d'un cabinet d'histoire naturelle, provenant de la succession de feu M. de Joubert, ancien trésorier des états de Languedoc, membre et correspondant de plusieurs académies. Paris: Imprimerie Nationale, [1792].

Desfontaines, René-Louiche. "Mémoire sur quelques nouvelles espèces d'oiseaux des côtes de Barbarie." *MARS* 1787/1789, 496–505.

———. "Observations sur l'irritabilité des organes sexuels d'un grand nombre de plantes." *MARS* 1787/1789, 468–480.

———. "Recherches sur un Arbrisseau connu des Anciens sous le nom de *Lotos de Lybie.*" *MARS* 1788/1790, 443–453.

———. "Histoire naturelle. Cours de botanique élémentaire et de physique végétale. Discours d'ouverture." *La Décade philosophique* 5 (1795): 449–461, 513–520; 6 (1796): 1–11, 129–143, 193–202, 321–330, 449–453.

Desperrières, Antoine. "Recherches sur les causes des maladies des gens de mer." *HSRM* 1786/1790, 105–113.

Desplaces de Montbron, Laurent-Benoît. *Preservatif contre l'agromanie, ou l'Agriculture réduite à ses vrais principes.* Paris: Hérissant, 1762.

Dezallier d'Argenville, Antoine-Joseph. *La Conchyliologie, ou Histoire Naturelle des Coquilles de mer, d'eau douce, terrestres et fossiles, avec un Traité de la Zoomorphose, ou représentation des Animaux qui les habitent.* 3d ed. 2 vols. Ed. Jacques Favanne de Montcervelle and Guillaume Favanne de Montcervelle. Paris: De Bure, 1780.

Diderot, Denis. "Cabinet d'Histoire naturelle." In *Encyclopédie, ou Dictionnaire raisonné des sciences, des arts et des métiers,* ed. Denis Diderot and Jean Le Rond d'Alembert. Paris: Briasson, 1751, 2:489–492.

[Diderot, Denis?]. "Histoire naturelle." In *Encyclopédie, ou Dictionnaire raisonné des sciences, des arts et des métiers,* ed. Denis Diderot and Jean Le Rond d'Alembert. Paris: Briasson, 1765, 8:225–230.

Diderot, Denis, and Jean Le Rond d'Alembert, eds. *Encyclopédie, ou Dictionnaire raisonné des sciences, des arts et des métiers, par une société de gens de lettres.* 35 vols. Paris: Briasson, 1751–1776. Facsimile reprint, Oxford: Pergamon, 1969.

Dourneau-Démophile. *Couplets civiques pour l'inauguration des Bustes de Francklin, Voltaire, Buffon, Jean-Jacques Rousseau, Marat et Pelletier, dans la salle de la Société populaire & Républicaine d'Avre-Libre (ci-devant Roye), Département de la Somme.* Paris: Moutard, year II/1793.

Dubois, Pierre-Marie-Auguste Broussonet, Antoine-Augustin Parmentier, Jean-Baptiste Dubois de Jancigny, and Jean-Laurent Lefebvre, eds. *Feuille du cultivateur.* 7 vols. Paris: Bureau de la Feuille du Cultivateur, 1790–1798.

Dubois, Dieudonné. *Rapport fait . . . au nom d'une commission, sur un Message du Directoire exécutif du 15 nivôse dernier, tendant au rapport de la loi du 9 du même mois, interprétative de celle du 17 prairial de l'an 4, concernant les terreins destinés à l'agrandissement du muséum d'histoire naturelle.* [Paris: n.p., year V/1797.]

Duchesne, Antoine-Nicolas. *Histoire naturelle des fraisiers, contenant les vues d'économie réunies à la botanique; & suivie de remarques particulières sur plusieurs points qui ont rapport à l'histoire naturelle générale.* Paris: P.-F. Didot le jeune, 1766.

Duhamel du Monceau, Henri-Louis, J.-P.-F. Guillot, Étienne Mignot de Montigny, Jean-Baptiste Le Roy, Jacques-René Tenon, Mathieu Tillet, and Antoine-Laurent de Lavoisier. "Rapport Fait à l'Académie Royale des Sciences, sur les Prisons, le 17 mars 1780." *MARS* 1780/1783, 409–424.

Duhamel du Monceau, Henri-Louis. *Traité sur la culture des terres, suivant les principaux de M. Tull, Anglais.* 6 vols. Paris: Guérin et Delatour, 1750–1761.

———. *Elémens de l'Agriculture.* Paris: Guérin, 1762.

———. *Du Transport, de la conservation et de la force des bois.* Paris: Delatour, 1767.

———. *Traité des Arbres Fruitiers, contenant leur figure, leur description, leur culture & c.* 2 vols. Paris: Saillant, 1768.

Du Pont de Nemours, Pierre-Samuel. *The Autobiography of Du Pont de Nemours.* Trans. Elizabeth Fox-Genovese. Wilmington, DE: Scholarly Resources, 1984.

Ellis, John. *Description du mangoustan et du fruit à pain.* Rouen: P. Machuel, 1779.

Eymery, Alexis. *Dictionnaire des Girouettes, ou Nos contemporains peints d'après eux-mêmes: ouvrage dans lequel sont rapportés des discours . . . écrits sous les gouvernemens qui ont eu lieu en France depuis vingt-cinq ans. . . .* Paris: A. Eymery, 1815.

Faujas de Saint-Fond, Barthélémi. *Observations de M. Faujas, Adjoint à la garde du Cabinet d'Histoire Naturelle du Roi, spécialement chargé des Correspondances des Jardins et du Cabinet. Sur le rapport du Comité des Finances de l'Assemblée Nationale, article "Jardin du Roi et Cabinet d'Histoire Naturelle," page 83.* Paris: Imprimerie de Chalon, 1790.

Favard de l'Anglade, Guillaume-Jean. *Rapport fait par Favard au nom d'une commission spéciale, sur des bâtimens et terreins réunis au Muséum d'histoire naturelle. Séance du 15 Brumaire an 6.* Paris: Imprimerie Nationale, year VI/1797.

Ferry de Saint-Constant, Jean-Louis, comte. *Génie de M. de Buffon, par M.****. Paris: Panckoucke, 1778.

Fougeroux de Bondaroy, Auguste-Denis. "Nouvelles Observations sur le Soufre." *MARS* 1780/1784, 105–110.

———. "Mémoire sur un Moyen proposé pour détruire le Méphitisme des Fosses d'aisance." *MARS* 1782/1785, 197–204.

———. "Observation sur le seigle ergoté." *MARS* 1783/1786, 101–103.

———. "Mémoire sur l'abricotier de Sibérie." *MARS* 1784/1787, 207–210.

———. "Mémoire sur une plante du Pérou, Nouvellement connu en France." *MARS* 1784/1787, 202–206.

———. "Sur les Etuves propres à la conservation des Grains." *MARS* 1786/1788, 423–439.

Fourcroy, Antoine-François de. "Mémoire sur la nature de la fibre charnue ou musculaire et sur la siège de l'irritabilité." *HSRM* 1783/1787, 502–513.

———. "Observation sur un nouveau moyen de se procurer facilement l'espèce de fluide élastique, connu sous le nom de *mofette atmosphérique,* & sur la production de ce gaz dans les Animaux." *MARS* 1786/1788, 546–549.

———. "Mémoire sur la coloration des matières végétales, par l'air vital, et sur une nouvelle préparation de couleurs solides pour la peinture." *MARS* 1789/1793, 335–342.

———. "Nouvelles expériences sur les matières animales, faites dans le laboratoire du lycée." *MARS* 1789/1793, 297–326.

———. "Extrait d'un Mémoire sur les propriétés médicinales de l'air vital." *Annales de Chimie* 4 (1790): 83–93.

———. *Eléments d'Histoire naturelle et de Chimie.* 4th ed. 5 vols. Paris: Cuchet, 1791.

———. *Discours sur l'Etat actuel des Sciences et des Arts dans la République Française. Prononcé à l'ouverture du Lycée des arts le dimanche 7 Avril 1793, l'an second de la République.* [Paris, 1793.]

Geoffroy Saint-Hilaire, Étienne. "Extrait d'un mémoire sur un nouveau genre de Quadrupèdes, de l'ordre des Rongeurs (*Glires* L.), lu à la Société d'Histoire Naturelle." *La Décade philosophique* 4 (1795): 193–206.

———. "Mémoire sur les rapports naturels des Makis Lémur." *Magazin Encyclopédique* 1 (1796): 20–36.

Geoffroy Saint-Hilaire, Étienne, and Georges Cuvier. "Histoire Naturelle des Orangs-Outangs, par Et. Geoffroy, professeur de zoologie, au Muséum d'Histoire naturelle, et G. Cuvier, professeur d'Histoire naturelle, aux Ecoles centrales de Paris." *Magazin Encyclopédique* 3 (1795): 451–463.

———. "Mémoire sur une nouvelle division des mammifères, et sur les principes qui doivent servir de base dans cette sorte de travail, lu à la Société d'Histoire naturelle, le premier Floréal de l'an troisième . . . ." *Magazin Encyclopédique* 6 (1795). Separately published.

Gersaint, Etienne-François. *Catalogue raisonné de coquilles et autres curiosités naturelles.* Paris: Flahault et Prault fils, 1736.

Gibbon, Edmund. *Histoire de la décadence et de la chûte de l'empire romain.* Paris: De Bure, 1776.

Grégoire, Henri. *Rapport et projet de décret présenté au nom du Comité d'Instruction publique, à la séance du 8 août.* Paris: Imprimerie Nationale, 1793.

———. *Rapport et projet de decret, sur les moyens d'améliorer l'agriculture en France, par l'établissement d'une maison d'économie rurale dans chaque département, présentés à la Séance du 13 du premier mois de l'an deuxième de la République Française, au nom des Comités d'Aliénation et d'Instruction publique, par le Citoyen Grégoire.* Paris: Imprimerie Nationale, year II/1793.

———. *Nouveaux développemens sur l'amélioration de l'Agriculture, par l'Etablissement de Maisons d'Economie rurale; présentés par le citoyen Grégoire à la Séance du 16 brumaire, l'an deuxième de la République une et indivisible.* Paris: Imprimerie Nationale, year II/1794.

———. *Oeuvres de l'abbé Grégoire.* 14 vols. Paris: Éditions d'Histoire Sociale; Nendeln: Kraus-Thomson Organisation, 1977.

———. *Mémoires de Grégoire, ancien évêque de Blois, député à l'Assemblée Constituante et à la Convention Nationale, Sénateur, membre de l'Institut; suivies de la Notice historique sur Grégoire d'Hippolyte Carnot.* Preface by Jean-Noël Jeanneney. Introduction by Jean-Michel Leniaud. Paris: Éditions de Santé, 1989.

Guillaume, James, ed. *Procès-verbaux du Comité d'Instruction Publique de la Convention Nationale.* 6 vols. Paris: Imprimerie Nationale, 1891.

Hales, Stephen. *La Statique des Végétaux et l'Analyse de l'Air, Expériences nouvelles lues à la Société Royale de Londres.* Trans. Georges-Louis Leclerc de Buffon. Paris: Jacques Vincent, 1735.

Haüy, René-Just. "Mémoire sur la structure de divers cristaux métalliques." *MARS* 1785/1788, 213–228.

———. "Observations sur la manière de faire les herbiers." *MARS* 1785/1788, 210–212.

———. "Mémoire où l'on expose une méthode analytique, pour résoudre les problèmes relatifs à la structure des Cristaux." *MARS* 1788/1790, 13–33.

Henry, Charles, ed. *Correspondance inédite de Condorcet et de Turgot, 1770–1779: Publiée avec des notes et une introduction d'après les autographes de la collection Minoret et les manuscrits de l'Institut.* Paris: Perrin, 1887.

Hérault de Séchelles, Marie-Jean. "Parallèle de Jean-Jacques Rousseau et le comte de Buffon." *Journal Encyclopédique* 3 (1786): 329–330.

———. *La visite à Buffon, ou Voyage à Montbard.* Paris: Solvet, 1801.

Hérissant, François-David. "Eclaircissemens sur l'organisation jusqu'ici inconnue d'une quantité considérable de productions animals, principalement des Coquilles des Animaux." *MARS* 1766/1769, 508–540.

Histoire et mémoires de l'Académie des Sciences. 1789. Paris: Du Pont, 1793.

Histoire et mémoires de l'Académie Royale des Sciences. 1760–1788. Paris: Imprimerie Royale, 1766–1791.

Histoire de la société royale de médecine . . . Avec les mémoires de médecine et de Physique médicale . . . Tirés des Registres de cette Société. 9 vols. Paris: Philippe-Denys Pierres, 1779–1790.

Holbach, Paul-Henri Thiry, baron d'. *Système social ou principes naturels de la morale et de la politique avec un examen de l'influence du gouvernement sur les moeurs.* 2 vols. London: n.p., 1773.

Ingen-Housz, Jan. *Expériences sur les végétaux, spécialement sur la Propriété qu'ils possèdent à un haut degré, soit d'améliorer l'Air quand ils sont au soleil, soit de la corrompre la nuit, ou lorsqu'ils sont à l'ombre; auxquelles on a joint une méthode nouvelle de juger du degré de salubrité de l'Atmosphère.* Trans. Jan Ingen-Housz. Paris: Didot, 1780.

Jaume Saint-Hilaire, Jean-Henri. *Notice des principaux objets d'histoire naturelle, conservés dans les galeries du Museum du Jardin des Plantes de Paris . . . On y a joint quelques Réfléxions sur la vie et les Ouvrages du Comte de Buffon.* Paris: Comminges, year IX/1801.

Johanneau, Éloi. *Tableau synoptique de la méthode botanique de B. et A. L. Jussieu.* Paris: Imprimerie de la République, year V/1797.

Journal des Sçavans. First series, Amsterdam, 1665–1781.

Journal Encyclopédique. Liège, 1756–1775. Facsimile reprint, Geneva: Slatkine, 1967.

Journal Encyclopédique ou Universel. Liège, 1775–1793. Facsimile reprint, Geneva: Slatkine, 1967.

Jussieu, Antoine de. *Traité des vertus des plantes.* Ed. Gandoger de Foigny. Nancy: H. Leclerc, 1771.

Jussieu, Antoine-Laurent de. "Examen de la famille des Renoncules." *MARS* 1773/1777, 214–240.

———. "Exposition d'un nouvel ordre de plantes adoptés dans les démonstrations du Jardin Royal." *MARS* 1774/1778, 175–197.

———. *Rapport de l'un des Commissaires chargés par le Roi de l'examen du Magnétisme animal.* Paris: Hérissant et Barrois, 1784.

———. "Mémoire sur les rapports existans entre les caractères des plantes, et leurs vertus." *HSRM* 1786/1790, 188–197.

———. *Genera plantarum secundum ordines naturales disposita juxta methodum in Horto Regio Parisiensi exatarum, anno M. D. CC. LXXIV.* Paris: Hérissant, 1789.

Kalm, Pehr. *Travels into North America; containing its natural history, and a circumstantial account of its plantations and agriculture in general, with the civil, ecclesiastical and commercial state of the country, the manners of inhabitants, & several curious & important remarks on various subjects.* Trans. J. R. Forster. 2 vols. London: Forster, 1770–1771.

Lacepède, Bernard-Germain-Étienne de La-Ville-sur-Illon, comte de. *Essai sur l'électricité naturelle et artificielle.* 2 vols. Paris: Imprimerie de Monsieur, 1781.

———. *Physique générale et particulière.* 2 vols. Paris: Imprimerie de Monsieur, 1782–1784.

———. *Histoire naturelle des Quadrupèdes Ovipares et des Serpens.* Paris: Hôtel de Thou, 1788

———. *Histoire naturelle des Serpens.* Paris: Hôtel de Thou, 1789.

———. *Extrait des registres de l'Assemblée Electorale du Departement de Paris . . . Discours prononcé le 15 Juin 1791 par M. Lacepède, Président, pour la clôture de la dernière session de l'Assemblée Electorale du Département de Paris, de 1790 et 1791. . . .* [Paris]: Prault, 1791.

———. "Introduction au Cours d'Ichthyologie, donné dans les galeries du Museum d'Histoire naturelle, par le citoyen Lacepède, et commencé le 13 floreal de l'an 3e." *Magazin Encyclopédique* 1 (year III/1795): 449.

———. *Discours d'ouverture et de clôture d'histoire naturelle des animaux vertebrés et a sang rouge, Donné dans le Muséum national d'Histoire Naturelle, l'an 6 de la république.* Paris: Plassan, [year VI/1797].

———. *Discours d'ouverture du cours d'histoire naturelle de l'homme, des quadrupèdes, des cétacées, des oiseaux, des quadrupèdes ovipares, des serpens et des poissons.* Paris: n.p., year VII/1799.

———. *Discours d'ouverture du cours de zoologie de l'an IX, par Lacepède.* Paris: Plassan, year IX/1800.

———. *Discours d'ouverture et de clôture du cours d'histoire naturelle donné dans le Muséum national d'Histoire naturelle, l'an VIII de la République.* Paris: Plassan, year VIII/1800.

———. *Discours de clôture du cours d'histoire naturelle de l'an IX.* Paris: n.p., year IX/1801.

———. *La poétique de la musique.* 2 vols. 1785. Facsimile reprint, Geneva: Slatkine, 1970.

Lacepède, Bernard-Germain-Étienne de La-Ville-sur-Illon, comte de, and Georges Cuvier. *La ménagerie du Muséum national d'Histoire naturelle, ou les animaux vivants, peints d'après nature . . . .* Paris: Miger, year X/1801.

Lalande, Joseph-Jerôme Lefrançois de. "Éloge de Vicq d'Azir." *La Décade philosophique* 3 (1794): 513–521; 4 (1795): 1–10.

Lamarck, Jean-Baptiste-Pierre-Antoine de Monet, chevalier de. *Dictionnaire de Botanique,* 3 vols. Paris: Panckoucke, 1783–1792. Continued by the abbé Jean-Louis-Marie Poiret. Part of *Encyclopédie méthodique, ou par ordre de matières. Par une Société de gens-de-lettres, de savans, et d'artistes.* 190 vols. Paris and Liège, 1782–1830.

———. "Mémoire sur les classes les plus convenables à établir parmi les végétaux, et sur l'analyse de leur nombre avec celles déterminées dans le règne animal, ayant égard de part et d'autre à la perfection graduée des organes." *MARS* 1785/1788, 437–453.

———. "Mémoire sur le genre du muscadier, *myristica.*" *MARS* 1788/1790, 148–169.

———. *Considérations en faveur du Chevalier de La Marck, ancien officier au régiment de Beaujolais, de l'Académie royale des Sciences, botaniste du Roi, attaché au cabinet d'histoire naturelle.* Paris: n.p., 1789.

———. *Mémoire sur le Projet du Comité des Finances, relatif à la suppression de la place de Botaniste attaché au Cabinet d'Histoire naturelle.* Paris: Gueffier, 1790.

———. *Mémoires de physique et d'histoire naturelle, établies sur des bases de raisonnement indépendantes de toute théorie; avec l'exposition de nouvelles considérations sur le cause générale des dissolutions; sur la matière du feu; sur la couleur des corps; sur la formation des composés; sur l'origine des minéraux; et sur l'organisation des corps vivans.* Paris: Lamarck, year VI/1797.

———. *Flore françoise ou Descriptions succinctes de toutes les plantes qui croissent naturellement en France, disposées selon une nouvelle méthode d'analyse, et précédées par un Exposé des Principes élémentaires de la Botanique.* 3d ed. 6 vols. Paris: Desaudray, 1815.

———. "Discours d'ouverture des cours donnés dans le Muséum d'Histoire naturelle, an VIII, an X, an XI et 1806." *Bulletin scientifique France-Belgique* 40 (1906): 21–157.

Lamarck, Jean-Baptiste-Pierre-Antoine de Monet, chevalier de, Jean-Guillaume Bruguière, Guillaume-Antoine Olivier, René-Just Haüy, and Bertrand Pelletier, eds. *Choix de mémoires sur divers sujets d'histoire naturelle par MM. Lamarck, Bruguière, Olivier, Hauy et Pelletier. Formant les Collections du Journal d'Histoire Naturelle.* Paris: Imprimerie du Cercle Social, 1792.

———. *Journal d'Histoire Naturelle, rédigé par MM. Lamarck, Bruguière, Olivier, Hauy et Pelletier.* Paris: Directeurs du Cercle Social, year II/1794.

Lanthenas, François-Xavier. *De l'influence de la liberté sur la santé, la morale, et le bonheur.* Paris: n.p., 1792.

Laplace, Pierre-Simon. "Sur les naissances, les mariages et les morts à Paris, depuis 1771 jusqu'en 1784; & dans toute l'étendue de la France, pendant les années 1781 & 1782." *MARS* 1783/1786, 693–702.

La Revellière-Lépeaux, Louis-Marie. *Mémoires de Larevellière-Lépeaux, membre du Directoire exécutif de la République française et de l'Institut national; publiés par son fils sur le manuscrit autographe de l'auteur et suivies des pièces justificatives et de correspondance inédits.* Ed. Ossian La Revellière-Lépeaux. 3 vols. Paris: E. Plon, Nourrit et Cie, 1895.

La Rochefoucauld, Louis-Alexandre de, Antoine Baumé, and Antoine-François de Fourcroy. "Examen d'un sable vert cuivreux, du Pérou." *MARS* 1786/1788, 465–477.

Lassone, Joseph-Marie-François de, Louis-Jean-Marie Daubenton, Jacques-René Tenon, Jean-Sylvain Bailly, Antoine-Laurent de Lavoisier, Pierre-Simon Laplace, Charles-Augustin Coulomb, and Jean D'Arcet. "Rapport des Commissaires chargés, par l'Académie, de l'examen du Projet d'un nouvel Hôtel-Dieu." *MARS* 1785/1788, 2–110; *HARS* 1786/1788, 1–24.

La Tour du Pin, Lucy de. *Escape from the Terror: The Journal of Madame de la Tour du Pin.* Ed. and trans. Felice Harcourt. London: Folio Society, 1979.

Lavater, Joseph Caspar. *Essai sur la Physiognomie, destiné à faire connoître l'homme et à le faire aimer.* 4 vols. La Haye: n.p., 1781–1803.

Lavoisier, Antoine-Laurent de. "Mémoires sur la Nature de l'Eau, et sur les Expériences par lesquelles on a prétendu prouver la possibilité de son changement en terre." *MARS* 1770/1773, 73–89, 90–107.

———. "Expériences sur la respiration des animaux, et sur les changemens qui arrivent à l'air en passant par leur poumon." *MARS* 1777/1780, 185–194.

———. "Considérations générales sur la nature des acides, et sur les Principes dont ils sont composés." *MARS* 1778/1781, 535–547.

———. "Mémoire sur la manière d'éclairer les salles de spectacle." *MARS* 1781/1784, 409–420.

———. "Mémoire sur les altérations qui arrivent à l'air dans plusieurs circonstances où se trouvent les hommes réunis en société." *HSRM* 1783/1787, 569–582.

———. "Réfléxions sur la décomposition de l'eau par les substances végétales et animales." *MARS* 1786/1788, 590–605.

———. "Observations générales, sur les couches modernes horizontales, qui ont été déposées par la mer, et sur les conséquences qu'on peut tirer de leurs dispositions, relativement à l'ancienneté du globe terrestre." *MARS* 1789/1793, 351–371.

Lavoisier, Antoine-Laurent de, and Armand Séguin. "Premier mémoire sur la respiration des animaux." *MARS* 1789/1793, 566–581.

Le Bègue de Presle, Achille-Guillaume, and C.-F.-A. de La Lauze. *Economie rurale et civile, ou moyens les plus économiques d'administrer & faire valoir les biens de campagne & de ville.* 2 vols. Paris: Buisson, 1789–1790.

Lebreton, François. *Manuel de botanique, à l'usage des amateurs et des voyageurs; Contenant les Principes de Botanique, l'Explication du Systême de Linné, un Catalogue des différens Végétaux étrangers, les moyens de transporter les Arbres & les Semences; la manière de former un Herbier, &c.* Paris: Prault, 1787.

Le Cat, Claude-Nicolas. *Traité de la couleur de la peau humaine en général, de celle des negres en particulier, et de la métamorphose d'une de ces couleurs en l'autre, soit de naissance, soit accidentellement.* Amsterdam: n.p., 1765.

Leroy, Charles-Georges. *Lettres sur les animaux.* Ed. Elizabeth Anderson. Oxford: Voltaire Foundation, 1994.

Le Roy, Jean-Baptiste. "Précis d'un Ouvrage sur les Hôpitaux, dans lequel on expose les principes résultant des observations de Physique et de Médecine qu'on doit avoir en vue dans la construction de ces édifices; avec un projet d'Hôpital disposé d'après ces principes." *MARS* 1787/1789, 588–600.

Le Roy, Jean-Baptiste, Mathieu Tillet, Mathurin-Jacques Brisson, Louis-Claude Cadet, Antoine-Laurent de Lavoisier, C. Bossut, Marie-Jean-Antoine-Nicolas Caritat, marquis de Condorcet, and Nicolas Desmarest. "Rapport fait à l'Académie des Sciences, sur la Machine Aérostatique, de Mrs. de Montgolfier." *HARS* 1783/1786, 4–7.

Linnaeus, Carolus. *Amoenitates academicae, seu Dissertationes variae physicae, medicae, botanicae, antehac seorsim editae nunc collectae et auctae cum tabulis aeneis.* 7 vols. Leiden: C. Haak, 1749–1769.

———. *The Elements of Botany: Containing the History of the Science.* Trans. H. Rose. London: T. Cadell and M. Hingston, 1775.

———. *Reflections on the Study of Botany.* Trans. J. E. Smith. 1785. Facsimile reprint, New York: Arno, 1965.

———. *A Dissertation on the sexes of plants.* Trans. James Edward Smith. London: James Edward Smith, 1786.

———. *Select Dissertations from the Amoenitates Academicae.* Trans F. J. Brand. Facsimile reprint, New York: Arno, 1977.

———. *L'équilibre de la nature.* Introduction by Camille Limoges. Paris: Vrin, 1987.

Locke, John. *An Essay Concerning Human Understanding.* Ed. Peter H. Nidditch. Oxford: Clarendon, 1991.

Magazin Encyclopédique, ou Journal des Sciences, des Lettres et des Arts, rédigé par Millin, Noel et Warens. Vols. 1–6. Paris: Imprimerie du Magazin Encyclopédique, year III/1795.

Malesherbes, Chrétien-Guillaume Lamoignon de. *Mémoire sur les moyens d'accélerer les progrès de l'économie rurale en France.* Paris: Ph.-D. Pierres, 1790.

———. *Observations de C.-G. Lamoignon de Malesherbes sur l'histoire naturelle générale et particulière de Buffon et Daubenton.* Ed. Paul Abeille. 2 vols. Paris: Pougens, year VI/1798.

Marmontel, Jean-François. *Mémoires: Édition critique établie par John Renwick.* 2 vols. Clermont-Ferrand: G. de Bussac, 1972.

Mauduyt de La Varenne, Pierre-Jean-Claude. "Mémoires sur l'électricité, considérée relativement à l'économie animale et à l'utilité dont elle peut être en médecine." *HSRM* 1776/1779, 461–528.

———. "Mémoire sur les effets généraux, la nature et l'usage du fluide électrique, considérée comme médicament." *HSRM* 1777–1778/1780 (2), 432–455.

———. "Mémoire sur le traitement électrique administré à quatre-vingt-deux malades." *HSRM* 1777–1778/1780 (2), 199–431.

———. "Mémoire sur les différentes manières d'administrer l'Electricité, et Observations sur les effets que ces divers moyens ont produits." *HSRM* 1780–1781/1785, 264–413.

———. "Ornithologie." In *Histoire naturelle des Animaux.* Paris: Panckoucke, 1782. Part of *Encyclopédie méthodique, ou par ordre de matières. Par une Société de gens-delettres, de savans, et d'artistes.* 190 vols. Paris and Liège, 1782–1830.

Mavidal, Jerôme, Emile Laurent, Marcel Reinhard, Georges Lefebvre, and Marc Bouloiseau, eds. *Archives parlementaires de 1789 à 1860: Recueil complet des débats législatifs et politiques des chambres françaises.* Series 1, vols. 1–87 (1787–1794). Paris: Imprimerie Nationale, 1867–1969.

Mémoires d'agriculture, d'économie rurale et domestique, publiés par la Société Royale d'Agriculture de Paris. 25 vols. Paris: Buisson/Cuchet/Bureau de la Feuille du Cultivateur, 1785–1791.

Mémoires pour l'Histoire des Sciences et des Beaux Arts [Journal de Trévoux]. Paris, 1700–1740.

Mercier, Louis-Sébastien. *L'An deux mille quatre cent quarante, rêve s'il en fut jamais.* London [Paris]: n.p., 1772.

———. *Tableau de Paris, critiqué par un solitaire du pied des Alpes.* 3 vols. Nyon: Natthey, 1783.

Mercure de France. Paris, January 1724–June 1781. Facsimile reprint, Geneva: Slatkine, 1967.

Mertrud, Antoine-Louis-François. *Discours prononcé à l'Amphitéâtre du Museum National d'Histoire Naturelle, pour la clôture du cours de l'Anatomie des animaux, le Primidi 11 Germinal l'an deuxième de la République Française une et indivisible.* [Paris: n.p., year II/1794.]

———. *Discours prononcé dans l'amphithéatre du Museum National d'histoire naturelle, à l'ouverture du cours de l'anatomie des animaux; le primidi 21 nivos, l'an deuxième de la République française, une et indivisible.* [Paris]: Imprimerie de la rue des Droits de l'Homme, [year II/1794].

Mesmer, Franz Anton. *Lettres de M. Mesmer à M. Vicq-d'Azyr, et à MM. les Auteurs du Journal de Paris.* Brussels: n.p., 1784.

Métra, Louis-François. *Correspondance secrete, politique et littéraire ou mémoires pour servir à l'histoire des Cours, des Sociétés et de la Littérature en France, depuis la mort de Louis XV.* 18 vols. London: John Adamson, 1788.

Millin de Grandmaison, Aubin-Louis. "Éloge de Linnaeus." *HSRM* 1777–1778/1780, 17–44.

———. *Rapport fait à la Société d'Histoire Naturelle, en sa séance du 21 prairial, Par A. L. Milin [sic], l'un de ses membres, sur l'ouvrage qui a pour titre: Calendario entomologico, etc. etc., de M. Giorna le fils, membre de la même société.* Paris: Imprimerie du Magazin Encyclopédique, year III/1795.

———, ed. *Actes de la Société d'Histoire Naturelle de Paris.* Paris: Imprimerie de la Société d'Histoire Naturelle, 1792.

Monge, Gaspard. "Mémoire sur le résultat de l'inflammation du Gaz inflammable & de l'Air déphlogistiqué, dans des vaisseaux clos." *MARS* 1783/1786, 78–88.

Montesquieu, Charles-Louis de Secondat, baron de La Brède et de. *Considérations sur les causes de la Grandeur des Romains, et de leur décadence.* Geneva: Jacques Desbordes, 1734.

———. *De l'Esprit des Lois.* Introduction by Victor Goldschmidt. 2 vols. Paris: Flammarion, 1979.

Morand. "Recherches sur quelques conformations monstrueuses des doigts dans l'homme." *MARS* 1770/1773, 137–147.

———. "Mémoire sur la population de Paris, et sur celle des Provinces de la France, avec des Recherches qui établissent l'accroissement de la Population de la Capitale & du reste du Royaume. Depuis le Commencement du siècle." *MARS* 1779/1782, 459–478.

[Morel, Jean-Marie]. *Tableau de l'École de Botanique du Jardin des Plantes de Paris, ou Catalogue général des Plantes qui y sont cultivées et rangées par classes, ordres, genres et espèces, d'après les principes de la Méthode naturelle de A. L. Jussieu. Suivi d'une Table alphabétique des Noms vulgaires des Plantes les plus fréquemment employées en Médecine, dans les Arts, la décoration des Jardins, etc.; avec les Noms des Genres et des Espèces auxquels elles se rapportent.* 1st ed. Paris: Didot le jeune, year VIII/1800. 2d ed. Paris: Méquignon l'aîné, year IX/1801.

Mourgue de Mont-Redon, Jacques-Antoine. "Recherches sur l'origine & sur la nature des Vapeurs qui ont régné dans l'Atmosphère pendant l'été de 1783." *MARS* 1781/1784, 754–773.

Necker, Jacques. *Sur la législation et le commerce des grains.* 2d ed. Paris: Prault, 1775.

Nollet, Jean-Antoine. *Leçons de physique expérimentale.* 6 vols. Paris: Frères Guerin, 1745–1768.

———. *Essai sur l'électricité des corps.* Paris: Frères Guerin, 1746.

———. "Sur les effets de la tonnerre, comparés avec ceux de l'electricité." *MARS* 1764/1767, 394–421.

Observations sur la physique, sur l'histoire naturelle et sur les arts. Series 1, vols. 1–2; series 2, vols. 1–43. Paris, 1771–1793.

Olivier, Guillaume-Antoine. "Observations sur la culture de l'arbre-à-Pain et des épiceries." *Journal d'Histoire naturelle* 2 (1794): 72–80.

Perrault, Claude. *Mémoires pour servir à l'histoire des Animaux.* Paris: Compagnie des Libraires, 1733.

Pétition des Propriétaires des Maisons et Terreins environnans le Jardin des Plantes, à la Convention Nationale. Paris: Démonville, [year III/1794–1795].

Pinel, Philippe. "Zoologie. Recherches sur une nouvelle méthode de classification des quadrupèdes, fondée sur la structure méchanique des parties osseuses qui servent à l'articulation de la mâchoire inférieure." In *Actes de la Société d'Histoire Naturelle de Paris.* Paris: Imprimerie de la Société d'Histoire Naturelle, 1792, 50–60.

Pluche, Noel-Antoine. *Le Spectacle de la Nature, ou entretien sur les particularités de l'histoire naturelle, qui ont paru les plus propres à rendre les jeunes gens curieux, et à leur former l'esprit.* 2d ed. 8 vols. Paris: Veuve Estienne, 1732–1751.

Poissonnier, Pierre-Isaac, Pierre-Jean-Claude Mauduyt de La Varenne, Caille, and Charles-Louis-François Andry. *Rapport des Commissaires de la Société royale de Médecine, nommés par le roi, pour faire l'examen du magnétisme animal.* Paris: Imprimerie Royale, 1784.

Ponce, Nicolas. *Les illustres Français, ou tableaux historiques des grands hommes de la France . . . par M. Ponce, . . . d'apres les dessins de M. Marillier.* Paris: Nicolas Ponce, [1790–1816].

Portal, Antoine. *Histoire de l'anatomie et de la chirurgie, contenant l'origine & les progrès de ces Sciences; avec un Tableau Chronologique des principales Découvertes, & un Catalogue des ouvrages d'Anatomie & de Chirurgie, des Mémoires Académiques, des Dissertations insérées dans les Journaux, & de la plupart des Theses qui ont été soutenues dans les Facultés de Médecine de l'Europe.* 6 vols. Paris: P. Fr. Didot le jeune, 1770–1773.

———. "Rapport sur le mort du Sieur Le Maire, & sur celle de son Epouse, marchands de Modes, à l'enseigne de la Corbeille Galante, rue Saint Honoré, causées par la vapeur du Charbon, le 3 Août 1774." *MARS* 1775/1778, 492–509.

———. "Observations sur la phthisie de Naissance." *MARS* 1781/1784, 631–644.

———. "Observations sur l'apoplexie." *MARS* 1781/1784, 623–630.

———. "Observations sur le traitement de la rage." *MARS* 1786/1788, 440–449.

———. *Observations sur les effets des Vapeurs méphitiques, dans l'Homme, sur les Noyés, sur les Enfans qui paroissent morts en naissant, et sur la Rage; avec un Précis du Traitement le mieux éprouvé en pareils cas. Sixème Edition, à laquelle on a joint des Observations sur les effets de plusieurs poisons dans le corps de l'homme, et sur les moyens d'en empêcher les suites funestes.* Paris: Imprimérie Royale, 1787.

———. "Observations sur les effets des vapeurs méphitiques, dans l'homme." *MARS* 1787/1789, 239–246.

Priestley, Joseph. *The History and Present State of Electricity, with Original Experiments.* 2d ed. London: J. Dodsley, 1769.

———. "Observations upon different Kinds of Air." *Philosophical Transactions* (1772): 156–247.

———. *Experiments and Observations on different kinds of Air.* 3 vols. London: J. Johnson, 1774–1777.

Quesnay, François. *Physiocratie, ou Constitution naturelle du gouvernement le plus avantageux au genre humain.* 2 vols. Leiden and Paris: Merlin, 1767–1768.

Raynal, Guillaume-Thomas-François. *Histoire politique et philosophique des établissemens et du commerce des Européens dans les deux Indes.* 6 vols. Amsterdam: n.p., 1770.

Reynier, Louis. "De l'influence du climat sur la forme et la nature des végétaux." *Journal d'Histoire naturelle* 1 (1792): 101–148.

Robespierre, Maximilien. "Rapport fait au nom du comité de Salut publique, sur les rapports des idées religieuses et morales avec les principes républicains et sur les fêtes nationales." *La Décade philosophique* 1 (1794): 177–191, 242–253.

Rollin, Charles. *Histoire ancienne des Egyptiens, des Carthaginois, des Assyriens, des Babyloniens, des Mèdes, et des Perses, des Macédoniens, des Grecs.* 13 vols. Paris: Frères Estienne, 1769–1778.

Rousseau, Jean-Jacques. *Discours sur les sciences et sur les arts. Édition critique avec une introduction et un commentaire par George R. Havens.* 1751. New York: Modern Language Association of America; London: Oxford University Press, 1946.

———. *Discours sur l'origine et les fondements de l'inégalité parmi les hommes.* 1755. Preface by Christian Delacampagne. Paris: Gallimard, 1965.

———. *Les Confessions de J. J. Rousseau (Les Rêveries du Promeneur solitaire.—Seconde partie des Confessions . . .) Edition enrichie d'un nouveau recueil de ses Lettres.* 10 vols. London and Neufchâtel: n.p., 1786–1790.

———. *Emile, ou de l'éducation.* Paris: Flammarion, 1966.

———. *Oeuvres complètes.* 3 vols. Paris: Seuil, 1971.

———. *Les Rêveries du promeneur solitaire.* Introduction by Jean Grenier. Paris: Gallimard, 1972.

———. *The Social Contract and Discourses.* Translated and introduced by G. D. H. Cole. London: Dent, 1973.

———. *Le Botaniste sans maître, ou manière d'apprendre seul la botanique: Fragments pour un dictionnaire des termes d'usage en botanique.* Paris: A. M. Métailié, 1983.

Royou, Thomas. *Le monde de verre reduite en poudre, ou Analyse et réfutation des Epoques de la nature de M. le comte de Buffon.* Paris: n.p., 1780.

Rozier, François, ed. *Cours complet d'Agriculture Théorique, Pratique, Économique, et de Médécine Rurale et Vêterinaire, suivi d'une Méthode pour étudier l'Agriculture par Principes; ou Dictionnaire universel d'Agriculture.* 10 vols. Paris: rue et hôtel Serpente [Cuchet], 1785–1797.

Rulvet, Joseph, Marie-Paule Depouhon-Ninnin, and Paul Servais, eds. *Lettres de Turgot à la Duchesse d'Enville, 1764–1774 et 1777–1780: Edition critique préparée par les etudiants en histoire de l'Université Catholique de Louvain.* Louvain: Bibliothèque de l'Université Louvain; Leiden: E. J. Brill, 1976.

Sage, Balthazar-Georges. *Observations sur un écrit, qui a pour titre, "Vues sur le Jardin royal des Plantes et le Cabinet d'Histoire naturelle."* Paris: Didot le jeune, 1790.

Seguin, Armand, and Antoine-Laurent de Lavoisier. "Premier mémoire sur la respiration des animaux." *MARS* 1789/1793, 566–583.

Senebier, Jean. *Action de la lumière sur la végétation.* Paris: Didot, 1780.

———. *Mémoires physico-chimiques, sur l'influence de la lumière solaire pour modifier les êtres des trois règnes de la nature, & surtout ceux du règne végétal.* 3 vols. Geneva: B. Chirol, 1782.

———. *Recherches sur l'influence de la lumière solaire pour métamorphoser l'air fixe en air pur pour la végétation. Avec des expériences & des considérations propres à faire connoître la nature des substances aëriformes.* Geneva: B. Chirol, 1783.

———. *Expériences sur l'action de la lumière solaire dans la végétation.* Geneva: Barde, Manget et Cie, 1788.

Sèze, Victor de. *Recherches physiologiques et philosophiques sur la sensibilité ou la vie animale.* Paris: Prault, 1786.

Sonnerat, Pierre. *Voyage à la Nouvelle Guinée, dans lequel on trouve la description des lieux, des observations physiques et morales, et des détails rélatifs à l'histoire naturelle dans le règne animal et le règne végétal.* Paris: Ruault, 1776.

Stork, William. *An Account of East-Florida: With remarks on its future importance to trade and commerce.* London: Woodfall, 1766.

Swift, Jonathan. *Voyages du capitaine Gulliver en divers pays éloignés . . . Nouvelle édition.* La Haye: Jean Swart, 1778.

Talleyrand, Charles-Maurice de. *Rapport sur l'instruction publique fait au nom du Comité de Constitution les 10, 11 et 19 septembre 1791.* Paris: Baudoin et Du Pont, 1791.

Tessier, Henri-Alexandre. "Expériences propres à développer les effets de la Lumière sur certaines Plantes." *MARS* 1783/1786, 133–156.

———. "Mémoire sur quelques particularités du *cupressus disticha* Lin. Appelé Cyprès *Chauve* par les Américains." *MARS* 1785/1788, 197–205.

———. "Mémoire sur la manière de parvenir à la connoissance exacte de tous les objets cultivés en grand dans l'Europe, et particulièrement dans la France." *MARS* 1786/1788, 574–589.

———. "Mémoire sur l'importation et les progrès des arbres à épicerie dans les colonies françoises." *MARS* 1789/1793, 585–596.

———. "Mémoire sur l'orage du dimanche 13 juillet 1788." *MARS* 1789/1793, 628–638.

Thibaudeau, Antoine-Clair. *Rapport fait au nom des comités d'instruction publique et des finances, sur le Muséum national d'histoire naturelle . . . à la séance du 21 frimaire, l'an 3.* Paris: Imprimerie Nationale, year III/1795.

Thouin, André. "Mémoire sur les avantages de la Culture des Arbres étrangers pour l'emploi de plusieurs terrains de différente nature abandonnés comme stériles." In *Mémoires d'agriculture, d'économie rurale et domestique, publiés par la Société Royale d'Agriculture de Paris.* 25 vols. Paris: Buisson, 1786, 3:43–95.

———. "Mémoire sur l'usage du Terreau de Bruyère dans la culture des arbrisseaux et arbustes étrangers, regardés jusqu'à présent comme délicats dans nos jardins." *MARS* 1787/1789, 481–495.

———. "Mémoire sur les avantages de la culture des arbres étrangers pour l'emploi de plusieurs terrains de différente nature, abandonnés comme steriles." *La Décade philosophique* 2 (1794): 129–132, 192–199.

Thouin, André, Charles-Germain Bourgeois, and Antoine-Augustin Parmentier. *Avis aux Cultivateurs dont les Récoltes ont été ravagées par la Grêle du 13 Juillet 1788. Rédigé par la Société Royale d'Agriculture, et publié par ordre du Roi.* Paris: Imprimerie Royale, 1788.

Thouret, Michel-Augustin. *Extrait de la Correspondance de la Société royale de médecine, relativement au magnétisme animal.* Paris: Imprimerie Royale, 1785.

———. "Rapport sur les Exhumations du Cimetière et de l'Eglise des Saints Innocens." *HSRM* 1786/1790, 227–271.

Thourry, de. "Mémoire qui a remporté le Prix proposé par l'Académie des Sciences, Belles-Lettres & Arts de Lyon, sur cette Question: L'Electricité de l'Athmosphère a-t-elle quelque influence sur le corps humain? quels sont les effets de cette influence?" *Observations sur la physique, sur l'histoire naturelle et sur les arts,* series 2, 9 (January–June 1777): 401–437.

Thunberg, Carl Petter. *Voyages de C. P. Thunberg au Japon, par le Cap de Bonne-Espérance, les isles de la Sonde, etc.* Trans. J. Langlès and Jean-Baptiste-Pierre-Antoine de Monet, chevalier de Lamarck. 2 vols. Paris: Dandré, year IV/1796.

Tillet, Mathieu. "Expériences et observations sur la végétation du Blé dans chacune des matières simples dont les terres labourables sont ordinairement composées, par lesquels on s'est rapproché de ceux qui constitutent ces mêmes terres à labourer." *MARS* 1772/1775, 99–156.

———. "Projet d'un tarif Propre à servir pour établir la valeur du Pain, proportionnément à celle du Blé & des Farines; avec des Observations sur la Mouture économique, comme base essentielle de ce Tarif; & sur les avantages du commerce des Farines, par préférence à celui du Blé en nature." *MARS* 1781/1784, 107–123.

Toscan, Georges-Louis. *Histoire du Lion de la Menagerie et de son chien, par le citoyen Toscan, bibliothécaire du Muséum.* Paris: n.p., year III/1794.

Tourneux, Maurice, ed. *Correspondance littéraire, philosophique et critique par Grimm, Diderot, Raynal, Meister, etc. revue sur les textes originaux, comprenant, outre ce qui a été publié à diverses époques, les fragments supprimés en 1813 par la censure; les parties inédites conservées à la Bibliothèque ducale de Gotha et à l'Arsenal à Paris.* 16 vols. Paris: Garnier frères, 1877–1882.

Tschudi, Jean-Baptiste-Louis-Théodore, baron de. *De la transplantation, de la naturalisation et du perfectionnement des végétaux.* London and Paris: Lambert and P. F. Didot le jeune, 1778.

Turgot, Anne-Robert-Jacques-Marie. *On the Progress of the Human Mind.* Trans. McQuillin De Grange. Hanover, NH: Sociological Press, 1929.

Venel, Gabriel-François. "Climat (Méd.)." In *Encyclopédie, ou Dictionnaire raisonné des Sciences, des arts et des métiers,* ed. Denis Diderot and Jean Le Rond d'Alembert. Paris: Briasson, 1754, 4:534–536.

Verdier, Delaune, and Picquenard. *Au Roi, et aux Représentants de la Nation. Dénonciation.* Paris: n.p., 1790.

Vicq d'Azyr, Félix. "Mémoires pour servir à l'anatomie des oiseaux." *MARS* 1772/1775 (2), 617–620; 1773/1777, 566–586, 489–521.

———. "Table pour servir à l'histoire anatomique et naturelle des corps vivans ou organiques." *Observations sur la physique, sur l'histoire naturelle et sur les arts,* series 2, 4 (1774): 479.

———. "Recherches sur la structure du Cerveau, du Cervelet, de la Moelle alongée, de la Moelle épinière; & sur l'origine des Nerfs de l'Homme et des Animaux." *MARS* 1781/1784, 495–622.

———. "Sur la structure du cerveau des Animaux comparé avec celui de l'Homme." *MARS* 1783/1786, 468–504.

———. *Traité d'Anatomie et de Physiologie, avec des Planches coloriées Representant au naturel les divers organes de l'Homme et des Animaux.* Vol. 1 (no more published). Paris: Didot l'aîné, 1786.

Volney, Constantin-François de Chassebeuf de. "Questions d'Economie Politique." *Magazin Encyclopédique* 1 (1795): 352–362.

Voltaire, François-Marie Arouet de. *Candide, ou l'optimisme, traduit de l'allemand de Mr. le docteur Ralph.* 1759. Ed. J. H. Brumfitt. Oxford: Oxford University Press, 1984.

Vues sur le Jardin Royal des Plantes et le Cabinet d'Histoire Naturelle. Paris: Baudoin, 1789.

Young, Arthur. *Travels in France, during the Years 1787, 1788, and 1789.* Ed. J. Kaplow. Garden City, NY: Anchor Books, 1969.

Secondary Sources

Adams, Percy G. *Travelers and Travel Liars, 1600–1800.* Berkeley: University of California Press, 1962.

Adams, Thomas McStay. *Bureaucrats and Beggars: French Social Policy in the Age of the Enlightenment.* New York: Oxford University Press, 1990.

Adams, William Howard. *The French Garden, 1500–1800.* London: Scolar Press, 1979.
Aftalion, Florin. *The French Revolution: An Economic Interpretation.* Trans. Martin Thom. Paris: Éditions de la Maison de l'Homme; Cambridge: Cambridge University Press, 1990.
Agnew, John A., and James S. Duncan, eds. *The Power of Place: Bringing Together Geographical and Sociological Imaginations.* Boston: Unwin Hyman, 1989.
Agulhon, Maurice. *Marianne into Battle: Republican Imagery and Symbolism in France, 1789–1880.* Paris: Éditions de la Maison des Sciences de l'Homme; Cambridge: Cambridge University Press, 1979.
Albertone, Manuela. "'Dans une république, nul n'est libre d'être ignorant': Révolution française et obligation scolaire." *Canadian Journal of History* 19 (1984): 17–38.
———. "Instruction et ordre naturel: Le point de vue physiocratique." *Revue d'histoire moderne et contemporaine* 33 (1986): 589–607.
Allan, D. G. C. "The Society of Arts and Government, 1754–1800: Public Encouragement of Arts, Manufactures, and Commerce in Eighteenth-Century England." *Eighteenth-Century Studies* 7 (1973–1974): 434–452.
Allen, David Elliston. *The Naturalist in Britain: A Social History.* Harmondsworth, Middlesex: Penguin, 1978.
———. "Arcana ex Multitudine: Prosopography as a Research Technique." *Archives of Natural History* 17 (1990): 349–359.
———. "Natural History in Britain in the Eighteenth Century." *Archives of Natural History* 20 (1993): 333–347.
Altman, Janet Gurkin. "Political Ideology in the Letter Manual (France, England, New England)." *Studies in Eighteenth-Century Culture* 18 (1988): 105–122.
———. "Teaching the 'People' to Write: The Formation of a Popular Civic Identity in the French Letter Manual." *Studies in Eighteenth-Century Culture* 22 (1992): 147–180.
Anderson, Wilda A. *Between the Library and the Laboratory: The Language of Chemistry in Eighteenth-Century France.* Baltimore: Johns Hopkins University Press, 1984.
———. "Scientific Nomenclature and Revolutionary Rhetoric." *Rhetorica* 7 (1989): 45–53.
Andia, Béatrice de, and Valérie Noëlle Jouffre, eds. *Fêtes et Révolution.* Dijon: Musée des Beaux-Arts, 1989–1990.
Appel, Toby A. *The Cuvier-Geoffroy Debate: French Biology in the Decades before Darwin.* New York: Oxford University Press, 1987.
Arasse, Daniel. *The Guillotine and the Terror.* Trans. Christopher Miller. Harmondsworth, Middlesex: Penguin, 1991.
Atran, Scott. *Cognitive Foundations of Natural History: Towards an Anthropology of Science.* Cambridge: Cambridge University Press, 1990.
Audelin, Louise. "Les Jussieu: Une dynastie de botanistes au XVIIIe siècle, 1680–1789." Dissertation, École des Chartes, Paris, 1987.
Baasner, Frank. "The Changing Meaning of 'Sensibilité,' 1654 till 1704." *Studies in Eighteenth-Century Culture* 14 (1985): 77–94.
Baczko, Bronislaw. "Lumières et utopie: Problèmes et recherches." *Annales: Économies, sociétés, civilisations* 26 (1971): 355–386.
Baehni, G. "Les grands systèmes botaniques depuis Linné." *Gesnerus* 14 (1957): 83–93.
Baker, Keith Michael. "Les débuts de Condorcet au secrétariat de l'Académie Royale des Sciences, 1773–1776." *Revue d'histoire des sciences et de leurs applications* 20 (1967): 229–280.
———. "Scientism, Elitism, and Liberalism: The Case of Condorcet." *Studies on Voltaire and the Eighteenth Century* 55 (1967): 129–165.
———. "Politics and Social Science in Eighteenth-Century France: The Société de

1789." In *French Government and Society, 1500–1850: Essays in Memory of Alfred Cobban,* ed. J. F. Bosher. London: Athlone, 1973, 208–230.

———. *Inventing the French Revolution: Essays on French Political Culture in the Eighteenth Century.* Cambridge: Cambridge University Press, 1990.

———, ed. *The French Revolution and the Creation of Modern Political Culture.* Vol. 1, *The Political Culture of the Old Regime.* Oxford: Pergamon, 1987.

Balan, Bernard. *L'ordre et le temps: L'anatomie comparée et l'histoire des vivants au XIXe siècle.* Paris: Vrin, 1979.

Bapst, G. "Histoire d'un cabinet minéralogique: Le cabinet d'histoire naturelle des princes de Condé." *Revue des deux mondes* 2 (1892): 437–449.

Barnes, Barry. *The Nature of Power.* Cambridge: Polity Press, 1989.

Barnes, Barry, David Bloor, and John Henry. *Scientific Knowledge.* London: Athlone, 1996.

Barnes, Barry, and Steven Shapin, eds. *Natural Order: Historical Studies of Scientific Culture.* Beverly Hills: Sage Publications, 1979.

Barnett, Dene. *The Art of Gesture: The Practices and Principles of Eighteenth-Century Acting.* Heidelberg: Carl Winter, 1987.

Barnouw, Jerry. "Feeling in Enlightenment Aesthetics." *Studies in Eighteenth-Century Culture* 18 (1988): 323–339.

Barsanti, Giulio. "Linné et Buffon: Deux images differentes de la nature et de l'histoire naturelle." *Studies on Voltaire and the Eighteenth Century* 216 (1983): 306–307.

Barthélémy-Madaule, Madeleine. *Lamarck, the Mythical Precursor: A Study of the Relations between Science and Ideology.* Cambridge: Harvard University Press, 1982.

Bassy, Alain M. "L'oeuvre de Buffon à l'Imprimerie Royale, 1749–1789." In *L'art du livre à l'Imprimerie Nationale.* Paris: Imprimerie Nationale, 1973, 171–189.

Baudrillard, Jean. "The System of Collecting." In *The Cultures of Collecting,* ed. John Elsner and Roger Cardinal. London: Reaktion Books, 1994, 7–24.

———. *The System of Objects.* Trans. James Benedict. London: Verso, 1996.

Bazin, Germain. *The Louvre.* London: Thames and Hudson, 1971.

Beale, Georgia R. "Early French Members of the Linnean Society of London, 1788–1802: From the Estates General to Thermidor." *Proceedings of the Annual Meeting of the Western Society for French History* 18 (1991): 272–282.

Beaune, Jean-Claude, Serge Benoît, Jean Gayon, and Denis Woronoff, eds. *Buffon 88: Actes du colloque international pour le bicentenaire de la mort de Buffon (Paris, Montbard, Dijon, 14–22 juin 1988).* Paris: Vrin and Institut Interdisciplinaire d'Études Épistémologiques, 1992.

Becker, G. *Tournefort.* Paris: Muséum National d'Histoire Naturelle, 1937.

Bedel, Charles. "L'enseignement des sciences pharmaceutiques." In *Enseignement et diffusion des sciences au XVIIIe siècle en France,* ed. René Taton. Paris: Hermann, 1986, 237–257.

Behrens, C. B. A. *Society, Government, and the Enlightenment: The Experiences of Eighteenth-Century France and Prussia.* London: Thames and Hudson, 1985.

Benjamin, Marina. "Elbow Room: Women Writers on Science, 1790–1840." In *Science and Sensibility: Gender and Scientific Enquiry, 1780–1845,* ed. Marina Benjamin. Oxford: Basil Blackwell, 1991, 27–59.

Bennett, Tony. *The Birth of the Museum: History, Theory, Politics.* London: Routledge, 1995.

Bernardi, Walter, and Antonello La Vergata, eds. *Lazzaro Spallanzani e la biologia del settecento: Teorie, esperimenti, istituzioni scientifiche.* Florence: Leo Olschki, 1982.

Bertin, Léon. "Buffon, homme d'affaires." In *Buffon,* ed. Roger Heim. Paris: Muséum National d'Histoire Naturelle, 1952, 87–104.

Bewell, Alan. "'On the Banks of the South Sea': Botany and Sexual Controversy in the Late Eighteenth Century." In *Visions of Empire: Voyages, Botany, and Representa-*

tions of Nature, ed. David Philip Miller and Peter Hanns Reill. Cambridge: Cambridge University Press, 1996, 173–193.

Biagioli, Mario. "Galileo's System of Patronage." *History of Science* 27 (1990): 1–62.

———. *Galileo, Courtier: The Practice of Science in the Culture of Absolutism.* Chicago: University of Chicago Press, 1993.

Bidal, Anne-Marie. *Inventaire des archives du Muséum National d'Histoire Naturelle: Première partie, série A, archives du Jardin du Roi.* Paris: Muséum National d'Histoire Naturelle, 1947.

Bien, David D. "Offices, Corps, and a System of State Credit: The Uses of Privilege under the Ancien Régime." In *The French Revolution and the Creation of Modern Political Culture,* vol. 1, *The Political Culture of the Old Regime,* ed. Keith Michael Baker. Oxford: Pergamon, 1987, 89–114.

———. "Manufacturing Nobles: The Chancellerie in France to 1789." *Journal of Modern History* 61 (1989): 445–486.

Birn, Raymond. "The French-Language Press and the Encyclopédie, 1750–1789." *Studies on Voltaire and the Eighteenth Century* 55 (1967): 263–286.

Biver, Marie-Louise. *Fêtes révolutionnaires à Paris.* Preface by Jean Tulard. Paris: Presses Universitaires de France, 1979.

Black, Jeremy. *Natural and Necessary Enemies: Anglo-French Relations in the Eighteenth Century.* London: Duckworth, 1986.

Blanckaert, Claude. "Buffon and the Natural History of Man: Writing History and the 'Foundational Myth' of Anthropology." *History of the Human Sciences* 6 (1993): 13–50.

Blanckaert, Claude, Claudine Cohen, Pietro Corsi, and Jean-Louis Fischer, eds. *Le Muséum au premier siècle de son histoire.* Paris: Muséum National d'Histoire Naturelle, 1997.

Bloch, Ernst. *Natural Law and Human Dignity.* Trans. D. Schmidt. Cambridge: MIT Press, 1986.

Bloch, Jean. *Rousseauism and Education in Eighteenth-Century France.* Oxford: Voltaire Foundation, 1995.

Bloch, Ruth H. *Visionary Republic: Millennial Themes in American Thought, 1756–1800.* Cambridge: Cambridge University Press, 1985.

Blunt, Wilfrid. *The Compleat Naturalist: A Life of Linnaeus.* London: Collins, 1971.

Bollème, Geneviève, ed. *Livre et société dans la France du XVIIIe siècle.* Paris: Mouton, 1965.

Bonnat, Jean-Louis, and Mireille Bossis, eds. *Ecrire, publier, lire: Les correspondances.* Nantes: Université de Nantes, 1983.

Bornet, Jean-Claude, ed. *La carmagnole des muses.* Paris: Armand Colin, 1986.

Bosher, J. F. *French Finances, 1770–1795: From Business to Bureaucracy.* Cambridge: Cambridge University Press, 1970.

———, ed. *French Government and Society, 1500–1850: Essays in Memory of Alfred Cobban.* London: Athlone, 1973.

Boulaine, Jean. "Les avatars de l'Académie d'agriculture sous la Révolution." In *Scientifiques et sociétés pendant la Révolution et l'Empire: Actes du 114e congrès national des sociétés savantes (Paris, 3–9 avril 1989), Section histoire des sciences et des techniques.* Paris: Éditions du CTHS, 1990, 211–227.

———. *Histoire de l'agronomie en France.* Paris: Lavoisier Tec et Doc, 1992.

Bouloiseau, Marc. *The Jacobin Republic, 1792–1794.* Trans. Jonathan Mandelbaum. Cambridge: Cambridge University Press; Paris: Éditions de la Maison des Sciences de l'Homme, 1987.

Bourde, André. *The Influence of the English on the French Agronomes, 1750–1789.* Cambridge: Cambridge University Press, 1953.

———. *Agronomie et agronomes en France au XVIIIe siècle.* 3 vols. Paris: SEVPEN, 1967.

Bourdier, Franck. "Buffon d'après ses portraits." In *Buffon,* ed. Roger Heim. Paris: Muséum National d'Histoire Naturelle, 1952, 167–180.

———. "Principaux aspects de la vie et de l'oeuvre de Buffon." In *Buffon,* ed. Roger Heim. Paris: Muséum National d'Histoire Naturelle, 1952, 15–86.

Bourguet, Marie-Noëlle. "L'image des terres incultes: La lande, la friche, le marais." In *La nature en Révolution, 1750–1800,* ed. Andrée Corvol. Paris: L'Harmattan, 1993, 15–29.

———. "Voyage, statistique, histoire naturelle: L'inventaire du monde au XVIIIe siècle." Rapport de synthèse, Université de Paris I, Panthéon, Sorbonne, 1993.

———. "La collecte du monde: Voyage et histoire naturelle, fin XVIIe siècle–début XIXe siècle." In *Le Muséum au premier siècle de son histoire,* ed. Claude Blanckaert et al. Paris: Muséum National d'Histoire Naturelle, 1997, 163–196.

Bowler, Peter J. "Bonnet and Buffon: Theories of Generation and the Problem of Species." *Journal for the History of Biology* 6 (1973): 259–281.

———. "Science and the Environment: New Agendas for the History of Science?" In *Science and Nature: Essays in the History of the Environmental Sciences,* ed. Michael Shortland. Oxford: Alden Press, 1993, 1–21.

Boyer, Ferdinand. *Le monde des arts en Italie et la France de la Révolution et de l'Empire: Études et recherches.* Torino: Società Editrice Internazionale, 1969.

———. "Le Muséum d'Histoire Naturelle et l'Europe des sciences sous la Convention." *Revue de l'histoire des sciences et de leurs applications* 20 (1973): 251–257.

Bradley, Margaret. "The Financial Basis of French Scientific Education and the Scientific Institutions of Paris, 1700–1815." *Annals of Science* 36 (1979): 451–492.

Brandenburg, David J. "Agriculture in the 'Encyclopédie': An Essay in French Intellectual History." *Agricultural History* 24 (1950): 96–108.

Broberg, Gunnar, ed. *Progress and Prospects in Linnaean Research.* Stockholm: Almqvist and Wiksell, 1980.

Broc, Numa. "Voyages et géographie au XVIIIe siècle." *Revue d'histoire des sciences et de leurs applications* 22 (1969): 137–154.

Brockliss, Laurence. *French Higher Education in the Seventeenth and Eighteenth Century: A Cultural History.* Oxford: Clarendon, 1987.

Brockliss, Laurence, and Colin Jones. *The Medical World of Early Modern France.* Oxford: Clarendon, 1997.

Brockway, Lucile H. *Science and Colonial Expansion: The Role of the British Royal Botanic Garden.* New York: Academic Press, 1979.

Brongniart, Adolphe. "Notice historique sur Antoine Laurent de Jussieu." *Annales des sciences naturelles* 7 (1837): 5–24.

Brown, A. W. "Some Political and Scientific Attitudes to Literature and the Arts in the Years Following the French Revolution." *Forum for Modern Language Studies* 2 (1966): 230–252.

Browne, Janet. "Botany for Gentlemen: Erasmus Darwin and *The Loves of the Plants.*" *Isis* 80 (1989): 593–621.

———. "Botany in the Boudoir and Garden: The Banksian Context." In *Visions of Empire: Voyages, Botany, and Representations of Nature,* ed. David Philip Miller and Peter Hanns Reill. Cambridge: Cambridge University Press, 1996, 153–172.

Brygoo, E. R. "Du Jardin et du Cabinet du Roi au Muséum d'Histoire Naturelle, en 1793: La continuité par les hommes." *Histoire et nature* 28–29 (1987–1988): 47–63.

Burke, Peter. *The Italian Renaissance: Culture and Society in Italy.* Cambridge: Polity Press, 1987.

———. *The Fabrication of Louis XIV.* New Haven: Yale University Press, 1992.

———. "Fables of the Bees: A Case-Study in Views of Nature and Society." In *Nature and Society in Historical Context,* ed. Mikuláš Teich, Roy S. Porter, and Bo Gustafsson. Cambridge: Cambridge University Press, 1997, 112–123.

———, ed. *New Perspectives on Historical Writing.* Cambridge: Polity Press, 1991.

Burkhardt, Richard W., Jr. "Lamarck, Evolution, and the Politics of Science." *Journal for the History of Biology* 3 (1970): 275–298.

———. "The Inspiration of Lamarck's Belief in Evolution." *Journal for the History of Biology* 5 (1972): 413–438.

———. *The Spirit of System: Lamarck and Evolutionary Biology.* Cambridge: Harvard University Press, 1977.

———. "Le comportement animal et l'idéologie de domestication chez Buffon et chez les éthologues modernes." In *Buffon 88: Actes du colloque international pour le bicentenaire de la mort de Buffon (Paris, Montbard, Dijon, 14–22 juin 1988),* ed. Jean-Claude Beaune, Serge Benoît, Jean Gayon, and Denis Woronoff. Paris: Vrin and Institut Interdisciplinaire d'Études Épistémologiques, 1992, 569–582.

———. "La ménagerie et la vie du Muséum." In *Le Muséum au premier siècle de son histoire,* ed. Claude Blanckaert et al. Paris: Muséum National d'Histoire Naturelle, 1997, 481–508.

Butel, Paul. "Revolution and the Urban Economy: Maritime Cities and Continental Cities." In *Reshaping France: Town, Country, and Region during the French Revolution,* ed. Alan Forrest and Peter Jones. Manchester: Manchester University Press, 1991, 37–51.

Buttoud, Gérard, and Yvonne Letouzey. "Les projets forestiers de la Révolution." *Revue forestière française* (1983): 9–20.

Callon, Michel. "Some Elements of a Sociology of Translation: Domestication of the Scallops and the Fishermen of St. Brieuc's Bay." In *Power, Action, and Belief: A New Sociology of Knowledge?* ed. John Law. London: Routledge and Kegan Paul, 1986, 196–233.

Canguilhem, Georges. *Études d'histoire et de philosophie des sciences.* Paris: Vrin, 1968.

Caron, Joseph. "'Biology' in the Life Sciences: A Historiographical Contribution." *History of Science* 26 (1988): 223–268.

Carter, Harold B. *His Majesty's Spanish Flock: Sir Joseph Banks and the Merinos of George III of England.* Sydney: Angus and Robertson, 1964.

———. *Sir Joseph Banks, 1743–1820.* London: British Museum (Natural History), 1988.

Castellani, Carlo. "The Problem of Generation in Bonnet and in Buffon: A Critical Comparison." In *Science, Medicine, and Society in the Renaissance: Essays to Honor Walter Pagel,* ed. Allan G. Debus. 2 vols. New York: Science History Publications, 1972, 2:265–288.

Censer, Jack R. "The Coming of a New Interpretation of the French Revolution?" *Journal of Social History* 21 (1987): 295–309.

Centenaire de la fondation du Muséum d'Histoire Naturelle, 10 juin 1793–10 juin 1893: Volume commémoratif publié par les professeurs du Muséum. Paris: Imprimerie Nationale, 1893.

Certeau, Michel de. *Heterologies: Discourse on the Other.* Trans. Brian Massumi. Minneapolis: University of Minnesota Press, 1986.

———. *The Practice of Everyday Life.* Vol. 1. Trans. Steven Rendall. Berkeley: University of California Press, 1988.

Chapin, Seymour L. "Scientific Profit from the Profit Motive: The Case of the La Pérouse Expedition." *Actes du XIIe congrès international d'histoire des sciences* 11 (1968–1971): 45–49.

———. "The Vicissitudes of a Scientific Institution: A Decade of Change at the Paris Observatory, 1789–1799." *Journal for the History of Astronomy* 21 (1990): 235–274.

Charlton, D. G. *New Images of the Natural in France: A Study in European Cultural History, 1750–1800.* Cambridge: Cambridge University Press, 1984.

Chartier, Roger. *Lectures et lecteurs dans la France d'ancien régime.* Paris: Seuil, 1987.

———. *Cultural History: Between Practices and Representations.* Trans. Lydia G. Cochrane. Cambridge: Polity Press, 1988.

———. "Texts, Printings, Readings." In *The New Cultural History,* ed. Lynn Hunt. Berkeley: University of California Press, 1989, 156–175.

———. *The Cultural Origins of the French Revolution: Bicentennial Reflections on the French Revolution.* Trans. Lydia G. Cochrane. Durham, NC: Duke University Press, 1991.

———. "Secrétaires for the People? Model Letters of the Ancien Régime: Between Court Literature and Popular Chapbooks." In Roger Chartier, Alain Boureau, and Cécile Dauphin, *Correspondence: Models of Letter-Writing from the Middle Ages to the Nineteenth Century,* trans. Christopher Woodall. Cambridge: Polity Press, 1997, 59–111.

———, ed. *A History of Private Life.* Vol. 3, *Passions of the Renaissance.* Trans. Arthur Goldhammer. Cambridge: Harvard University Press, Belknap Press, 1987.

Chartier, Roger, Alain Boureau, and Cécile Dauphin. *Correspondence: Models of Letter-Writing from the Middle Ages to the Nineteenth Century.* Trans. Christopher Woodall. Cambridge: Polity Press, 1997.

Chaussinand-Nogaret, Guy. *The French Nobility in the Eighteenth Century: From Feudalism to Enlightenment.* Trans William Doyle. Cambridge: Cambridge University Press, 1985.

Cherni, Amor. "Dégénération et dépravation: Rousseau chez Buffon." In *Buffon 88: Actes du colloque international pour le bicentenaire de la mort de Buffon (Paris, Montbard, Dijon, 14–22 juin 1988),* ed. Jean-Claude Beaune, Serge Benoît, Jean Gayon, and Denis Woronoff. Paris: Vrin and Institut Interdisciplinaire d'Études Épistémologiques, 1992, 143–154.

Chevalier, Auguste. *La vie et l'oeuvre de René Desfontaines, fondateur de l'herbier du Muséum: La carrière d'un savant sous la Révolution.* Paris: Muséum National d'Histoire Naturelle, 1939.

Chisick, Harvey. *The Limits of Reform in the Enlightenment: Attitudes toward the Education of the Lower Classes in Eighteenth-Century France.* Princeton: Princeton University Press, 1984.

Clark, William, Jan V. Golinski, and Simon Schaffer, eds. *The Sciences in Enlightened Europe.* Chicago: University of Chicago Press, 1999.

Cobban, Alfred. *The Social Interpretation of the French Revolution.* Cambridge: Cambridge University Press, 1965.

Coleman, William. *Georges Cuvier, Zoologist: A Study in the History of Evolution Theory.* Cambridge: Harvard University Press, 1964.

Collins, Harry, and Trevor Pinch. *The Golem: What Everyone Should Know about Science.* Cambridge: Cambridge University Press, 1993.

Corbin, Alain. *The Foul and the Fragrant: Odor and the French Social Imagination.* Trans. Jonathan Mandelbaum. Leamington Spa: Berg, 1986.

Corsi, Pietro. *The Age of Lamarck: Evolutionary Theories in France, 1790–1830.* Revised ed. Trans. Jonathan Mandelbaum. Berkeley: University of California Press, 1988.

Corvol, Andrée. *L'homme et l'arbre sous l'ancien régime.* Preface by Pierre Chaunu. Paris: Economica, 1984.

———. "L'arbre et la nature, XVIIe–XXe siècle." *Histoire economie et société* 6 (1987): 67–82.

———. "The Transformation of a Political Symbol: Tree Festivals in France from the Eighteenth to the Twentieth Centuries." *French History* 4 (1990): 455–486.

———, ed. *La nature en Révolution, 1750–1800.* Paris: L'Harmattan, 1993.

Corvol, Andrée, and Isabelle Richefort, eds. *Nature, environnement, et paysage: L'héritage du XVIIIe siècle: Guide de recherche archivistique et bibliographique.* Paris: L'Harmattan, 1995.

Cotgrove, Denis, and Stephen Daniels, eds. *The Iconography of Landscape: Essays on the Symbolic Representation, Design, and Use of Past Environments.* Cambridge: Cambridge University Press, 1988.

Coutura, Johel. "Le Musée de Bordeaux." *Dix-huitième siècle* 19 (1987): 149–164.

Cowell, F. R. *The Garden as Fine Art from Antiquity to Modern Times.* London: Weidenfeld and Nicolson, 1978.

Cowling, Mary. *The Artist as Anthropologist: The Representation of Type and Character in Victorian Art.* Cambridge: Cambridge University Press, 1989.

Crocker, Lester G. *Nature and Culture: Ethical Thought in the French Enlightenment.* Baltimore: Johns Hopkins University Press, 1963.

Crosland, Maurice P. "The Development of a Professional Career in Science in France." In *The Emergence of Science in Western Europe,* ed. Maurice P. Crosland. London: Macmillan, 1975, 139–160.

———. *Science under Control: The French Academy of Sciences, 1795–1914.* Cambridge: Cambridge University Press, 1992.

———. *Studies in the Culture of Science in France and Britain since the Enlightenment.* Aldershot: Variorum, 1995.

———, ed. *The Emergence of Science in Western Europe.* London: Macmillan, 1975.

Crow, Thomas Eugene. *Painters and Public Life in Eighteenth-Century Paris.* New Haven: Yale University Press, 1985.

———. *Emulation: Making Artists for Revolutionary France.* New Haven: Yale University Press, 1995.

Cunningham, Andrew, and Roger French, eds. *The Medical Enlightenment of the Eighteenth Century.* Cambridge: Cambridge University Press, 1990.

Cunningham, Andrew, and Nicholas Jardine, eds. *Romanticism and the Sciences.* Cambridge: Cambridge University Press, 1990.

Dagognet, François. "Valentin Haüy, Étienne Geoffroy Saint-Hilaire, Augustin-Pyramus de Candolle." *Revue d'histoire des sciences et de leurs applications* 25 (1972): 327–336.

———. "L'animal selon Condillac." In Étienne Bonnot de Condillac, *Traité des animaux,* ed. François Dagognet. Paris: Vrin, 1987, 10–131.

Dance, S. Peter. *A History of Shell Collecting.* Leiden: E. J. Brill, 1986.

Darnton, Robert. *Mesmerism and the End of the Enlightenment in France.* Cambridge: Harvard University Press, 1968.

———. "Le lieutenant de police J.-P. Lenoir, la guerre des farines, et l'approvisionnement de Paris à la veille de la Révolution." *Revue d'histoire moderne et contemporaine* 16 (1969): 611–624.

———. "In Search of the Enlightenment: Recent Attempts to Create a Social History of Ideas." *Journal of Modern History* 43 (1971): 113–132.

———. "The Encyclopédie Wars of Prerevolutionary France." *American Historical Review* 78 (1973): 1331–1352.

———. *The Business of Enlightenment: A Publishing History of the Encyclopédie, 1775–1800.* Cambridge: Harvard University Press, Belknap Press, 1979.

———. *The Literary Underground of the Old Régime.* Cambridge: Harvard University Press, 1982.

———. "The Epistemological Strategy of the Encyclopédie." In *Gelehrte Bücher vom Humanismus bis zur Gegenwart,* ed. Bernhard Fabian and Paul Raabe. Wiesbaden: Harrassowitz, 1983, 119–136.

———. *The Great Cat Massacre and Other Episodes in French Cultural History.* Harmondsworth, Middlesex: Penguin, 1985.

———. "The Literary Revolution of 1789." *Studies in Eighteenth-Century Culture* 21 (1991): 3–26.

Darnton, Robert, and Daniel Roche, eds. *Revolution in Print: The Press in France, 1775–1800.* Berkeley: University of California Press, 1989.

Daston, Lorraine. "Nature by Design." In *Picturing Science: Producing Art,* ed. Caroline A. Jones and Peter Galison. New York: Routledge, 1998, 232–253.

Daston, Lorraine, and Katharine Park. *Wonders and the Order of Nature, 1150–1750.* New York: Zone Books, 1998.

Daudin, Henri. *Les méthodes de la classification et l'idée de série en botanique et en zoologie de Linné à Lamarck, 1740–1790.* Paris: Félix Alcan, 1926.

———. *Cuvier et Lamarck: Les classes zoologiques et l'idée de série animale, 1790–1830.* 2 vols. Facsimile reprint, Paris: Belles-Lettres, 1988.

———. *De Linné à Jussieu: Méthode de la classification et idée de série en botanique et zoologie, 1740–1790.* Facsimile reprint, Paris: Belles-Lettres, 1988.

Daumas, Maurice. *Lavoisier: Théoricien et expérimentateur.* Paris: Presses Universitaires de France, 1955.

———. "Manuels épistolaires et identité sociale, XVIe–XVIIIe siècles." *Revue d'histoire moderne et contemporaine* 40 (1993): 529–556.

Dawson, Virginia P. *Nature's Enigma: The Problem of the Polyp in the Letters of Bonnet, Trembley, and Réaumur.* Philadelphia: American Philosophical Society, 1987.

Dawson, Warren R. *Catalogue of the Manuscripts in the Library of the Linnean Society of London.* London: Linnean Society, 1934.

Delaporte, François. *Nature's Second Kingdom: Explorations of Vegetality in the Eighteenth Century.* Trans. Arthur Goldhammer. Cambridge: MIT Press, 1982.

Denby, David J. *Sentimental Narrative and the Social Order in France, 1760–1820.* Cambridge: Cambridge University Press, 1994.

Desaive, Jean-Paul, Jacques-Philippe Goubert, and Emmanuel Le Roy Ladurie, eds. *Médecins, climat, et epidémies à la fin du XVIIIe siècle.* Paris: Mouton, 1972.

Desmond, Adrian. *The Politics of Evolution: Morphology, Medicine, and Reform in Radical London.* Chicago: University of Chicago Press, 1989.

Desmond, Ray. *Dictionary of British and Irish Botanists and Horticulturalists Including Plant Collectors and Botanical Artists.* London: Taylor and Francis, 1977.

Dettelbach, Michael. "Humboldtian Science." In *Cultures of Natural History,* ed. Nicholas Jardine, James A. Secord, and E. C. Spary. Cambridge: Cambridge University Press, 1996, 287–304.

Dhombres, Jean. "L'enseignement des mathématiques par la 'méthode révolutionnaire': Les leçons de Laplace à l'École Normale de l'an III." *Revue de l'histoire des sciences* 33 (1980): 316–348.

Dhombres, Jean, and Nicole Dhombres. *Naissance d'un nouveau pouvoir: Sciences et savants en France, 1793–1824.* Paris: Bibliothèque Historique Payot, 1989.

Dhombres, Nicole. *Les savants en Révolution, 1789–1799.* Paris: Cité des Sciences et de l'Industrie, 1989.

Dowd, David Lloyd. *Pageant-Master of the Republic: Jacques-Louis David and the French Revolution.* Lincoln: University of Nebraska Studies, 1948.

Driver, Felix, and Gillian Rose, eds. *Nature and Science: Essays in the History of Geographical Knowledge.* Special issue of *Historical Geography Research Series* 28 (1992).

Drouin, Jean-Marc. "Linné et l'économie de la nature." In *Sciences, techniques, et encyclopédies,* ed. D. Hue. Paris: Association Diderot, 1991, 147–158.

———. "L'histoire naturelle à travers un périodique: *La Décade philosophique.*" In *La nature en Révolution, 1750–1800,* ed. Andrée Corvol. Paris: L'Harmattan, 1993, 175–181.

———. *Réinventer la nature: L'écologie et son histoire.* Paris: Flammarion, 1993.

———. "L'histoire naturelle: Problèmes scientifiques et engouement mondain." In

Nature, environnement, et paysage: L'héritage du XVIIIe siècle: Guide de recherche archivistique et bibliographique, ed. Andrée Corvol and Isabelle Richefort. Paris: L'Harmattan, 1995, 19–27.

———. "Le Jardin des plantes à travers *La Décade philosophique.*" In *Les jardins entre science et représentation,* ed. Jean-Louis Fischer. Paris: CTHS, forthcoming.

Dubois, Jean, ed. *Dictionnaire de la langue française.* Paris: Larousse, 1987.

Duchet, Michèle. *Anthropologie et histoire au siècle des Lumières: Buffon, Voltaire, Rousseau, Helvétius, Diderot.* Paris: Maspero, 1971.

Dulieu, Louis. "Antoine Gouan, 1733–1821." *Revue d'histoire des sciences et de leurs applications* 20 (1967): 33–48.

Dumont, Martine. "Le succès mondain d'une fausse science: La physiognomonie de Johann Kaspar Lavater." *Actes de la recherche en sciences sociales* 54 (1984): 3–30.

Duris, Pascal. *Linné et la France, 1780–1850.* Geneva: Droz, 1993.

Duveen, Denis I., and Herbert S. Klickstein. *A Bibliography of the Works of Antoine-Laurent Lavoisier, 1743–1794.* With supplement. London: Dawsons, 1965.

Eddy, John H., Jr. "Buffon, Organic Alterations, and Man." *Studies in the History of Biology* 7 (1984): 1–45.

———. "Buffon's Histoire Naturelle: History? A Critique of Recent Interpretations." *Isis* 85 (1994): 644–661.

Edwards, Paul, ed. *The Encyclopedia of Philosophy.* New York: Macmillan, 1967.

Ehrard, Jean. *L'idée de nature en France dans la première moitié du XVIIIe siècle.* Reprint. Paris: Albin Michel, 1994.

Ehrard, Jean, and Paul Viallaneix, eds. *Les fêtes de la Révolution: Colloque de Clermont-Ferrand, juin 1974.* Paris: Société des Études Robespierristes, 1977.

Eisinger, Chester E. "The Influence of Natural Rights and Physiocratic Doctrines on American Agrarian Thought during the Revolutionary Period." *Agricultural History* 21 (1947): 13–23.

Elias, Norbert. *The Civilising Process.* Vol. 1, *The History of Manners.* Vol. 2, *Power and Civility.* Trans. Edmund Jephcott. Oxford: Blackwell, 1978–1982.

———. *The Court Society.* Trans. Edmund Jephcott. Oxford: Blackwell, 1983.

Ellis, Steven G. *Tudor Ireland: Crown, Community, and the Conflict of Cultures, 1470–1603.* London: Longman, 1985.

Elsner, John, and Roger Cardinal, eds. *The Cultures of Collecting.* London: Reaktion Books, 1994.

Emerson, Roger L. "The Edinburgh Society for the Importation of Foreign Seeds and Plants, 1764–1773." *Eighteenth-Century Life* 7 (1982): 73–95.

Eriksson, Gunnar. "Linnaeus the Botanist." In *Linnaeus: The Man and His Work,* ed. Tore Frängsmyr. Canton, MA: Science History Publications, 1994, 63–109.

Ewan, Joseph. "Fougeroux de Bondaroy (1732–1789) and His Projected Revision of Duhamel du Monceau's 'Traité' (1755) on Trees and Shrubs." *Proceedings of the American Philosophical Society* 103 (1959): 807–818.

Falls, William P. "Buffon et l'agrandissement du Jardin du Roi à Paris." *Archives du Muséum d'Histoire Naturelle,* series 6, 10 (1933): 131–200.

———. "Buffon et les premières bêtes du Jardin du Roi: Histoire ou légende?" *Isis* 30 (1939): 491–494.

Farber, Paul L. "Buffon and the Concept of Species." *Journal for the History of Biology* 5 (1972): 259–284.

———. "Buffon and Daubenton: Divergent Traditions within the *Histoire Naturelle.*" *Isis* 66 (1975): 63–74.

———. "Research Traditions in Eighteenth-Century Natural History." In *Lazzaro Spallanzani e la biologia del settecento: Teorie, esperimenti, istituzioni scientifiche,* ed. Walter Bernardi and Antonello La Vergata. Florence: Leo Olschki, 1982, 397–403.

Faugue, Danielle. "Il y a deux cents ans: L'expédition Lapérouse." *Revue d'histoire des sciences* 38 (1985): 149–160.

Fayet, Joseph. *La Révolution française et la science, 1789–1795.* Paris: Marcel Rivière, 1960.

Feldman, Theodore S. "Late Enlightenment Meteorology." In *The Quantifying Spirit in the Eighteenth Century,* ed. Tore Frängsmyr, John L. Heilbron, and Robin E. Rider. Berkeley: University of California Press, 1990, 143–177.

———. "The Ancient Climate in the Eighteenth and Early Nineteenth Century." In *Science and Nature: Essays in the History of the Environmental Sciences,* ed. Michael Shortland. Oxford: Alden Press, 1993, 23–40.

Festy, Octave. *L'agriculture pendant la Révolution française.* Vol. 1, *Les conditions de production et de récolte des céréales: Étude d'histoire économique, 1789–1795.* Vol. 2, *L'utilisation des jachères.* Paris: Gallimard, 1947–1950.

Findlen, Paula. *Possessing Nature: Museums and Collecting in Early Modern Italy.* Berkeley: University of California Press, 1994.

Fischer, Jean-Louis. "Étienne Geoffroy Saint-Hilaire." *Revue d'histoire des sciences et de leurs applications* 25 (1972): 293–390.

Fish, Stanley. *Is There a Text in This Class? The Authority of Interpretative Communities.* Cambridge: Harvard University Press, 1980.

Fisher, John. *The Origins of Garden Plants.* London: Constable, 1982.

Fliegelman, Jay. *Prodigals and Pilgrims: The American Revolution against Patriarchal Authority, 1750–1800.* Cambridge: Cambridge University Press, 1982.

Flourens, Pierre. *Des manuscrits de Buffon.* Paris: Garnier, 1860.

Forgan, Sophie. "The Architecture of Display: Museums, Universities, and Objects in Nineteenth-Century Britain." *History of Science* 32 (1994): 139–162.

Forrest, Alan, and Peter Jones, eds. *Reshaping France: Town, Country, and Region during the French Revolution.* Manchester: Manchester University Press, 1991.

Forster, Robert. "Obstacles to Agricultural Growth in Eighteenth-Century France." *American Historical Review* 75 (1970): 1600–1615.

Fortunet, Françoise, Philippe Jobert, and Denis Woronoff. "Buffon en affaires." In *Buffon 88: Actes du colloque international pour le bicentenaire de la mort de Buffon (Paris, Montbard, Dijon, 14–22 juin 1988),* ed. Jean-Claude Beaune, Serge Benoît, Jean Gayon, and Denis Woronoff. Paris: Vrin and Institut Interdisciplinaire d'Études Épistémologiques, 1992, 13–28.

Foucault, Michel. *Madness and Civilisation: A History of Insanity in the Age of Reason.* Trans. Richard Howard. London: Tavistock, 1967.

———. *The Birth of the Clinic: An Archaeology of Medical Perception.* Trans. Alan Sheridan. London: Tavistock, 1973.

———. "Cuvier's Position in the History of Biology." *Critical Anthropology* 4 (1979): 125–130.

———. *Power/Knowledge.* Brighton: Harvester Press, 1980.

———. *The Order of Things: An Archaeology of the Human Sciences.* Trans. Alan Sheridan. London: Tavistock, 1985.

———. *Discipline and Punish: The Birth of the Prison.* Trans. Alan Sheridan. London: Penguin, 1991.

Fox, Robert. "The Rise and Fall of Laplacian Physics." *Historical Studies of the Physical Sciences* 4 (1974): 89–136.

———. "Scientific Enterprise and the Patronage of Research in France, 1800–1900." In *The Patronage of Science in the Nineteenth Century,* ed. G. L'Estrange Turner. Leiden: Noordhoff, 1976, 9–51.

———. "Learning, Politics, and Polite Culture in Provincial France: The Sociétés Savantes in the Nineteenth Century." *Historical Reflections* 7 (1980): 543–564.

Fox, Robert, and George Weisz, eds. *The Organisation of Science and Technology in France, 1808–1914.* Cambridge: Cambridge University Press, 1980.

Fox-Genovese, Elizabeth. *The Origins of Physiocracy: Economic Revolution and Social Order in Eighteenth-Century France.* Ithaca, NY: Cornell University Press, 1976.

François, Yves. "Notes pour l'histoire du Jardin des Plantes: Sur quelques projets d'amenagement du Jardin au temps de Buffon." *Bulletin du Muséum d'Histoire Naturelle,* series 2, 22 (1950): 675–693.

———. "Buffon au Jardin du Roi, 1739–1788." In *Buffon,* ed. Roger Heim. Paris: Muséum National d'Histoire Naturelle, 1952, 105–124.

Frängsmyr, Tore, ed., *Linnaeus: The Man and His Work.* Canton, MA: Science History Publications, 1994.

Frängsmyr, Tore, John L. Heilbron, and Robin E. Rider, eds. *The Quantifying Spirit in the Eighteenth Century.* Berkeley: University of California Press, 1990.

Frei, H. W. *The Eclipse of Biblical Narrative.* New Haven: Yale University Press, 1974.

French Caricature and the French Revolution, 1789–1799. Chicago: University of Chicago Press, 1988.

Fried, Michael. *Absorption and Theatricality: Painting and Beholder in the Age of Diderot.* Chicago: University of Chicago Press, 1980.

Friguglietti, James. "Interpreting vs. Understanding the Revolution: François Furet and Albert Soboul." *Consortium on Revolutionary Europe, 1750–1850: Proceedings* 17 (1987): 23–36.

Fruchtman, Jack, Jr. *The Apocalyptic Politics of Richard Price and Joseph Priestley: A Study in Late Eighteenth-Century English Republican Millennialism.* Philadelphia: American Philosophical Society, 1983.

Furet, François. *Interpreting the French Revolution.* Trans. Elborg Forster. Paris: Éditions de la Maison des Sciences de l'Homme; Cambridge: Cambridge University Press, 1988.

Furet, François, and Denis Richet. *French Revolution.* London: Weidenfeld and Nicolson, 1970.

Galliano, Paul, Robert Philippe, and Philippe Suissel. *La France des Lumières, 1715–1789: Histoire de la France.* Paris: Culture, Art, Loisirs, 1970.

Garrett, Clarke. *Respectable Folly: Millenarians and the French Revolution in France and England.* Baltimore: Johns Hopkins University Press, 1975.

Gartrell, Ellen G. *Electricity, Magnetism, and Animal Magnetism: A Checklist of Printed Sources, 1600–1850.* Wilmington, DE: Scholarly Resources, 1975.

Gascoigne, John. *Joseph Banks and the English Enlightenment: Useful Knowledge and Polite Culture.* Cambridge: Cambridge University Press, 1994.

Gasking, Elizabeth. *Investigations into Generation, 1651–1828.* London: Hutchinson, 1967.

Gay, Peter. *The Enlightenment: An Interpretation.* 2 vols. London: Weidenfeld and Nicolson, 1967.

———. "The Enlightenment as Medicine and as Cure." In *The Age of the Enlightenment: Studies Presented to Theodore Besterman,* ed. W. H. Barber, J. H. Brumfitt, R. A. Leigh, R. Shackleton, and S. S. B. Taylor. Edinburgh: Oliver and Boyd, 1967, 375–386.

———, ed. *Eighteenth-Century Studies Presented to Arthur M. Wilson.* Hanover, NH: University Press of New England, 1972.

Gaziello, Catherine. *L'expédition de Lapérouse, 1785–1788: Replique française aux voyages de Cook.* Paris: CTHS, 1984.

Géhin, Étienne. "Rousseau et l'histoire naturelle de l'homme social." *Revue française de sociologie* 22 (1981): 15–31.

Geison, Gerald L. *Professions and the French State, 1700–1900.* Philadelphia: University of Pennsylvania Press, 1984.

Gelbart, Nina Rattner. "The French Revolution as Medical Event: The Journalistic Gaze." *History of European Ideas* 10 (1989): 417–427.

Geoffroy Saint-Hilaire, Isidore. *Vie, travaux, et doctrine scientifique d'Étienne Geoffroy Saint-Hilaire.* Paris: Bertrand, 1847.

Gierke, Otto. *Natural Law and the Theory of Society, 1500–1800.* Trans. E. Barker. Boston: Beacon Press, 1957.

Gillespie, Neal C. "Natural History, Natural Theology, and Social Order: John Ray and the 'Newtonian Ideology.'" *Journal of the History of Biology* 20 (1987): 1–49.

Gillispie, Charles Coulston. "The Formation of Lamarck's Evolutionary Theory." *Archives internationales d'histoire des sciences* 9 (1956): 323–338.

———. "The *Encyclopédie* and the Jacobin Philosophy of Science: A Study in Ideas and Consequences." In *Critical Problems in the History of Science,* ed. Marshall Clagett. New York: Madison, 1959, 255–289.

———. "Science in the French Revolution." *Behavioural Sciences* 4 (1959): 69–73.

———. "Probability and Politics: Laplace, Condorcet, and Turgot." *Proceedings of the American Philosophical Society* 116 (1972): 1–20.

———. *Science and Polity in France at the End of the Old Regime.* Princeton: Princeton University Press, 1980.

———. "Science and Secret Weapons Development in Revolutionary France, 1792–1804: A Documentary History." *Historical Studies in the Physical and Biological Sciences* 23 (1992): 35–152.

———, ed. *Dictionary of Scientific Biography.* 16 vols. New York: Charles Scribner, 1970–1980.

Glacken, Clarence J. *Traces on the Rhodian Shore: Nature and Culture in Western Thought from Ancient Times to the End of the Eighteenth Century.* Berkeley: University of California Press, 1967.

Glass, Bentley. "Heredity and Variation in the Eighteenth Century Concept of Species." In *Forerunners of Darwin,* ed. Bentley Glass, Owsei Temkin, and William L. Straus Jr. Baltimore: Johns Hopkins University Press, 1959, 144–172.

Glass, Bentley, Owsei Temkin, and William L. Straus Jr., eds. *Forerunners of Darwin.* Baltimore: Johns Hopkins University Press, 1959.

Goldgar, Anne. *Impolite Learning: Conduct and Community in the Republic of Letters, 1680–1750.* New Haven: Yale University Press, 1995.

Goldstein, Jan, ed., *Foucault and the Writing of History.* Oxford: Blackwell, 1994.

Golinski, Jan V. "Utility and Audience in Eighteenth-Century Chemistry: Case Studies of William Cullen and Joseph Priestley." *British Journal for the History of Science* 21 (1988): 1–32.

———. *Science as Public Culture: Chemistry and Enlightenment in Britain, 1760–1820.* Cambridge: Cambridge University Press, 1992.

Gooding, David. *Experiment and the Making of Meaning: Human Agency in Scientific Observation and Experiment.* Dordrecht: Kluwer Academic, 1990.

Gooding, David, Trevor Pinch, and Simon Schaffer, eds. *The Uses of Experiment: Studies in the Natural Sciences.* Cambridge: Cambridge University Press, 1989.

Goodman, David. *Buffon's Natural History.* Milton Keynes: Open University Press, 1980.

Goodman, Dena. *The Republic of Letters: A Cultural History of the French Enlightenment.* Ithaca, NY: Cornell University Press, 1994.

Gould, Cecil. *Trophy of Conquest: The Musée Napoléon and the Creation of the Louvre.* London: Faber and Faber, 1965.

Goupil, Michelle. *Du flou au clair? Histoire de l'affinité chimique de Cardan à Prigogine.* Paris: Éditions du CTHS, 1991.

Greene, John C. *The Death of Adam: Evolution and Its Impact on Western Thought.* 2d ed. Ames: Iowa State University Press, 1996.

Grenon, Michel. "Science ou vertu? L'idée de progrès dans le débat sur l'instruction publique, 1789–1795." *Études françaises* 25 (1989): 177–190.

Grinevald, Paul-Marie. "Les éditions de l'*Histoire naturelle*." In *Buffon 88: Actes du colloque international pour le bicentenaire de la mort de Buffon (Paris, Montbard, Dijon, 14–22 juin 1988),* ed. Jean-Claude Beaune, Serge Benoît, Jean Gayon, and Denis Woronoff. Paris: Vrin and Institut Interdisciplinaire d'Études Épistémologiques, 1992, 631–637.

Grove, Richard H. *Green Imperialism: Colonial Expansion, Tropical Island Edens, and the Origins of Environmentalism, 1600–1860.* Cambridge: Cambridge University Press, 1995.

Guedès, Michel. "La méthode taxonomique d'Adanson." *Revue d'histoire des sciences et de leurs applications* 20 (1967): 361–386.

———. "Jussieu's Natural Method." *Taxon* 22 (1973): 211–219.

Guerlac, Henry. *Lavoisier: The Crucial Year: The Background and Origins of His First Experiments on Combustion in 1772.* Ithaca, NY: Cornell University Press, 1961.

Guery, Alain. "Le roi dépensier: Le don, la contrainte, et l'origine du système financier de la monarchie française d'ancien régime." *Annales: Économies, sociétés, civilisations* 39 (1984): 1241–1269.

Guillaume, A., and V. Chaudun. "La collection de modèles réduits d'instruments agricoles et horticoles du muséum à propos d'une lettre inédite de A. Thouin." *Bulletin du Muséum National d'Histoire Naturelle,* series 2, 16 (1944): 137–141.

Guillaumin, André. "André Thouin et l'enrichissement des collections de plantes vivantes du Muséum aux dépens des jardins de la liste civile, des émigrés, et condamnés: D'après ses notes manuscrits." *Bulletin du Muséum National d'Histoire Naturelle,* series 2, 16 (1944): 483–490.

Guillerme, André. "Network: Birth of a Category in Engineering Thought during the French Restoration." *History and Technology* 8 (1992): 151–166.

Gunn, J. D. W. *Beyond Liberty and Property: The Process of Self-Recognition in Eighteenth-Century Political Thought.* Kingston, Ontario: McGill–Queen's University Press, 1983.

Gusdorf, Georges. *Dieu, la nature, l'homme au siècle des Lumières.* Paris: Payot, 1972.

———. *La conscience révolutionnaire: Les idéologues.* Paris: Payot, 1979.

Habermas, Jürgen. *The Structural Transformation of the Public Sphere: An Inquiry into a Category of Bourgeois Society.* Trans. T. Burger and F. Lawrence. Cambridge: MIT Press, 1989.

Hahn, Roger. "The Problems of the French Scientific Community, 1793–1795." *Actes du XIIe congrès international d'histoire des sciences* 3 (1968): 37–40.

———. "Elite scientifique et démocratie politique dans la France révolutionnaire." *Dix-huitième siècle* 1 (1969): 229–235.

———. *The Anatomy of a Scientific Institution: The Paris Academy of Sciences, 1666–1803.* Berkeley: University of California Press, 1971.

———. "Sur les débuts de la carrière scientifique de Lacepède." *Revue d'histoire des sciences* 27 (1974): 347–353.

———. "L'autobiographie de Lacepède retrouvée." *Dix-huitième siècle* 7 (1975): 49–85.

———. "Scientific Careers in Eighteenth-Century France." In *The Emergence of Science in Western Europe,* ed. Maurice P. Crosland. London: Macmillan, 1975, 127–138.

———. "Scientific Research as an Occupation in Eighteenth-Century Paris." *Minerva* 13 (1975): 501–513.

———. "Science and the Arts in France: The Limitations of an Encyclopedic Ideology." *Studies in Eighteenth-Century Culture* 10 (1981): 77–93.

———. "The Triumph of Scientific Activity: From Louis XVI to Napoleon." *Proceed-*

ings of the Annual Meeting of the Western Society for French History 16 (1989): 204–211.

Haigh, Elizabeth L. "Vitalism, the Soul, and Sensibility: The Physiology of Théophile Bordeu." *Journal for the History of Medicine and Allied Sciences* 31 (1976): 30–44.

———. *Xavier Bichat and the Medical Theory of the Eighteenth Century.* London: Wellcome Institute, 1984.

Hampson, Norman. *Will and Circumstance: Montesquieu, Rousseau, and the French Revolution.* London: Duckworth, 1983.

Hamy, Ernest-Théodore. "Les derniers jours du Jardin du Roi et la fondation du Muséum d'Histoire Naturelle." In *Centenaire de la fondation du Muséum d'Histoire naturelle, 10 juin 1793–10 juin 1893: Volume commémoratif publié par les professeurs du Muséum.* Paris: Imprimerie Nationale, 1893, 1–162.

Hanks, Lesley. *Buffon avant l'*Histoire naturelle. Paris: Presses Universitaires de France, 1966.

Hanley, William. "The Policing of Thought: Censorship in Eighteenth-Century France." *Studies on Voltaire and the Eighteenth Century* 183 (1980): 265–295.

Hannaway, Caroline C. "The Société Royale de Médecine and Epidemics in the *Ancien Régime.*" *Bulletin for the History of Medicine* 46 (1972): 257–273.

Hannaway, Owen, and Caroline C. Hannaway. "La fermeture du cimetière des Innocents." *Dix-huitième siècle* 2 (1977): 181–191.

Harding, Robert. *Anatomy of a Power Elite: The Provincial Governors of Early Modern France.* New Haven: Yale University Press, 1978.

Harten, Elke. *Museen und Museumsprojekte der französischen Revolution: Ein Beitrag zur Entstehungsgeschichte einer Institution.* Kunstgeschichte, Band 24. Münster: Lit, 1989.

Harten, Hans-Christian, and Elke Harten. *Die Versöhnung mit der Natur: Gärten, Freiheitsbäume, republikanische Wälder, heilige Berge, und Tugendparks in der französischen Revolution.* Reinbek: Rowohlt, 1989.

Harvey, John. *Early Nurserymen: With Reprints of Documents and Lists.* London: Phillimore and Co., 1974.

Hawthorn, Geoffrey. *Plausible Worlds.* Princeton: Princeton University Press, 1992.

Heilbron, John L. *Electricity in the Seventeenth and Eighteenth Centuries: A Study of Early Modern Physics.* Berkeley: University of California Press, 1979.

Heim, Roger. "Preface à Buffon." In *Buffon,* ed. Roger Heim. Paris: Muséum Nationale d'Histoire Naturelle, 1952.

———, ed. *Buffon.* Paris: Muséum National d'Histoire Naturelle, 1952.

Heller, John Lewis. *Studies in Linnaean Method and Nomenclature.* Frankfurt am Main: Peter Lang, 1983.

Herbert, R. L. *David, Voltaire, Brutus, and the French Revolution: An Essay in Art and Politics.* London: Allen Lane, 1972.

Hilts, Victor. "Enlightenment Views on the Genetic Perfectibility of Man." In *Transformation and Tradition in the Sciences: Essays in Honour of I. Bernard Cohen,* ed. Everett Mendelsohn. Cambridge: Cambridge University Press, 1984, 255–271.

Hoffman, Philip T. "Institutions and Agriculture in Old Regime France." *Politics and Society* 16 (1988): 241–264.

Holdengräber, Paul. "'A Visible History of Art': The Forms and Preoccupations of the Early Museum." *Studies in Eighteenth-Century Culture* 17 (1987): 107–117.

Holub, Robert C. *Reception Theory: A Critical Introduction.* London: Methuen, 1983.

Hondt, Jean-Loup d'. "Louis-Auguste-Guillaume Bosc (1759–1828), conventionnel et naturaliste, premier systématicien français des bryozoaires actuels." In *Scientifiques et sociétés pendant la Révolution et l'Empire: Actes du 114e congrès national des sociétés savantes (Paris 3–9 avril 1989), Section histoire des sciences et des techniques.* Paris: Éditions du CTHS, 1990, 241–258.

Hooper-Greenhill, Eilean. "The Museum in the Disciplinary Society." In *Museum Studies in Material Culture,* ed. Susan M. Pearce. Leicester: Leicester University Press, 1989, 61–72.

———. *Museums and the Shaping of Knowledge.* London: Routledge, 1992.

Horn, Jeffrey. "The Revolution as Discourse." *History of European Ideas* 13 (1991): 623–632.

Horowitz, Asher. *Rousseau, Nature, and History.* Buffalo: University of Toronto Press, 1987.

———. "'Laws and Customs Thrust Us Back into Infancy': Rousseau's Historical Anthropology." *Review of Politics* 52 (1990): 215–241.

Howard, Rio Cecily. "Guy de La Brosse: The Founder of the Jardin des Plantes in Paris." Ph.D. dissertation, Cornell University, 1974.

Howse, Derek, ed. *Background to Discovery: Pacific Exploration from Dampier to Cook.* Berkeley: University of California Press, 1990.

Huard, Pierre. "L'enseignement médico-chirurgical." In *Enseignement et diffusion des sciences en France au XVIIIe siècle,* ed. René Taton. Reprint. Paris: Hermann, 1986, 171–236.

Huard, Pierre, and Ming Wong. "Les enquêtes scientifiques françaises et l'exploration du monde éxotique aux XVIIe et XVIIIe siècles." *Bulletin de l'école française d'Extrême-Orient* 52 (1964): 143–155.

Hublard, Emile. *Le naturaliste hollandois Pierre Lyonet: Sa vie et ses oeuvres (1706–1789) d'après des lettres inédites.* Brussels: J. Lebègue et Cie, 1910.

Huisman, Philippe. *French Watercolours of the Eighteenth Century.* Trans. D. Imber. London: Thames and Hudson, 1969.

Hunt, Lynn. *Politics, Culture, and Class in the French Revolution.* Berkeley: University of California Press, 1984.

———. "The Unstable Boundaries of the French Revolution." In *A History of Private Life,* vol. 4, *From the Fires of Revolution to the Great War,* ed. Michelle Perrot, trans. Arthur Goldhammer. Cambridge: Harvard University Press, Belknap Press, 1990, 13–45.

———. *The Family Romance of the French Revolution.* London: Routledge, 1992.

———, ed. *The New Cultural History,* Berkeley: University of California Press, 1989.

Impey, Oliver, and Arthur MacGregor, eds. *The Origins of Museums: The Cabinet of Curiosities in Sixteenth- and Seventeenth-Century Europe.* Oxford: Clarendon, 1985.

Iriye, Masumi. "Le Vau's Menagerie and the Rise of the Animalier: Enclosing, Dissecting, and Representing the Animal in Early Modern France." Ph.D. dissertation, University of Michigan, 1994.

Jackson, Myles. "Natural and Artificial Budgets: Accounting for Goethe's Economy of Nature." *Science in Context* 7 (1994): 409–431.

Jacob, Margaret C. *The Radical Enlightenment: Pantheists, Freemasons, and Republicans.* London: Allen and Unwin, 1981.

Jardine, Nicholas. *The Scenes of Enquiry.* Oxford: Clarendon, 1991.

———. "The Laboratory Revolution in Medicine as Rhetorical and Aesthetic Accomplishment." In *Romanticism and the Sciences,* ed. Andrew Cunningham and Nicholas Jardine. Cambridge: Cambridge University Press, 1990, 304–323.

Jardine, Nicholas, James A. Secord, and E. C. Spary, eds. *Cultures of Natural History.* Cambridge: Cambridge University Press, 1996.

Jardine, Nicholas, and E. C. Spary. "Introduction: The Natures of Cultural History." In *Cultures of Natural History,* ed. Nicholas Jardine, James A. Secord, and E. C. Spary. Cambridge: Cambridge University Press, 1996, 3–13.

Jaume, Lucien. "Le public et le privé chez les Jacobins, 1789–1794." *Revue française de science politique* 37 (1987): 230–248.

Johns, Adrian. "Dolly's Wax: The Historical Physiology of Interpretation in Early

Modern England." In *The Practice and Representation of Reading in England,* ed. James Raven, Helen Small, and Naomi Tadmor. Cambridge: Cambridge University Press, 1996, 138–161.

———. "Science and the Book in Modern Cultural Historiography." *Studies in History and Philosophy of Science* 35 (1997): 23–59.

———. *The Nature of the Book: Print and Knowledge in the Making.* Chicago: University of Chicago Press, 1998.

Jones, Colin. "Bourgeois Revolution Revivified: 1789 and Social Change." In *Rewriting the French Revolution,* ed. Colin Lucas. Oxford: Clarendon, 1991, 69–118.

Jordan, David P. *The Revolutionary Career of Maximilien Robespierre.* New York: Macmillan, 1985.

Jordanova, Ludmilla J. "The Natural Philosophy of Lamarck in Its Historical Context." Ph.D. dissertation, University of Cambridge, 1976.

———. "Earth Science and Environmental Medicine: The Synthesis of the Late Enlightenment." In *Images of the Earth: Essays in the History of the Environmental Sciences,* ed. Ludmilla J. Jordanova and Roy S. Porter. Chalfont St. Giles: British Society for the History of Science, 1979, 119–146.

———. "Policing Public Health in France, 1780–1815." In *Public Health: Proceedings of the 5th International Symposium on the Comparative History of Medicine—East and West,* ed. Teizo Ogawa. Tokyo: Saikon Publishing, 1981, 12–32.

———. "Urban Health in the French Enlightenment." *Society for the Social History of Medicine Bulletin* 32 (1983): 31–33.

———. *Lamarck.* Oxford: Oxford University Press, 1984.

———. "Environmentalism in the Eighteenth Century." In *Nature and Science: Essays in the History of Geographical Knowledge,* ed. Felix Driver and Gillian Rose. Special issue of *Historical Geography Research Series* 28 (1992).

———, ed. *Languages of Nature: Critical Essays on Science and Literature.* London: Free Association Press, 1986.

Jordanova, Ludmilla J., and Roy S. Porter, eds. *Images of the Earth: Essays in the History of the Environmental Sciences.* Chalfont St. Giles: British Society for the History of Science, 1979.

Joyce, Patrick, ed. *The Historical Meanings of Work.* Cambridge: Cambridge University Press, 1987.

Julia, Dominique. *Les trois couleurs du tableau noir: La Révolution.* Paris: Belin, 1981.

Kaplan, Steven Laurence. *Bread, Politics, and Political Economy in the Reign of Louis XV.* 2 vols. The Hague: Nijhoff, 1976.

Kaplan, Steven Laurence, and Cynthia J. Koepp, eds. *Work in France: Representations, Meaning, Organisation, and Practice.* Ithaca, NY: Cornell University Press, 1986.

Kennedy, R. Emmet, Jr. "François Furet: Post-Patriot Historian of the French Revolution." *Proceedings of the 11th Annual Meeting of the Western Society for French History* 11 (1984): 194–200.

———. *A Cultural History of the French Revolution.* New Haven: Yale University Press, 1989.

Kenyon, Timothy. "Utopia in Reality: 'Ideal' Societies in Social and Political Theory." *History of Political Thought* 3 (1982): 123–155.

Kersaint, Georges. "Antoine-François de Fourcroy, 1755–1809: Sa vie et son oeuvre." *Mémoires du Muséum d'Histoire Naturelle,* series D, 2 (1966): 1–296.

Kettering, Sharon. *Patrons, Brokers, and Clients in Seventeenth Century France.* New York: Oxford University Press, 1986.

Knafla, Louis A., Martin S. Staum, and T. H. E. Travers, eds. *Science, Technology, and Culture in Historical Perspective.* Calgary: University of Calgary Press, 1976.

Knorr-Cetina, Karin D., and Michael Mulkay, eds. *Science Observed: Perspectives on the Social Study of Science.* London: Sage, 1983.

Koerner, Lisbet. "Carl Linnaeus in His Time and Place." In *Cultures of Natural History,* ed. Nicholas Jardine, James A. Secord, and E. C. Spary. Cambridge: Cambridge University Press, 1996, 145–162.

———. "Purposes of Linnaean Travel: A Preliminary Research Report." In *Visions of Empire: Voyages, Botany, and Representations of Nature,* ed. David Philip Miller and Peter Hanns Reill. Cambridge: Cambridge University Press, 1996, 117–152.

Kors, Alan Charles. *D'Holbach's Coterie: An Enlightenment in Paris.* Princeton: Princeton University Press, 1976.

Koselleck, Reinhard. *Critique and Crisis: Enlightenment and the Pathogenesis of Modern Society.* Oxford: Berg, 1988.

Kury, Lorelai. "Les instructions de voyage dans les expéditions scientifiques françaises, 1750–1830." *Revue d'histoire des sciences* 51 (1998): 65–91.

Lacroix, Jean-Bernard. "L'approvisionnement des ménagéries et les transports d'animaux sauvages par le Compagnie des Indes au XVIIIe siècle." *Revue française d'histoire d'outre-mer* 65 (1978): 153–179.

LaFreniere, Gilbert F. "Rousseau and the European Roots of Environmentalism." *Environmental History Review* 14 (1990): 41–72.

Laissus, Joseph. "La succession de Le Monnier au Jardin du Roi: Antoine-Laurent de Jussieu et René-Louiche Desfontaines." *Comptes rendus du 91e congrès national des sociétés savantes, 1966, Section des sciences* 1 (1967): 137–152.

Laissus, Yves. "Les voyageurs naturalistes du Jardin du Roi et du Muséum d'Histoire Naturelle: Essai de portrait-robot." *Revue d'histoire des sciences* 34 (1981): 259–317.

———. "Le Jardin du Roi." In *Enseignement et diffusion des sciences en France au XVIIIe siècle,* ed. René Taton. Reprint, Paris: Hermann, 1986, 287–341.

———. "Les cabinets d'histoire naturelle." In *Enseignement et diffusion des sciences en France au XVIIIe siècle,* ed. René Taton. Reprint, Paris: Hermann, 1986, 342–384.

Lamadon, A. "Fêtes en Révolution, 1789–1794." *Revue d'Auvergne* 103 (1989): 59–82.

Lamande, Pierre. "La mutation de l'enseignement scientifique en France (1750–1810) et le rôle des écoles centrales: L'exemple de Nantes." Special issue of *Sciences et techniques en perspective* 15 (1988–1989).

Lang, Catherine. "Joseph Dombey (1742–1794), un botaniste au Pérou et au Chili: Présentation des sources." *Revue d'histoire moderne et contemporaine* 35 (1988): 262–274.

Langins, Janis. *La République avait besoin de savants: Les débuts de l'École polytechnique: L'École centrale des travaux publics et les cours révolutionnaires de l'an III.* Paris: Belin, 1987.

———. "Sur l'enseignement et les examens à l'école polytechnique sous le Directoire: À propos d'une lettre inédite de Laplace." *Revue d'histoire des sciences* 40 (1987): 145–177.

Larrère, Catherine. *L'invention de l'économie au XVIIIe siècle: Du droit naturel à la physiocratie.* Paris: Presses Universitaires de France, 1992.

Larson, James L. *Reason and Experience: The Representation of Natural Order in the Work of Carl Linnaeus.* Berkeley: University of California Press, 1971.

———. "Not without a Plan: Geography and Natural History in the Late Eighteenth Century." *Journal of the History of Biology* 19 (1986): 447–488.

———. *Interpreting Nature: The Science of Living Form from Linnaeus to Kant.* Baltimore: Johns Hopkins University Press, 1994.

Latour, Bruno. "The Powers of Association." In *Power, Action, and Belief: A New Sociology of Knowledge?* ed. John Law. London: Routledge and Kegan Paul, 1986, 264–280.

———. "Visualization and Cognition: Thinking with Eyes and Hands." *Knowledge and Society* 6 (1986): 1–40.

———. *Science in Action: How to Follow Scientists and Engineers through Society.* Milton Keynes: Open University Press, 1987.

———. *The Pasteurisation of France.* Trans. Alan Sheridan and John Law. Cambridge: Harvard University Press, 1988.

———. *We Have Never Been Modern.* Trans. Catherine Porter. New York: Harvester Wheatsheaf, 1993.

Latour, Bruno, and Steve Woolgar. *Laboratory Life: The Social Construction of Scientific Facts.* London: Sage, 1979.

Laurent, Goulven. *Paléontologie et évolution en France, 1800–1860: Cuvier-Lamarck à Darwin.* Paris: Éditions du CTHS, 1987.

Law, John, ed. *Power, Action, and Belief: A New Sociology of Knowledge?* Sociological Review Monographs, 32. London: Routledge and Kegan Paul, 1986.

Law, John, Michel Callon, and Arie Rip, eds. *Mapping the Dynamics of Science and Technology.* Basingstoke: Macmillan, 1986.

Lawrence, Chris. "The Nervous System and Society in the Scottish Enlightenment." In *Natural Order: Historical Studies of Scientific Culture,* ed. Barry Barnes and Steven Shapin. Beverly Hills: Sage Publications, 1979.

Lawrence, George H. M., ed. *Adanson: The Bicentennial of Michel Adanson's Familles des Plantes.* 2 vols. Pittsburgh: Hunt Botanical Library and Carnegie Institute of Technology, 1963–1964.

Lefebvre, Georges. *La Révolution française.* 3d ed. Paris: Presses Universitaires de France, 1963.

Légée, Georgette. "Le Muséum sous la Révolution, l'Empire, et la Restauration." *Comptes rendus du 95e congrès national des sociétés savantes, Reims, 1970, Section histoire moderne* 1 (1974): 747–760.

———. "Étienne-Pierre Ventenat (1757–1806), botaniste limousin, face aux problèmes de classification et de sexualité végétales." *Comptes rendus du 102e congrès national des sociétés savantes, Limoges, 1977, Section des sciences* 3 (1977): 33–46.

Légée, Georgette, and Michel Guédès. "Lamarck botaniste et évolutionniste." *Histoire et nature* 17–18 (1981): 19–31.

Le Huray, Peter, and James Day, eds. *Music and Aesthetics in the Eighteenth and Early-Nineteenth Centuries.* Abridged ed. Cambridge: Cambridge University Press, 1988.

Leith, James A. *The Idea of Art as Propaganda in France, 1750–1799: A Study in the History of Ideas.* Toronto: University of Toronto Press, 1965.

———. *Media and Revolution: Moulding a New Citizenry in France during the Terror.* Toronto: University of Toronto Press, 1968.

———. *Space and Revolution: Projects for Monuments, Squares, and Public Buildings in France, 1789–1799.* Montreal: McGill–Queen's University Press, 1991.

Lemoine, Robert. "L'enseignement scientifique dans les collèges bénédictins." In *Enseignement et diffusion des sciences en France au XVIIIe siècle,* ed. René Taton. Reprint, Paris: Hermann, 1986, 101–123.

Lenardon, A. *Index du Journal Encyclopédique, 1756–1793.* Geneva: Slatkine, 1976.

Lenoble, R. *Histoire de l'idée de nature.* Paris: Albin Michel, 1969.

Lepenies, Wolf. *Das Ende der Naturgeschichte: Wandel kultureller Selbstverständlichkeiten in den Wissenschaften des 18. und 19. Jahrhunderts.* Munich: Hanser Verlag, 1976.

———. "De l'histoire naturelle à l'histoire de la nature." *Dix-huitième siècle* 11 (1979): 175–182.

———. *Gefährliche Wahlverwandtschaften: Essays zur Wissenschaftsgeschichte.* Stuttgart: Philipp Reclam, 1989.

Le Rougetel, Hazel. "Encouragement Given by the Society of Arts to Tree Planters: John Buxton's Work at Shadwell, Norfolk." *Journal of the Royal Society of Arts* 129 (1981): 678–681.

———. *The Chelsea Gardener: Philip Miller, 1691–1771.* London: Natural History Museum Publications, 1990.

Leroy, Jean François. "Adanson dans l'histoire de la pensée scientifique." *Revue d'histoire des sciences et de leurs applications* 20 (1967): 349–360.

Letouzey, Yvonne. *Le Jardin des Plantes à la croisée des chemins avec André Thouin, 1747–1824.* Paris: Muséum National d'Histoire Naturelle, 1989.

Levin, Miriam R. "'Ideology' and Neoclassicism: The Problem of Creating a Natural Society through Artificial Means." *Consortium on Revolutionary Europe, 1750–1850: Proceedings* (1981): 177–187.

Licoppe, Christian. *La formation de la pratique scientifique: Le discours de l'expérience en France et en Angleterre, 1630–1820.* Paris: Éditions La Découverte, 1996.

Liedman, Sven-Eric. "Utilitarianism and the Economy." In *Science in Sweden: The Royal Swedish Academy of Sciences, 1739–1989,* ed. Tore Frängsmyr. Canton, MA: Science History Publications, 1989, 23–44.

Limoges, Camille. "Économie de la nature et idéologie juridique chez Linné." *Actes du XIIIe congrès international d'histoire des sciences* 9 (1971–1974): 25–30.

———. "The Development of the Muséum d'Histoire Naturelle of Paris, c. 1800–1914." In *The Organisation of Science and Technology in France, 1808–1914,* ed. Robert Fox and George Weisz. Cambridge: Cambridge University Press, 1980, 211–240.

Lindroth, Sten. "The Two Faces of Linnaeus." In *Linnaeus: The Man and His Work,* ed. Tore Frängsmyr. Canton, MA: Science History Publications, 1994, 1–62.

Lough, John. *The Contributors to the Encyclopédie.* London: Grant and Cutler, 1973.

———. *France on the Eve of Revolution: British Travellers' Observations, 1763–1788.* London: Croom Helm, 1987.

Lovejoy, A. O. "Buffon and the Problem of Species." In *Forerunners of Darwin, 1745–1859,* ed. Bentley Glass, Owsei Temkin, and William L. Straus Jr. Baltimore: Johns Hopkins University Press, 1959, 84–113.

———. *The Great Chain of Being.* Cambridge: Harvard University Press, 1970.

Lucas, Colin, ed. *The French Revolution and the Creation of Modern Political Culture.* Vol. 2, *The Political Culture of the French Revolution.* Oxford: Pergamon, 1988.

———. *Rewriting the French Revolution.* Oxford: Clarendon, 1991.

Lux, David S. *Patronage and Royal Power in Seventeenth-Century France: The Académie de Physique in Caen.* Ithaca, NY: Cornell University Press, 1989.

Lynch, Michael. *Art and Artifact in Laboratory Science: A Study of Shop Work and Shop Talk in a Research Laboratory.* London: Routledge and Kegan Paul, 1985.

———. *Scientific Practice and Ordinary Action: Ethnomethodology and Social Studies of Science.* Cambridge: Cambridge University Press, 1993.

Lyon, John, and Phillip R. Sloan. *From Natural History to the History of Nature: Readings from Buffon and His Critics.* Notre Dame: University of Notre Dame Press, 1981.

Ly-Tio-Fane, Madeleine. *The Triumph of J. N. Ceré and His Isle Bourbon Collaborators.* Preface by Y. Perotin. Paris: Mouton, 1970.

———. *Pierre Sonnerat, 1748–1814: An Account of His Life and Work.* [Reduit, Mauritius]: n.p., 1976.

———. "A Reconnaissance of Tropical Resources during Revolutionary Years: The Role of the Paris Museum d'Histoire Naturelle." *Archives of Natural History* 18 (1991): 333–362.

MacDougall, Elisabeth B., and F. Hamilton Hazlehurst, eds. *The French Formal Garden.* Dumbarton Oaks: Harvard University Press, 1974.

Mackay, David. *In the Wake of Cook: Exploration, Science, and Empire, 1780–1801.* London: Croom Helm, 1985.

———. "Agents of Empire: The Banksian Collectors and Evaluation of New Lands." In *Visions of Empire: Voyages, Botany, and Representations of Nature,* ed. David Philip

Miller and Peter Hanns Reill. Cambridge: Cambridge University Press, 1996, 38–47.
Maniquis, Robert M. "The Puzzling Mimosa: Sensitivity and Plant Symbols in Romanticism." *Studies in Romanticism* 8 (1969): 129–155.
Mannheim, Karl. *Ideology and Utopia: An Introduction to the Sociology of Knowledge.* Preface by Louis Wirth and Bryan Turner. London: Routledge, 1991.
Marouby, Christian. "From Early Anthropology to the Literature of the Savage: The Naturalisation of the Primitive." *Studies in Eighteenth-Century Culture* 14 (1985): 289–298.
Mathias, Peter. "Concepts of Revolution in England and France in the Eighteenth Century." *Studies in Eighteenth-Century Culture* 14 (1985): 29–45.
Mauskopf, Seymour M. "Minerals, Molecules, and Species." *Archives internationales d'histoire des sciences* 23 (1970): 185–206.
Mauss, Marcel. *The Gift: The Form and Reason for Exchange in Archaic Societies.* Trans. W. D. Halls. London: Routledge, 1990.
Mauzi, Robert. *L'idée du bonheur au XVIIIe siècle.* Reprint. Paris: Albin Michel, 1994.
Maza, Sarah. "Luxury, Morality, and Social Change: Why There Was No Middle-Class Consciousness in Prerevolutionary France." *Journal of Modern History* 69 (1997): 199–229.
McClellan, Andrew. *Inventing the Louvre: Art, Politics, and the Origins of the Modern Museum in Eighteenth-Century Paris.* Cambridge: Cambridge University Press, 1994.
McClellan, James E., III. "The Scientific Press in Transition: Rozier's Journal and the Scientific Societies in the 1770s." *Annals of Science* 36 (1979): 425–449.
———. "The Académie Royale des Sciences, 1699–1793: A Statistical Portrait." *Isis* 72 (1981): 541–567.
———. *Colonialism and Science: Saint Domingue in the Old Regime.* Baltimore: Johns Hopkins University Press, 1992.
Médecine et sciences de la vie au XVIIIe siècle. Special issue of *Revue de synthèse* 105, nos. 113–114 (1984).
Meek, Ronald L. *The Economics of Physiocracy: Essays and Translations.* London: Allen and Unwin, 1962.
Melot, Michael. "Caricature and the French Revolution: The Situation in France in 1789." In *French Caricature and the French Revolution.* Chicago: University of Chicago Press, 1988.
Men/Women of Letters. Special issue of *Yale French Studies* 71 (1986).
Merrick, Jeffrey. "Patriarchalism and Constitutionalism in Eighteenth-Century Parlementary Discourse." *Studies in Eighteenth-Century Culture* 20 (1990): 317–330.
Méthivier, Hubert. *L'ancien régime en France, XVIe-XVIIe-XVIIIe siècles.* Paris: Presses Universitaires de France, 1981.
Metzger, Hélène. *Les doctrines chimiques en France du début du XVIIe à la fin du XVIIIe siècle.* Paris: Presses Universitaires de France, 1923.
Michel, André, and Gaston Migeon. *Les grandes institutions de France. Le Musée du Louvre: Sculptures et objets d'art du moyen age, de la renaissance, et des temps modernes.* Paris: Renouard, 1912.
Michon, Georges, ed. *Correspondance de Maximilien et Augustin Robespierre.* Paris: Félix Alcan, 1926.
Middleton, Robin. "The Château and Gardens of Mauperthuis: The Formal and the Informal." In *Garden History: Issues, Approaches, Methods.* Washington, DC: Dumbarton Oaks Research Library and Collection, 1992, 219–242.
Miller, David Philip. "Joseph Banks, Empire, and 'Centers of Calculation' in Late Hanoverian London." In *Visions of Empire: Voyages, Botany, and Representations of Nature,* ed. David Philip Miller and Peter Hanns Reill. Cambridge: Cambridge University Press, 1996, 21–37.

Miller, David Philip, and Peter Hanns Reill, eds. *Visions of Empire: Voyages, Botany, and Representations of Nature.* Cambridge: Cambridge University Press, 1996.

Moran, Bruce T., ed. *Patronage and Institutions: Science, Technology, and Medicine at the European Court, 1500–1750.* Woodbridge, Suffolk: Boydell, 1991.

Moran, Francis, III. "Between Primates and Primitive: Natural Man as the Missing Link in Rousseau's Second Discourse." *Journal of the History of Ideas* 54 (1993): 37–58.

Morange, Michel. "Condorcet et les naturalistes de son temps." In *Sciences à l'époque de la Révolution française,* ed. Roshdi Rashed. Paris: Albert Blanchard, 1988, 445–464.

Moravia, Serge. "From Homme Machine to Homme Sensible: Changing Eighteenth Century Models of Man's Image." *Journal of the History of Ideas* 39 (1978): 45–60.

Mornet, Daniel. *Le sentiment de la nature en France de J.-J. Rousseau à Bernardin de Saint-Pierre: Essai sur les rapports de la littérature et des moeurs.* Paris: Hachette, 1907.

———. "Les enseignements des bibliothèques privées, 1750–1780." *Revue d'histoire littéraire de la France* 17 (1910): 449–496.

———. *Les sciences de la nature en France au XVIIIe siècle: Un chapitre de l'histoire des idées.* New York: Franklin, 1971.

Mousnier, Roland. *The Institutions of France under the Absolute Monarchy, 1598–1789.* Trans. Brian Pearce and Arthur Goldhammer. 2 vols. Chicago: University of Chicago Press, 1979–1984.

Mukerji, Chandra. *Territorial Ambitions and the Gardens of Versailles.* Cambridge: Cambridge University Press, 1997.

Mullan, John. "Hypochondria and Hysteria: Sensibility and the Physicians." *Eighteenth Century: Theory and Interpretations* 25 (1984): 141–174.

Müller-Wille, Staffan. *Botanik und weltweiter Handel: Zur Begründung eines Natürlichen Systems der Pflanzen durch Carl von Linné, 1707–1778.* Studien zur Theorie der Biologie, Band 3. Berlin: Verlag für Wissenschaft und Bildung, 1999.

Müntz, Eugène. "Les annexations de collections d'art et de bibliothèques et leur rôle dans les rélations internationales, principalement pendant la Révolution française." Part 2. *Revue d'histoire diplomatique* 9 (1895): 375–393.

Nathans, Benjamin. "Habermas's 'Public Sphere' in the Era of the French Revolution." *French Historical Studies* 16 (1990): 620–644.

Nicolas, J.-P. "Adanson: Ses travaux sur les blés, ses observations sur l'orge miracle." *Journal d'agriculture tropicale et de botanique appliquée* 11 (1964): 231–249.

Nouvelle biographie universelle. 46 vols. Paris: Firmin Didot, 1853–1866.

O'Brian, Patrick. *Joseph Banks: A Life.* London: Collins Harvill, 1987.

Ophir, Adi, and Steven Shapin. "The Place of Knowledge: A Methodological Survey." *Science in Context* 4 (1991): 3–22.

Osborne, Michael A. "Applied Natural History and Utilitarian Ideals: 'Jacobin Science' at the Muséum d'Histoire Naturelle, 1789–1870." In *Recreating Authority in Revolutionary France,* ed. Bryant T. Raglan Jr. and Elizabeth A. Williams. New Brunswick: Rutgers University Press, 1992, 125–143.

———. *Nature, the Exotic, and the Science of French Colonialism.* Bloomington: Indiana University Press, 1994.

Outram, Dorinda. "The Language of Natural Power: The Éloges of Georges Cuvier and the Public Language of Nineteenth-Century Science." *History of Science* 16 (1978): 153–178.

———. Introduction to *The Letters of Georges Cuvier: A Summary Calendar of Manuscript and Printed Materials Preserved in Europe, the United States of America, and Australasia,* ed. Dorinda Outram. Chalfont St. Giles: British Society for the History of Science, 1980, 1–11.

———. "The Ordeal of Vocation: The Paris Academy of Sciences and the Terror, 1793–1795." *History of Science* 21 (1983): 251–273.

———. *Georges Cuvier: Vocation, Science, and Authority in Post-Revolutionary France.* Manchester: Manchester University Press, 1984.

———. "Uncertain Legislator: Georges Cuvier: Laws of Nature in Their Intellectual Context." *Journal of the History of Biology* 19 (1986): 323–368.

———. *The Body and the French Revolution: Sex, Class, and Political Culture.* New Haven: Yale University Press, 1989.

———. "New Spaces in Natural History." In *Cultures of Natural History,* ed. Nicholas Jardine, J. A. Secord, and E. C. Spary. Cambridge: Cambridge University Press, 1996, 249–265.

Ozouf, Mona. "L'opinion publique." In *The Political Culture of the Old Regime,* ed. Keith Michael Baker. Oxford: Pergamon, 1987, 419–434.

———. *La fête révolutionnaire, 1789–1799.* Paris: Gallimard, 1976. Translated by Alan Sheridan under the title *Festivals and the French Revolution.* Cambridge: Harvard University Press, 1988.

———. "Public Opinion at the End of the Old Regime." *Journal of Modern History,* supplement, 60 (1988): 1–21.

———. "La Révolution française et l'idée de l'homme nouveau." In *The French Revolution and the Creation of Modern Political Culture,* vol. 2, *The Political Culture of the French Revolution,* ed. Colin Lucas. Oxford: Pergamon, 1988, 213–232.

———. "La Révolution française au tribunal de l'utopie." In *The French Revolution and the Creation of Modern Political Culture,* vol. 3, *The Transformation of Political Culture, 1789–1848,* ed. François Furet and Mona Ozouf. Oxford: Pergamon, 1989, 561–574.

Pagden, Anthony R. *The Fall of Natural Man: The American Indians and the Origins of Comparative Ethnology.* Cambridge: Cambridge University Press, 1982.

Pagès, Alain. "La communication circulaire." In *Ecrire, publier, lire: Les correspondances,* ed. Jean-Louis Bonnat and Mireille Bossis. Nantes: Université de Nantes, 1983, 343–361.

Palmer, Robert R. *The Improvement of Humanity: Education and the French Revolution.* Princeton: Princeton University Press, 1985.

Pappas, John. "Buffon vu par Berthier, Feller, et les *Nouvelles ecclésiastiques.*" *Studies on Voltaire and the Eighteenth Century* 216 (1983): 26–28.

Parker, Noel. *Portrayals of Revolution: Images, Debates, and Patterns of Thought on the French Revolution.* New York: Harvester Wheatsheaf, 1990.

Paul, Charles. *Science and Immortality: The Éloges of the Paris Academy of Sciences.* Berkeley: University of California Press, 1980.

Paulson, Ronald. *Representations of Revolution, 1789–1820.* New Haven: Yale University Press, 1983.

Paulson, William R. *Enlightenment, Romanticism, and the Blind in France.* Princeton: Princeton University Press, 1987.

Pearce, Susan M., ed. *Museum Studies in Material Culture.* Leicester: Leicester University Press, 1989.

———. *Objects of Knowledge.* London: Athlone, 1990.

———. *Interpreting Objects and Collections.* London: Routledge, 1994.

Pearl, J. L. "The Role of Personal Correspondence in the Exchange of Scientific Information in Early Modern France." *Renaissance and Reformation* 8 (1984): 106–113.

Pera, Marcello. *The Ambiguous Frog: The Galvani-Volta Controversy on Animal Electricity.* Princeton: Princeton University Press, 1992.

Peronnet, Michel. "L'invention de l'*ancien régime* en France." *History of European Ideas* 14 (1992): 49–58.

Petersen, Susanne. *Lebensmittelfrage und revolutionäre Politik in Paris, 1792–1793: Stu-*

dien zum Verhältnis von revolutionärer Bourgeoisie und Volksbewegung bei Herausbildung der Jakobinerdiktatur. Munich: Oldenbourg, 1979.

Pick, Daniel. *Faces of Degeneration: A European Disorder, c. 1848–c. 1918.* Cambridge: Cambridge University Press, 1989.

Pickering, Andrew, ed. *Science as Practice and Culture.* Chicago: University of Chicago Press, 1992.

Pickstone, John V. "Museological Science? The Place of the Analytical/Comparative in Nineteenth-Century Science, Technology, and Medicine." *History of Science* 32 (1994): 111–138.

Pieters, Florence F. J. M. "Note on the Menagerie and Zoological Cabinet of Stadholder William V of Holland, Directed by Aernout Vosmaer." *Journal of the Society for the Bibliography of Natural History* 9 (1980): 539–563.

Pinault, Madeleine. *The Painter as Naturalist from Dürer to Redouté.* Trans. Philip Sturgess. Paris: Flammarion, 1991.

Pomian, Krszysztof. *Collectionneurs, amateurs, et curieux: Paris, Venise, XVIe–XVIIIe siècle.* Paris: Gallimard, 1987.

Pommier, Edouard. *Le problème du musée à la veille de la Révolution.* Montargis: Musée Girodet, 1989.

———. "La fête de thermidor an VI." In *Fêtes et Révolution,* ed. Béatrice de Andia and Valérie Noëlle Jouffre. Dijon: Musée des Beaux-Arts, 1989–1990, 178–215.

———. "La Révolution française et l'origine des musées." Paper presented at Department of History of Art and Architecture, Cambridge, April 1992.

Popkin, Jeremy D. "Les journaux républicains, 1795–1799." *Revue d'histoire moderne et contemporaine* 31 (1984): 143–157.

Porter, Roy S. "Making Faces: Physiognomy and Fashion in Eighteenth-Century England." *Études anglaises* 38 (1985): 385–396.

Poulot, Dominique. "Naissance du monument historique." *Revue d'histoire moderne et contemporaine* 32 (1985): 418–450.

———. "Musée et société dans l'Europe moderne." *Mélanges de l'école française de Rome, moyen âge–temps modernes* 98 (1986): 991–1096.

———. "Le musée entre l'histoire et ses légendes." *Débat* 49 (1988): 69–83.

———. "Le Louvre imaginaire: Essai sur le statut du musée en France, des Lumières à la République." *Historical Reflections* 17 (1991): 171–204.

Powell, Dulcie. "The Voyage of the Plant Nursery, H.M.S. Providence, 1791–1793." *Economical Botany* 31 (1977): 387–431.

Pratt, Mary Louise. *Imperial Eyes: Travel Writing and Transculturation.* London: Routledge, 1992.

Prevost, M., Roman d'Amat, R. Limouzin-Lamothe, J.-P. Lobies, and H. Tribout de Morembert, eds. *Dictionnaire de biographie française.* Paris: Letouzey et Ané, 1933–.

Price, Roger. *The Economic Modernisation of France, 1730–1875.* London: Macmillan, 1975.

Pueyo, Guy. "Duhamel du Monceau, précurseur des études climatiques et microclimatiques." *Actes du XIIe congrès international d'histoire des sciences, 1968* 12 (1971): 63–68.

Py, Gilbert. *Rousseau et les éducateurs: Étude sur la fortune des idées pédagogiques de Jean-Jacques Rousseau en France et en Europe au XVIIIe siècle.* Oxford: Voltaire Foundation, 1997.

Rabreau, Daniel. "Architecture et fêtes dans la nouvelle Rome." In *Les fêtes de la Révolution: Colloque de Clermont-Ferrand, juin 1974,* ed. Jean Ehrard and Paul Viallaneix. Paris: Société des Études Robespierristes, 1977, 355–376.

Rabreau, Daniel, and Monique Mosser. "Paris en 1778: L'architecture en question." *Dix-huitième siècle* 11 (1979): 141–164.

Raglan, Bryant T., Jr., and Elizabeth A. Williams, eds. *Recreating Authority in Revolutionary France.* New Brunswick: Rutgers University Press, 1992.

Raitières, Anna. "Lettres à Buffon dans les registres de l'ancien régime, 1739–1788." *Histoire et nature* 17–18 (1980–1981): 85–148.

Ramsey, Matthew. *Professional and Popular Medicine in France, 1770–1830.* Cambridge: Cambridge University Press, 1988.

Rappaport, Rhoda. "Revolutions, Accidents, and 'Bouleversements.'" *Histoire et nature* 19–20 (1981–1982): 57–58.

Rashed, Roshdi, ed. *Sciences à l'époque de la Révolution française.* Paris: Albert Blanchard, 1988.

Ravitch, Norman. "On François Furet and the French Revolution." *Proceedings of the 11th Annual Meeting of the Western Society for French History* 11 (1984): 201–206.

Reddy, William. *Money and Liberty in Modern Europe: A Critique of Historical Understanding.* New York: Cambridge University Press, 1987.

Regourd, François. "La Société Royale d'Agriculture de Paris face à l'espace colonial, 1761–1793." *Bulletin du centre d'histoire des espaces atlantiques,* new series, 8 (1998): 155–194.

———. "Maîtriser la nature: Un enjeu colonial: Botanique et agronomie en Guyane et aux Antilles, XVII–XVIIIe siècles." *Revue française d'histoire d'outre-mer* 86 (1999): 39–63.

Rheinberger, Hans-Jörg. "Buffon: Zeit, Veränderung, und Geschichte." *History and Philosophy of the Life Sciences* 12 (1990): 203–223.

Rice, Howard C., Jr. "Jefferson's Gift of Fossils to the Museum of Natural History in Paris." *Proceedings of the American Philosophical Society* 95 (1951): 597–627.

Richefort, Isabelle. "Métaphores et représentations de la nature sous la Révolution." In *Nature, environnement, et paysage: L'héritage du XVIIIe siècle: Guide de recherche archivistique et bibliographique,* ed. Andrée Corvol and Isabelle Richefort. Paris: L'Harmattan, 1995, 3–17.

Riley, James C. *The Seven Years War and the Old Regime in France: The Economic and Financial Toll.* Princeton: Princeton University Press, 1986.

———. *The Eighteenth-Century Campaign to Avoid Disease.* New York: St. Martin's Press, 1987.

Ritvo, Harriet. *The Animal Estate: The English and Other Animals in the Victorian Age.* Cambridge: Harvard University Press, 1987.

———. "At the Edge of the Garden: Nature and Domestication in 18th- and 19th-Century Britain." *Huntington Library Quarterly* 55 (1992): 363–378.

Roberts, Lissa. "The Death of the Sensuous Chemist: The 'New' Chemistry and the Transformation of Sensuous Technology." *Studies in History and Philosophy of Science* 26 (1995): 503–530.

Robins, William J., and Mary Christine Howson. "André Michaux's New Jersey Garden and Pierre Paul Saunier, Journeyman Gardener." *Proceedings of the American Philosophical Society* 102 (1958): 351–370.

Roche, Daniel. *Le siècle des Lumières en province: Académies et académiciens provinciaux, 1680–1789.* 2 vols. Paris: Mouton, 1978.

———. *The People of Paris: An Essay in Popular Culture in the Eighteenth Century.* Trans. Marie Evans and Gwynne Lewis. Leamington Spa: Berg, 1987.

———. *La culture des apparences: Une histoire du vêtement, XVIIe–XVIIIe siècles.* Paris: Fayard, 1989.

Roe, Shirley A. *Matter, Life, and Generation: Eighteenth Century Embryology and the Haller-Wolff Dispute.* Cambridge: Cambridge University Press, 1981.

Roe, Shirley A., and Renato G. Mazzolini, eds. *Science against the Unbelievers: The Correspondence of Bonnet and Needham, 1760–1780.* Oxford: Voltaire Foundation, 1986.

Roger, Jacques. *Les sciences de la vie dans la pensée française du XVIIIe siècle.* Paris: Armand Colin, 1963. Translated by Robert Ellrich under the title *The Life Sciences in 18th Century French Thought,* ed. Keith R. Benson. Stanford: Stanford University Press, 1997.

———. "Buffon et la théorie de l'anthropologie." In *Enlightenment Essays in Honour of Lester G. Crocker,* ed. Alfred J. Bingham and Virgil W. Topazio. Oxford: Voltaire Foundation, 1979, 253–261.

———. *Buffon: Un philosophe au Jardin du Roi.* Paris: Fayard, 1989. Translated by Sarah Lucille Bonnefoi under the title *Buffon: A Life in Natural History,* ed. L. Pearce Williams. Ithaca, NY: Cornell University Press, 1997.

Rookmaaker, L. C. "Histoire du rhinocéros de Versailles, 1770–1793." *Revue d'histoire des sciences* 36 (1983): 307–318.

Rosen, George. *From Medical Police to Social Medicine: Essays on the History of Health Care.* New York: Science History Publications, 1974.

Rosenthal, Jean-Laurent. *The Fruits of Revolution: Property Rights, Litigation, and French Agriculture, 1700–1860.* Cambridge: Cambridge University Press, 1992.

Roule, Louis. *L'histoire de la nature vivante d'après l'oeuvre des grands naturalistes français.* 6 vols. Paris: Ernest Flammarion, 1924–1932.

Rousseau, George S., and Roy S. Porter. *The Ferment of Knowledge: Studies in the Historiography of Eighteenth-Century Science.* Cambridge: Cambridge University Press, 1980.

Rudé, George. *The Crowd in the French Revolution.* Westport, CT: Greenwood Press, 1986.

———. *The French Revolution.* London: Weidenfeld and Nicolson, 1988.

Rupke, Nicolaas. *Richard Owen: Victorian Naturalist.* New Haven: Yale University Press, 1994.

Saisselin, R. G. "The French Garden in the Eighteenth Century: From Belle Nature to the Landscape of Time." *Journal of Garden History* 5 (1985): 284–297.

Salomon-Bayet, Claire. *L'institution de la science et l'expérience du vivant: Méthode et expérience à l'Académie Royale des Sciences, 1666–1793.* Paris: Flammarion, 1978.

Savage, Henry, Jr., and Elizabeth J. Savage. *André and François André Michaux.* Charlottesville: University Press of Virginia, 1986.

Schaffer, Simon. "Natural Philosophy and Public Spectacle in the Eighteenth Century." *History of Science* 21 (1983): 1–43.

———. "Astronomers Mark Time: Discipline and the Personal Equation." *Science in Context* 2 (1986): 115–145.

———. "Scientific Discoveries and the End of Natural Philosophy." *Social Studies of Science* 16 (1986): 387–420.

———. "Authorized Prophets: Comets and Astronomers after 1759." *Studies in Eighteenth-Century Culture* 17 (1988): 45–74.

———. "Measuring Virtue; Eudiometry, Enlightenment, and Pneumatic Medicine." In *The Medical Enlightenment of the Eighteenth Century,* ed. Andrew Cunningham and Roger French. Cambridge: Cambridge University Press, 1990, 281–318.

———. "The Eighteenth Brumaire of Bruno Latour." Essay review. *Studies in History and Philosophy of Science* 22 (1992): 174–192.

———. "Self Evidence." *Critical Inquiry* 8 (1992): 328–362.

———. "The Earth's Fertility as a Social Fact in Early Modern England." In *Nature and Society in Historical Context,* ed. Mikuláš Teich, Roy S. Porter, and Bo Gustafsson. Cambridge: Cambridge University Press, 1997, 124–147.

Schaffer, Simon, and Steven Shapin. *Leviathan and the Airpump: Hobbes, Boyle, and the Experimental Life.* Princeton: Princeton University Press, 1985.

Schama, Simon. *Citizens: A Chronicle of the French Revolution.* Harmondsworth, Middlesex: Penguin, 1988.

Scheler, Lucien. "Antoine-Laurent Lavoisier et le *Journal d'Histoire Naturelle.*" *Revue d'histoire des sciences et de leurs applications* 14 (1961): 1–9.

———. "Antoine-Laurent Lavoisier, Michel Adanson, et la rédaction des programmes de prix de l'Académie Royale des Sciences." *Revue d'histoire des sciences et de leurs applications* 14 (1961): 257–284.

Schiebinger, Londa. "Skeletons in the Closet: The First Illustrations of the Female Skeleton in Nineteenth-Century Anatomy." In *The Making of the Modern Body: Sexuality and Society in the Nineteenth Century,* ed. Thomas Laqueur and Catherine Gallagher. Berkeley: University of California Press, 1987, 1–38.

———. "The Private Life of Plants: Sexual Politics in Carl Linnaeus and Erasmus Darwin." In *Science and Sensibility: Gender and Scientific Enquiry, 1780–1845,* ed. Marina Benjamin. Oxford: Basil Blackwell, 1991, 121–143.

———. *Nature's Body: Sexual Politics and the Making of Modern Science.* London: Pandora, 1993.

Schiller, Joseph. "Les laboratoires d'anatomie et de botanique à l'Académie des Sciences au XVIIe siècle." *Revue d'histoire des sciences et de leurs applications* 17 (1964): 97–114.

Schmidt, Adolf. *Tableaux de la révolution française publiés sur les papiers inédits du département et de la police secrète de Paris.* 3 vols. Leipzig: Veit, 1867–1870.

———. *Paris pendant la révolution d'après les rapports de la police secrète, 1789–1800.* Vols. 1–4. Trans. P. Viollet. Paris: Champion, 1880–1894.

Scientifiques et sociétés pendant la Révolution et l'Empire: Actes du 114e congrès national des sociétés savantes (Paris 3–9 avril 1989), Section histoire des sciences et des techniques. Paris: Éditions du CTHS, 1990.

Secord, Anne. "Corresponding Interests: Artisans and Gentlemen in Natural History Exchange Networks." *British Journal for the History of Science* 27 (1994): 383–408.

Secord, James A. "Nature's Fancy: Charles Darwin and the Breeding of Pigeons." *Isis* 72 (1981): 162–186.

———. *Controversy in Victorian Geology: The Cambrian-Silurian Dispute.* Princeton: Princeton University Press, 1986.

———. "The Discovery of a Vocation: Darwin's Early Geology." *British Journal for the History of Science* 24 (1991): 133–157.

Sennett, Richard. *The Fall of Public Man.* New York: Alfred Knopf, 1977.

Shapin, Steven. "The Audience for Science in Eighteenth-Century Edinburgh." *History of Science* 12 (1974): 95–121.

———. "Property, Patronage, and the Politics of Science: The Founding of the Royal Society of Edinburgh." *British Journal for the History of Science* 7 (1974): 1–41.

———. "Pump and Circumstance: Robert Boyle's Literary Technology." *Social Studies of Science* 14 (1984): 481–520.

———. "The House of Experiment in Seventeenth-Century England." *Isis* 79 (1988): 373–404.

———. "'A Scholar and a Gentleman': The Problematic Identity of the Scientific Practitioner in Early Modern England." *History of Science* 29 (1991): 279–327.

———. *A Social History of Truth: Civility and Science in Seventeenth-Century England.* Chicago: University of Chicago Press, 1994.

Shaw, George Bernard. *Back to Methuselah: A Metabiological Pentateuch.* Leipzig: Bernhard Tauchnitz, 1922.

Sherman, Daniel. *Worthy Monuments: Art Museums and the Politics of Culture in Nineteenth-Century France.* Cambridge: Harvard University Press, 1989.

Shklar, Judith N. "The Political Theory of Utopia: From Melancholy to Nostalgia." *Daedalus* 94 (1965): 367–381.

Shortland, Michael, ed. *Science and Nature: Essays in the History of the Environmental Sciences.* Oxford: Alden Press, 1993.

Shulim, Joseph I. *Liberty, Equality, and Fraternity: Studies in the Era of the French Revolution and Napoleon.* New York: Peter Lang, 1989.
Simon, Jonathan. "The Alchemy of Identity: Pharmacy and the Chemical Revolution, 1777–1809." Ph.D. dissertation, University of Pittsburgh, 1997.
Sloan, Phillip R. "The Idea of Racial Degeneracy in Buffon's Histoire Naturelle." In *Racism in the Eighteenth Century,* ed. Harold E. Pagliaro. Cleveland: Press of Case Western Reserve University, 1973.
———. "The Buffon-Linnaeus Controversy." *Isis* 67 (1976): 356–375.
———. "Buffon, German Biology, and the Historical Interpretation of Biological Species." *British Journal for the History of Science* 12 (1979): 109–153.
———. "Organic Molecules Revisited." In *Buffon 88: Actes du colloque international pour le bicentenaire de la mort de Buffon (Paris, Montbard, Dijon, 14–22 juin 1988),* ed. Jean-Claude Beaune, Serge Benoît, Jean Gayon, and Denis Woronoff. Paris: Vrin and Institut Interdisciplinaire d'Études Épistémologiques, 1992, 162–187.
———. "Buffon Studies Today." *History of Science* 32 (1994): 469–477.
Smeaton, W. A. "The Early Years of the Lycée and the Lycée des Arts: A Chapter in the Lives of A. L. Lavoisier and A. F. de Fourcroy." *Annals of Science* 11 (1955): 257–267, 309–319.
———. "Lavoisier's Membership of the Société Royale d'Agriculture and the Comité d'Agriculture." *Annals of Science* 12 (1956): 267–277.
———. *Fourcroy: Chemist and Revolutionary, 1755–1809.* Cambridge: Heffer and Sons, 1962.
———. "Monsieur and Madame Lavoisier in 1789: The Chemical Revolution and the French Revolution." *Ambix* 36 (1989): 1–4.
Smith, Bernard. *European Vision in the South Pacific.* 2d ed. New Haven: Yale University Press, 1985.
Smith, Jay M. *The Culture of Merit: Nobility, Royal Service, and the Making of Absolute Monarchy in France, 1600–1789.* Ann Arbor: University of Michigan Press, 1996.
Soboul, Albert. *La Révolution française, 1789–1799.* Paris: Éditions Sociales, 1948.
———. *Les sans-culottes parisiens en l'an II: Histoire politique et sociale des sections de Paris, 2 juin 1793–9 thermidor an II.* 2d ed. Paris: Clavreuil, 1960.
Spadafora, David. *La France à la veille de la Révolution.* 2 vols. Paris: Société d'Enseignement Supérieur, 1966.
———. *The Idea of Progress in Eighteenth-Century Britain.* New Haven: Yale University Press, 1990.
Spary, E. C. "Political, Natural, and Bodily Economies." In *Cultures of Natural History,* ed. Nicholas Jardine, James A. Secord, and E. C. Spary. Cambridge: Cambridge University Press, 1996, 178–196.
———. "Le spectacle de la nature: Contrôle du public et vision républicaine dans le Muséum jacobin." In *Le Muséum au premier siècle de son histoire,* ed. Claude Blanckaert, Claudine Cohen, Pietro Corsi, and Jean-Louis Fischer. Paris: Muséum National d'Histoire Naturelle, 1997, 457–479.
———. "L'invention de l' 'expédition scientifique': Histoire naturelle, empire, et Égypte." In *L'invention scientifique de la Méditerranée,* ed. Marie-Noëlle Bourguet, Bernard Lepetit, Daniel Nordmann, and Maroula Sinarellis. Paris: Éditions de l'École des Hautes Études en Sciences Sociales, 1998, 119–138.
———. "Codes of Passion: Natural History Specimens as a Polite Language in Late Eighteenth-Century France." In *Wissenschaft als kulturelle Praxis, 1750–1900,* ed. Hans Erich Bödeker, Peter Hanns Reill, and Jürgen Schlumbohm. Göttingen: Vanderhoek and Ruprecht, 1999, 105–135.
———. "The 'Nature' of Enlightenment." In *The Sciences in Enlightened Europe,* ed.

William Clark, Jan V. Golinski, and Simon Schaffer. Chicago: University of Chicago Press, 1999, 272–304.

———. "Rococo Readings of the Book of Nature." In *History of the Book/History of the Sciences,* ed. Marina Frasca Spada and Nicholas Jardine. Cambridge: Cambridge University Press, forthcoming.

———. "Forging Nature at the Republican Muséum." Unpublished paper.

———. "The Nut and the Orange: Natural History, Natural Religion, and Republicanism in Late Eighteenth-Century France." Unpublished paper.

Sprigath, Gabriele. "Sur le vandalisme révolutionnaire, 1792–1794." *Annales historiques de la Révolution française* 52 (1980): 510–535.

Stafford, Barbara Maria. *Voyage into Substance: Art, Science, Nature, and the Illustrated Travel Account, 1760–1840.* Cambridge: MIT Press, 1984.

———. *Body Criticism: Imaging the Unseen in Enlightenment Art and Medicine.* Cambridge: MIT Press, 1991.

———. *Artful Science: Enlightenment Entertainment and the Eclipse of Visual Education.* Cambridge: MIT Press, 1994.

Stafleu, Frans A. "L'Héritier de Brutelle: The Man and His Work." In Charles-Louis L'Héritier de Brutelle, *Sertum Anglicum 1788: Facsimile with Critical Studies and a Translation.* Pittsburgh: Hunt Botanical Library, 1963, xiii–xliii.

———. "Adanson and His 'Familles des Plantes.'" In *Adanson: The Bicentennial of Michel Adanson's Familles des Plantes,* 2 vols., ed. George H. M. Lawrence. Pittsburgh: Hunt Botanical Library and Carnegie Institute of Technology, 1963–1964, 1:123–264.

———. *Linnaeus and the Linnaeans: The Spreading of Their Ideas in Systematic Botany.* Utrecht: Oosthoek, 1971.

Starobinski, Jean. *Jean-Jacques Rousseau: Transparency and Obstruction.* Trans. Arthur Goldhammer. Introduction by Robert J. Morrissey. Chicago: University of Chicago Press, 1988.

Staum, Martin S. "Science and Government in the French Revolution." In *Science, Technology, and Culture in Historical Perspective,* ed. Louis A. Knafla, Martin S. Staum, and T. H. E. Travers. Calgary: University of Calgary Press, 1976.

———. *Cabanis: Enlightenment and Medical Philosophy in the French Revolution.* Princeton: Princeton University Press, 1980.

Stemerding, Dirk. *Plants, Animals, and Formulae: Natural History in the Light of Latour's* Science in Action *and Foucault's* The Order of Things. Enschede: Universiteit Twente, 1991.

Stettler, Antoinette. "Sensation und Sensibilität: Zu John Lockes Einfluß auf das Konzept der Sensibilität im 18. Jahrhundert." *Gesnerus* 45 (1988): 445–460.

Stevens, Peter F. "Haüy and A.-P. de Candolle: Crystallography, Botanical Systematics, and Comparative Morphology." *Journal of the History of Biology* 17 (1984): 49–82.

———. *The Development of Biological Systematics: Antoine-Laurent de Jussieu, Nature, and the Natural System.* New York: Columbia University Press, 1994.

Stewart, Larry. *The Rise of Public Science: Rhetoric, Technology, and Natural Philosophy in Newtonian Britain, 1660–1750.* Cambridge: Cambridge University Press, 1992.

Stocking, George W., Jr. "Essays on Museums and Material Culture." In *Objects and Others: Essays on Museums and Material Culture,* ed. George W. Stocking, Jr. Madison: University of Wisconsin Press, 1985, 3–14.

———, ed. *Objects and Others: Essays on Museums and Material Culture.* Madison: University of Wisconsin Press, 1985.

Strandberg, Runar. "The French Formal Garden after Le Nostre." In *The French Formal Garden,* ed. Elisabeth B. MacDougall and F. Hamilton Hazlehurst. Washington, DC: Dumbarton Oaks and Harvard University Press, 1974, 43–67.

Strauss, Leo. *Natural Right and History.* Chicago: University of Chicago Press, 1953.

Strong, Roy. *Art and Power: Renaissance Festivals, 1450–1650.* Woodbridge, Suffolk: Boydell, 1984.

Stroup, Alice. *Royal Funding of the Parisian Académie Royale des Sciences during the 1690s.* Philadelphia: American Philosophical Society, 1987.

———. *A Company of Scientists: Botany, Patronage, and Community at the Seventeenth-Century Parisian Royal Academy of Sciences.* Berkeley: University of California Press, 1990.

Sturdy, David J. *Science and Social Status: The Members of the Académie des Sciences, 1666–1750.* Woodbridge, Suffolk: Boydell, 1995.

Suleiman, Susan R., and Inge Crosman, eds. *The Reader in the Text: Essays on Audience and Interpretation.* Princeton: Princeton University Press, 1980.

Sutton, Geoffrey V. "Electric Medicine and Mesmerism." *Isis* 72 (1981): 375–392.

———. *Science for a Polite Society: Gender, Culture, and the Demonstration of Enlightenment.* Boulder, CO: Westview Press, 1995.

Szyfman, Léon. *Jean-Baptiste Lamarck et son époque.* Paris: Masson, 1982.

Taton, René, ed. *Enseignement et diffusion des sciences en France au XVIIIe siècle.* Reprint. Paris: Hermann, 1986.

Te Heesen, Anke. *Der Weltkasten: Die Geschichte einer Bildenzyklopädie aus dem 18. Jahrhundert.* Göttingen: Wallstein Verlag, 1997.

Teich, Mikulás. "Circulation, Transformation, Conservation of Matter, and the Balancing of the Biological World in the Eighteenth Century." In *Lazzaro Spallanzani e la biologia del settecento: Teorie, esperimenti, istituzioni scientifiche,* ed. Walter Bernardi and Antonello La Vergata. Florence: Leo Olschki, 1982, 363–380.

Terrall, Mary. "Salon, Academy, and Boudoir: Generation and Desire in Maupertuis's Science of Life." *Isis* 87 (1996): 217–229.

Théodoridès, Jean. "Le comte de Lacepède, 1756–1825: Naturaliste, musicien, et homme politique." *Comptes rendus du 96e congrès national des sociétés savantes, Toulouse, 1971, Section scientifique* 1 (1974): 47–61.

Thomas, Keith. *Man and the Natural World: Changing Attitudes in England, 1500–1800.* Harmondsworth, Middlesex: Penguin, 1984.

Thomas, Nicholas. *Entangled Objects.* Cambridge: Harvard University Press, 1991.

———. "Licensed Curiosity: Cook's Pacific Voyages." In *The Cultures of Collecting,* ed. John Elsner and Roger Cardinal. London: Reaktion Books, 1994, 116–136.

Thompson, E. P. "The Moral Economy of the English Crowd in the Eighteenth Century." *Past and Present* 50 (1971): 76–136.

Thompson, J. M. *The French Revolution.* Oxford: Basil Blackwell, 1986.

Todericiu, Doru. "L'Académie Royale des Sciences de Paris et la mise en valeur des richesses minérales du Roussillon au XVIIIe siècle." *Comptes rendus du 106e congrès national des sociétés savantes, Perpignan, 1981, Section des sciences* 4 (1982): 153–159.

Tomaselli, Sylvana. "The Enlightenment Debate on Women." *History Workshop* 20 (1985): 101–124.

Torlais, Jean. "Un cabinet d'histoire naturelle français datant du XVIIIe siècle." *Revue d'histoire des sciences et de leurs applications* 14 (1961): 87–88.

———. *Un esprit encyclopédique en dehors de 'l'Encyclopédie': Réaumur: D'après des documents inédits.* 2d ed. Paris: Albert Blanchard, 1961.

———. "La physique expérimentale." In *Enseignement et diffusion des sciences en France au XVIIIe siècle,* ed. René Taton. Reprint, Paris: Hermann, 1986, 342–384.

Torrens, Hugh. "Under Royal Patronage: The Early Work of John Mawe (1766–1829) in Geology and the Background to His Travels in Brazil in 1807–1810." Unpublished paper.

Trahard, Pierre. *La sensibilité révolutionnaire, 1789–1794.* Paris: Boivin, 1936.
Turner, G. L'Estrange, ed. *The Patronage of Science in the Nineteenth Century.* Leiden: Noordhoff, 1976.
Tuveson, Ernest L. *Millennium and Utopia: A Study in the Development of the Idea of Progress.* Berkeley: University of California Press, 1949.
Tytler, Graeme. *Physiognomy in the European Novel.* Princeton: Princeton University Press, 1982.
Vachon, Max. "Lamarck et son enseignement au Muséum." *Histoire et nature* 17–18 (1980–1981): 7–17.
Van Kley, Dale. "Church, State, and the Ideological Origins of the French Revolution: The Debate over the General Assembly of the Gallican Clergy in 1765." *Journal of Modern History* 51 (1979): 629–666.
———. *The Damiens Affair and the Unravelling of the Ancien Regime, 1750–1770.* Princeton: Princeton University Press, 1984.
Van Tieghem, Paul. *Le sentiment de la nature dans le pré-romantisme européen.* Paris: A. G. Nizet, 1960.
Vartanian, Aram. "Trembley's Polyp, La Mettrie, and Eighteenth-Century French Materialism." *Journal for the History of Ideas* 11 (1950): 259–286.
———. "Diderot and the Technology of Life." *Studies in Eighteenth-Century Culture* 15 (1986): 11–32.
Venturi, Franco. *Utopia and Reform in the Enlightenment.* Cambridge: Cambridge University Press, 1971.
Vergo, Peter, ed. *The New Museology.* London: Reaktion Books, 1989.
Vess, David M. *Medical Revolution in France, 1789–1796.* Gainesville: University Presses of Florida, 1975.
Viard, J. *Le tiers espace: Essai sur la nature.* Paris: Méridiens Klincksieck, 1990.
Vidler, Anthony. "Grégoire, Lenoir, et les 'monuments parlants.'" In *La carmagnole des muses,* ed. Jean-Claude Bornet. Paris: Armand Colin, 1986, 131–159.
———. *The Writing on the Walls: Architectural Theory in the Late Enlightenment.* London: Butterworth Architecture, 1989.
Viel, Claude. "Duhamel du Monceau: Naturaliste, physicien, et chimiste." *Revue d'histoire des sciences* 38 (1985): 55–71.
Vincent-Buffault, Anne. *The History of Tears: Sensibility and Sentimentality in France.* Basingstoke: Macmillan, 1991.
Viroli, Maurizio. *Jean-Jacques Rousseau and the "Well-Ordered Society."* Trans. Derek Hanson. Cambridge: Cambridge University Press, 1988.
Vovelle, Michel. *The Fall of the French Monarchy, 1787–1792.* Trans. Susan Burke. Paris: Éditions de la Maison des Sciences de l'Homme; Cambridge: Cambridge University Press, 1984.
Vyverberg, Henry. *Human Nature, Cultural Diversity, and the French Enlightenment.* Oxford: Oxford University Press, 1989.
Wade, Ira O. *The Intellectual Origins of the French Enlightenment.* Princeton: Princeton University Press, 1971.
Walker, W. Cameron. "Animal Electricity before Galvani." *Annals of Science* 2 (1937): 84–113.
Wallis, P. J., R. V. Wallis, T. D. Whittet, and J. G. L. Burnby. *Eighteenth Century Medics: Subscriptions, Licences, Apprenticeships.* 2d ed. Newcastle: Project for Historical Bibliography, 1986.
Webster, Andrew. *Science, Technology, and Society: New Directions.* Basingstoke: Macmillan, 1991.
Wellman, Kathleen. "Medicine as a Key to Defining Enlightenment Issues: The Case of Julien Offray de La Mettrie." *Studies in Eighteenth-Century Culture* 17 (1988): 35–89.

———. *La Mettrie: Medicine, Philosophy, and Enlightenment.* Durham: Duke University Press, 1992.

Weulersse, Georges. *Le mouvement physiocratique en France de 1756 à 1770.* Paris: Félix Alcan, 1910.

———. *La physiocratie sous les ministères de Turgot et de Necker, 1774–1781.* Paris: Presses Universitaires de France, 1950.

———. *Les physiocrates à l'aube de la révolution, 1781–1792.* Revised by Corinne Beutler. Paris: EHESS, 1985.

Wiebenson, Dora. *The Picturesque Garden in France.* Princeton: Princeton University Press, 1978.

Williams, Elizabeth A. *The Physical and the Moral: Anthropology, Physiology, and Philosophical Medicine in France, 1750–1850.* Cambridge: Cambridge University Press, 1994.

Williams, L. Pearce. "Science, Education, and the French Revolution." *Isis* 44 (1953): 311–330.

Wilson, Adrian. *The Making of Man-Midwifery: Childbirth in England, 1660–1770.* London: UCL Press, 1995.

Wise, M. Norton. "Work and Waste: Political Economy and Natural Philosophy in Nineteenth Century Britain." Part 1. *History of Science* 27 (1989): 263–301.

Wokler, Robert. "Tyson and Buffon on the Orang-Utan." *Studies on Voltaire and the Eighteenth Century* 155 (1976): 2301–2319.

Wood, Paul B. "The Natural History of Man in the Scottish Enlightenment." *History of Science* 27 (1989): 89–123.

Woodbridge, Kenneth. *Princely Gardens: The Origins and Development of the French Formal Style.* New York: Rizzoli, 1986.

Woolgar, Steve. *Science: The Very Idea.* Chichester: Ellis Horwood, 1988.

Woronoff, Denis. *The Thermidorean Regime and the Directory, 1794–1799.* Trans. Julian Jackson. Paris: Éditions de la Maison des Sciences de l'Homme; Cambridge: Cambridge University Press, 1984.

Zirkle, Conway. "An Overlooked Eighteenth-Century Contribution to Plant Breeding and Plant Selection." *Journal of Heredity* 59 (1968): 195–198.

INDEX

(Italicized entries denote page numbers for references in figure captions.)